Springer Proceedings in Mathematics & Statistics

Volume 297

Springer Proceedings in Mathematics & Statistics

This book series features volumes composed of selected contributions from workshops and conferences in all areas of current research in mathematics and statistics, including operation research and optimization. In addition to an overall evaluation of the interest, scientific quality, and timeliness of each proposal at the hands of the publisher, individual contributions are all refereed to the high quality standards of leading journals in the field. Thus, this series provides the research community with well-edited, authoritative reports on developments in the most exciting areas of mathematical and statistical research today.

More information about this series at http://www.springer.com/series/10533

Melvyn B. Nathanson
Editor

Combinatorial and Additive Number Theory III

CANT, New York, USA, 2017 and 2018

Springer

Editor
Melvyn B. Nathanson
Department of Mathematics
Lehman College and the Graduate Center
City University of New York
New York, NY, USA

ISSN 2194-1009 ISSN 2194-1017 (electronic)
Springer Proceedings in Mathematics & Statistics
ISBN 978-3-030-31108-7 ISBN 978-3-030-31106-3 (eBook)
https://doi.org/10.1007/978-3-030-31106-3

Mathematics Subject Classification (2010): 03H15, 11B05, 11B13, 11B75, 11D07, 11D25, 11E04, 14H05, 15A12, 15B51

This Springer imprint is published by the registered company Springer Nature Switzerland AG
The registered company address is: Gewerbestrasse 11, 6330 Cham, Switzerland

Preface

Workshops on Combinatorial and Additive Number Theory (CANT) have been organized at the CUNY Graduate Center in New York every year since 2003. The 4-day CANT conferences are held in May, usually from Tuesday to Friday of the week immediately preceding or immediately following Memorial Day. They have become a fixed point in the number theory calendar.

These workshops are arranged by the New York Number Theory Seminar. The seminar was started in 1981 by David and Gregory Chudnovsky, Harvey Cohn, and Melvyn B. Nathanson, and for 38 years has been meeting at the CUNY Graduate Center every Thursday afternoon during the academic year, and also in the summer.

This volume contains papers presented at the CANT 2017 and CANT 2018 workshops. There are 17 papers on important topics in number theory and related parts of mathematics. These topics include sumsets, partitions, convex polytopes and discrete geometry, Ramsey theory, commutative algebra and arithmetic geometry, and applications of logic and nonstandard analysis to number theory.

I thank the Number Theory Foundation, Springer, and the Journal of Number Theory (Elsevier) for their support of CANT.

I am grateful to Springer and to mathematics editor Dahlia Fisch for making possible the publication of the proceedings of the CANT 2017 and CANT 2018 workshops.

Previous volumes are [1] and [2].

New York, USA — Melvyn B. Nathanson

References

1. M. B. Nathanson, editor, *Combinatorial and Additive Number Theory–CANT 2011 and 2012*, Springer Proc. Math. Stat., vol. 101, Springer, New York, 2014.
2. M. B. Nathanson, editor, *Combinatorial and Additive Number Theory II–CANT 2015 and 2016*, Springer Proc. Math. Stat., vol. 220, Springer, New York, 2017.

Contents

Weighted Zero-Sums for Some Finite Abelian Groups of Higher Ranks

S. D. Adhikari, Bidisha Roy and Subha Sarkar

Abstract In this article, we consider the study of the Davenport constant with weight $\{\pm 1\}$ for some finite abelian groups of higher ranks and take up some related questions. For instance, we show that for an odd prime p, any sequence over $G = (\mathbb{Z}/p\mathbb{Z})^3$ of length $4p - 3$ which contains at least five zero-elements, there is a $\{\pm 1\}$-weighted zero-sum subsequence of length p. We also show that for an odd prime p and for a positive even integer $k \geq 2$ which divides $p - 1$, if θ is an element of order k of the multiplicative group $(\mathbb{Z}/p\mathbb{Z})^*$ and A is the subgroup of $(\mathbb{Z}/p\mathbb{Z})^*$ generated by θ, then any sequence over $(\mathbb{Z}/p\mathbb{Z})^{k+1}$ of length $4p + \frac{p-1}{k} - 1$ contains an A-weighted zero-sum subsequence of length $3p$. In the introduction, we give a small expository account of the area and mention some relevant expository articles.

1 Introduction

Let G be a finite abelian group (written additively) and let $\exp(G)$ be the *exponent* of the group G. By a sequence over G we mean a finite sequence of elements from G in which repetition of terms is allowed. In this way we can view a sequence as an element of the free abelian monoid $\mathcal{F}(G)$ with multiplicative notation.

We call a sequence $S = g_1 g_2 \ldots g_k \in \mathcal{F}(G)$ to be a *zero-sum sequence* if $g_1 + g_2 + \cdots + g_k = 0$ where 0 is the identity element of G.

For an abelian group G, the *Davenport constant* $D(G)$ is defined to be the least positive integer ℓ such that if we take any sequence of length ℓ from G, there is a

S. D. Adhikari (✉)
(Formerly at Harish-Chandra Research Institute) Department of Mathematics, Ramakrishna Mission Vivekananda Educational and Research Institute, Belur 711202, India
e-mail: adhikari@hri.res.in

B. Roy · S. Sarkar
Harish-Chandra Research Institute, HBNI, Jhunsi, Allahabad, India
e-mail: bidisharoy@hri.res.in

S. Sarkar
e-mail: subhasarkar@hri.res.in

M. B. Nathanson (ed.), *Combinatorial and Additive Number Theory III*, Springer Proceedings in Mathematics & Statistics 297,
https://doi.org/10.1007/978-3-030-31106-3_1

non-empty zero-sum subsequence. The early motivation for the study of this constant [33] was factorization in algebraic number fields. Later this constant found important roles in graph theory (see for instance, [13] or [19]) and in the proof of the infinitude of Carmichael numbers by Alford et al. [10].

Given a finite abelian group $G = (\mathbb{Z}/n_1\mathbb{Z}) \times (\mathbb{Z}/n_2\mathbb{Z}) \times \cdots \times (\mathbb{Z}/n_d\mathbb{Z})$ with $n_1|n_2|\cdots|n_d$, writing $M(G) = 1 + \sum_{i=1}^{d}(n_i - 1)$, it is trivial to see that $M(G) \leq D(G) \leq |G|$. The equality $D(G) = |G|$ holds if and only if $G = \mathbb{Z}/n\mathbb{Z}$, the cyclic group of order n. Olson [30, 31] proved that $D(G) = M(G)$ for all finite abelian groups of rank 2 and for all p-groups. It is also known that $D(G) > M(G)$ for infinitely many finite abelian groups of rank $d > 3$ (see [21], for instance).

The best known bound is due to van Emde Boas and Kruyswijk [12] who proved that

$$D(G) \leq n\left(1 + \log\frac{|G|}{n}\right), \tag{1}$$

where n is the exponent of G. This was again proved by Alford et al. [10].

We state the following conjectures:

1. We have $D(G) = M(G)$ for all G with rank $d = 3$ or $G = (\mathbb{Z}/n\mathbb{Z})^d$ [20] and [18].
2. For $G = (\mathbb{Z}/n_1\mathbb{Z}) \times (\mathbb{Z}/n_2\mathbb{Z}) \times \cdots \times (\mathbb{Z}/n_d\mathbb{Z})$ with $n_1|n_2|\cdots|n_d$, $D(G) \leq \sum_{i=1}^{d} n_i$ [28].

For an abelian group G with $\exp(G) = n$, the *Erdős–Ginzburg–Ziv constant* $\mathsf{s}(G)$ is defined to be the least positive integer ℓ such that if we take any sequence of length ℓ over G there is a zero-sum subsequence of length n.

The name Erdős–Ginzburg–Ziv constant is after the prototype of zero-sum result [17] by Erdős, Ginzburg and Ziv, where it was proved that $\mathsf{s}(\mathbb{Z}/n\mathbb{Z}) \leq 2n - 1$. The example of the sequence $(\underbrace{0, 0, \ldots, 0}_{n-1}, \underbrace{1, 1, \ldots, 1}_{n-1})$ of length $(2n - 2)$ having no zero-sum subsequence of length n, establishes that $\mathsf{s}(\mathbb{Z}/n\mathbb{Z}) = 2n - 1$.

For the group $G = (\mathbb{Z}/n\mathbb{Z})^2$, Kemnitz [27] had conjectured that $\mathsf{s}(G) = 4n - 3$. In 2000, Rónyai [34] came very close to it by proving that $\mathsf{s}((\mathbb{Z}/p\mathbb{Z})^2) \leq 4p - 2$, for a prime p and finally the conjecture was confirmed by Reiher [32] in 2007.

Till now the exact value of the constant $\mathsf{s}(G)$ where $G = (\mathbb{Z}/n\mathbb{Z})^d$ and $d \geq 3$ is unknown. For all odd integers n, Elsholtz [16] proved a lower bound as

$$\mathsf{s}((\mathbb{Z}/n\mathbb{Z})^d) \geq (1 \cdot 125)^{\lfloor \frac{d}{3} \rfloor}(n - 1)2^d + 1.$$

In the other direction, Alon and Dubiner [11] proved that there is an absolute constant $c > 0$ so that for all n,

$$\mathsf{s}((\mathbb{Z}/n\mathbb{Z})^d) \leq (cd\log_2 d)^d n.$$

For further readings in this direction we refer to the following articles [8, 11, 13, 15, 19].

The present article is related to a particular weighted generalization of the above zero-sum constants, first considered (see the early papers [3, 4, 9, 35]) about twelve years back, which became popular (see [22, 23, 25, 36–38]) and the results here have found some applications (see [24, 26]) as well.

Let G be a finite abelian group of exponent n and $A \subset [1, n-1]$. We call a sequence $S = g_1 g_2 \dots g_k \in \mathcal{F}(G)$ to be an *A-weighted zero-sum sequence* if there exist $a_1, \dots, a_k$ in A such that

$$\sum_{i=1}^{k} a_i g_i = 0.$$

The *Davenport constant of G with weight A*, denoted by $D_A(G)$, is then defined to be the least positive integer ℓ such that any sequence $(x_1, x_2, \dots, x_\ell)$ over G of length ℓ, has a non-empty A-weighted zero-sum subsequence.

Notice that when $A = \{1\}$, we get the classical Davenport constant $D(G)$.

Similarly, for a finite abelian group G of exponent n and a non-empty subset A of $[1, n-1]$, one defines $\mathsf{s}_A(G)$ (as introduced in [2]; the notation used here being the standard one at present) to be the least integer k such that any sequence S of length k of elements in G has an A-weighted zero-sum subsequence of length n. Once again, taking $A = \{1\}$, one recovers the classical Erdős–Ginzburg–Ziv constant $\mathsf{s}(G)$.

For $G = (\mathbb{Z}/n\mathbb{Z})$, exact values and good bounds of $D_A(G)$ and $\mathsf{s}_A(G)$ are known for several subsets $A \subset [1, n-1]$ of weights (see for instance [1, 2, 4, 9, 14]).

When $A = \{\pm 1\}$, it was proved in [4] that $\mathsf{s}_{\{\pm 1\}}(\mathbb{Z}/n\mathbb{Z}) = n + \lfloor \log_2 n \rfloor$ and for an odd integer n, it was proved in [2] that $\mathsf{s}_{\{\pm 1\}}((\mathbb{Z}/n\mathbb{Z})^2) = 2n - 1$.

When $\exp(G)$ is even, the following asymptotic behavior of $\mathsf{s}_{\{\pm 1\}}(G)$ was established in [5]:

For finite abelian groups of even exponent and fixed rank, we have

$$\mathsf{s}_{\{\pm 1\}}(G) = \exp(G) + \log_2 |G| + O(\log_2 \log_2 |G|) \quad \text{as} \quad \exp(G) \to \infty.$$

In Sect. 2, we take up the first open case of odd exponent and discuss the problem of determining the exact value of the constant $\mathsf{s}_{\{\pm 1\}}((\mathbb{Z}/p\mathbb{Z})^3)$.

The combinatorial constant $\mathsf{s}_A(G)$ can be further generalized as follows. For any integer $m \geq 1$, the constant $\mathsf{s}_{mn,A}(G)$ is the least positive integer ℓ such that any sequence S over G of length ℓ has an A-weighted zero-sum subsequence of length mn. When $m = 1$, we get $\mathsf{s}_{n,A}(G) = \mathsf{s}_A(G)$.

Recently, Adhikari and Mazumdar [6] considered the rank 3 case and by a method of Rónyai in [34], they proved the following result.

Theorem 1 *For an odd prime p, we have $\mathsf{s}_{3p,\{\pm 1\}}((\mathbb{Z}/p\mathbb{Z})^3) \leq \frac{9p-3}{2}$.*

In another paper [7] the above authors proved the following result for elementary abelian p-groups of even rank.

Theorem 2 *Let p be an odd prime and let $k \geq 3$ be a divisor of $p-1$. Let θ be an element of order k of the multiplicative group $(\mathbb{Z}/p\mathbb{Z})^*$ and A be the subgroup of $(\mathbb{Z}/p\mathbb{Z})^*$ generated by θ. Then, we have*

$$\mathsf{s}_{3p,A}((\mathbb{Z}/p\mathbb{Z})^{2k}) \leq 5p - 2.$$

Here in Sect. 3, we prove a result similar to Theorem 2 for elementary abelian p-groups of odd rank. More precisely, we shall prove the following.

Theorem 3 *Let p be an odd prime and let $k \geq 2$ be an even integer which divides $p-1$. Let θ be an element of order k of the multiplicative group $(\mathbb{Z}/p\mathbb{Z})^*$ and A be the subgroup of $(\mathbb{Z}/p\mathbb{Z})^*$ generated by θ. Then, we have*

$$\mathsf{s}_{3p,A}((\mathbb{Z}/p\mathbb{Z})^{k+1}) \leq 4p + \frac{p-1}{k} - 1.$$

Remark 1 If $(e_1, e_2, \ldots, e_{k+1})$ be a basis of $(\mathbb{Z}/p\mathbb{Z})^{k+1}$, then observing that the sequence

$$\mathbf{0}^{3p-1}\prod_{i=1}^{k+1} e_i$$

has no A-weighted zero-sum subsequence of length $3p$, we see that

$$\mathsf{s}_{3p,A}((\mathbb{Z}/p\mathbb{Z})^{k+1}) \geq 3p + k + 1.$$

So, for general k, there is a gap between this lower bound and the upper bound given by the theorem above. If $k = p-1$, then the lower bound obtained above is $3p + k + 1 = 4p$ and the upper bound obtained in the theorem is $4p + \frac{p-1}{k} - 1 = 4p$.

Remark 2 Since p is an odd prime, $2 \mid (p-1)$ and by putting $k = 2$ in Theorem 3, one obtains Theorem 1.

2 The Group $(\mathbb{Z}/p\mathbb{Z})^3$

Before taking up the particular group $(\mathbb{Z}/p\mathbb{Z})^3$, we determine the weighted Davenport constant $D_{\{\pm 1\}}((\mathbb{Z}/n\mathbb{Z})^d)$, where n and d are positive integers.

We begin by the following Lemma.

Lemma 1 *For given positive integers n and d, consider the group $G = (\mathbb{Z}/n\mathbb{Z})^d$ and let $(y_1, \ldots, y_\ell)$ be a sequence over G of length $\ell > d \log_2 n$.*

Then there exists a non-empty $J \subset [1, \ell]$ and $\epsilon_j \in \{\pm 1\}$ for each $j \in J$ such that

$$\sum_{j \in J} \epsilon_j y_j = 0.$$

Proof Consider the sequence of 2^ℓ elements $\left(\sum_{j\in I} y_j\right)_{I\subset[1,\ell]}$ over G. Since $2^\ell > n^d$, there exist $I_1, I_2 \subset [1,\ell]$ with $I_1 \neq I_2$ such that

$$\sum_{j\in I_1} y_j = \sum_{j\in I_2} y_j.$$

Set $J = I_1 \cup I_2 \setminus I_1 \cap I_2$ and

$$\epsilon_j = \begin{cases} 1 & \text{when } j \in I_1 \cap J; \\ -1 & \text{when } j \in I_2 \cap J. \end{cases}$$

Then it is clear that J is non-empty and $\sum_{j\in J} \epsilon_j y_j = 0$. □

Lemma 2 *For given positive integers n and d, there exists a sequence of length $d\lfloor \log_2 n\rfloor$ over $G = (\mathbb{Z}/n\mathbb{Z})^d$ such that it has no non-empty $\{\pm 1\}$-weighted zero-sum subsequence. That is, $D_{\{\pm 1\}}((\mathbb{Z}/n\mathbb{Z})^d) \geq d\lfloor \log_2 n\rfloor + 1$.*

Proof Let us define $r \in \mathbb{N}$ by $2^{r+1} \leq n < 2^{r+2}$ and consider the following sequence of length $d(r+1)$ over G.

$$(1,0,0,\ldots,0), (2,0,0,\ldots,0), (2^2,0,0,\ldots,0), \ldots, (2^r,0,0,\ldots,0),$$

$$(0,1,0,\ldots,0), (0,2,0,\ldots,0), (0,2^2,0,\ldots,0), \ldots, (0,2^r,0,\ldots,0),$$

$$(0,0,1,\ldots,0), (0,0,2,\ldots,0), (0,0,2^2,\ldots,0), \ldots, (0,0,2^r,\ldots,0),$$

$$\cdots\cdots\cdots\cdots\cdots\cdots$$

$$(0,0,0,\ldots,1), (0,0,0,\ldots,2), (0,0,0,\ldots,2^2), \ldots, (0,0,0,\ldots,2^r).$$

This sequence has $(r+1)d = d\lfloor \log_2 n\rfloor$ elements and a $\{\pm 1\}$-weighted sum of any non-empty subsequence of it gives rise to an element whose absolute value in each co-ordinate is $\leq 2^{r+1} - 1$ and hence is not the zero element of G by the uniqueness of binary representation of a number. □

Theorem 4 *For given positive integers n and d,*

$$D_{\{\pm 1\}}((\mathbb{Z}/n\mathbb{Z})^d) = d\lfloor \log_2 n\rfloor + 1.$$

Proof The result follows from Lemmas 1 and 2. □

Lemma 3 ([6]) *Let p be an odd prime and G be the group $(\mathbb{Z}/p\mathbb{Z})^3$. Then there exists a sequence of length $4p - 4$ such that there is no $\{\pm 1\}$-weighted zero-sum subsequence of length p. In other words $\mathsf{s}_{\{\pm 1\}}(G) \geq 4p - 3$.*

Remark 3 Let p be an odd prime. From Theorem 4, $D_{\{\pm 1\}}((\mathbb{Z}/p\mathbb{Z})^3) = 3\lfloor \log_2 p \rfloor + 1$. For $p > 3\lfloor \log_2 p \rfloor$, given a sequence S of length p over $(\mathbb{Z}/p\mathbb{Z})^3$, if T is a maximal $\{\pm 1\}$-weighted zero-sum subsequence of S, then $|S| - |T| \leq 3\lfloor \log_2 p \rfloor$. So, given a sequence over $(\mathbb{Z}/p\mathbb{Z})^3$, of length $p + 3\lfloor \log_2 p \rfloor$ with $3\lfloor \log_2 p \rfloor$ zeros, there must be a $\{\pm 1\}$-weighted zero-sum subsequence of length p.

We expect that $\mathsf{s}_{\{\pm 1\}}(G) = 4p - 3$; that is, given a sequence of length $4p - 3$ over $(\mathbb{Z}/p\mathbb{Z})^3$, there is a zero-sum subsequence of length p. We observe the following conditional result.

Theorem 5 *Let p be an odd prime and $G = (\mathbb{Z}/p\mathbb{Z})^3$. Then, any sequence $S = x_1 \cdots x_{4p-3}$ of length $4p - 3$ over G with at least five zero-elements has a subsequence $x_{i_1}, \ldots, x_{i_p}$ of length p such that*

$$\omega_1 x_{i_1} + \cdots + \omega_p x_{i_p} = 0$$

in G where $\omega_i \in \{\pm 1\}$.

Proof Let $z_1, \ldots, z_5$ be 5 zero-elements in the sequence. We denote the remaining $4p - 8$ elements by $g_1, \ldots g_{4p-8}$.

If $p \leq 5$, then we trivially get a required zero-sum subsequence of length p with the z_i's. So, we assume that $p > 5$.

Key step. Consider the $3p$ elements $g_1, \ldots, g_{3p-3}, z_1, z_2, z_3$ (which is possible, since for $p > 5$, $4p - 8 > 3p - 3$) and rewrite them as

$$a_1, b_1, c_1, \ldots, a_p, b_p, c_p,$$

where a_p, b_p, c_p are the elements z_1, z_2, z_3.

If the sums $a_i + b_j + c_k$ corresponding to the distinct triples (i, j, k) are all distinct, they give us all the p^3 elements of the group G. In that case, adding the subsequence of the remaining elements of length $p - 3$ to these three-element sequences, we get subsequences of length p whose sums will run over all the elements of the group G and hence giving a zero-sum subsequence of length p.

If the sums $a_i + b_j + c_k$ are not all distinct, two 3-sums will be the same and they will produce a non-empty $\{\pm 1\}$-weighted zero-sum subsequence not involving z_1, z_2, z_3. (For instance, if $a_1 + b_1 + c_p = a_2 + b_3 + c_4$, we have $a_1 + b_1 - a_2 - b_3 - c_4 = 0$ as $c_p = z_3 = 0$.) We denote it by T_1 and observe that $1 \leq |T_1| \leq 6$.

Next, we remove elements of this sequence T_1 from the sequence $g_1 \ldots g_{3p-3}$ and replace them by the same number of elements from $g_{3p-2}, \ldots, g_{4p-8}$ (which are $p - 5$ in number).

After that we repeat the above mentioned "key-step" and stop when we reach the stage when $p \geq |T_1 \cup T_2 \cup \cdots \cup T_r| \geq p - 5$.

Thus we adjoin some elements from $z_1, \ldots, z_5$ with $T_1 \cup T_2 \cup \cdots \cup T_r$ to get a p-length $\{\pm 1\}$-weighted zero-sum subsequence of S. □

Remark 4 As is easy to observe, the proof of the above theorem goes through if, instead of five zero-elements, the sequence $S = x_1 \cdots x_{4p-3}$ has three zero-elements and a pair of elements x_i, x_j such that either $x_i = x_j$ or $x_i = -x_j$.

Also, we can generalize Theorem 5 as follows:

Let $d \geq 3$ be an integer and $p \geq d\lfloor \log_2 n \rfloor$ be any prime. (We note that for odd d, the counter example leading to Lemma 3 can be modified to give $\mathsf{s}_{\{\pm 1\}}((\mathbb{Z}/p\mathbb{Z})^d) \geq (d+1)p - d$).

Then any sequence $S = a_1 \cdots a_{(d+1)p-d}$ of length $(d+1)p - d$ over $G = (\mathbb{Z}/p\mathbb{Z})^d$ with at least $2d - 1$ zero-elements contains a subsequence $a_{i_1} \ldots a_{i_p}$ of length p such that

$$\epsilon_1 a_{i_1} + \cdots + \epsilon_p a_{i_p} = 0$$

in G where $\epsilon_i \in \{\pm 1\}$.

3 Proof of Theorem 3

We start with some lemmas.

The following lemma has been proved in [7]; we record it here.

Lemma 4 *Let p be an odd prime and let k be a divisor of $p - 1$. Let θ be an element of order k of $(\mathbb{Z}/p\mathbb{Z})^*$ and $D = \{0, \theta, \theta^2, \ldots, \theta^k\}$. For a positive integer m, let us consider the vector space*

$$\mathcal{C} = \left\{ \text{functions } f : D^m \to (\mathbb{Z}/p\mathbb{Z}) \right\}$$

over the field $\mathbb{Z}/p\mathbb{Z}$. Then the monomials $\prod_{1 \leq i \leq m} x_i^{r_i}$, $r_i \in [0, k]$ constitute a basis of $\mathcal{C}$ over $\mathbb{Z}/p\mathbb{Z}$.

Now, we state the well known theorem of Chevalley-Warning (see for instance, [29]).

Theorem 6 *Let p be a prime number and F a finite field of characteristic p. For $i = 1, 2, \ldots, m$, let $f_i \in F[x_1, x_2, \ldots, x_n]$ be a non-zero polynomial of degree d_i in n-variables over the field F. Let N denote the number of n-tuples $(x_1, x_2, \ldots, x_n)$ of elements of F such that*

$$f_i(x_1, x_2, \ldots, x_n) = 0,$$

for all $i = 1, 2, \ldots, m$. If $d_1 + d_2 + \cdots + d_m < n$, then

$$N \equiv 0 \pmod{p}.$$

In particular, if $N \geq 1$, then there is a non-zero simultaneous solution over F.

Lemma 5 *Let p be an odd prime and let $k \geq 2$ be an even integer which divides $p-1$. Let $A = \{\theta, \theta^2, \ldots, \theta^k = 1\}$ be the subgroup of $(\mathbb{Z}/p\mathbb{Z})^*$ generated by θ which is of order k. Let $S = \prod_{i=1}^{t} w_i \in \mathcal{F}((\mathbb{Z}/p\mathbb{Z})^{k+1})$ be a sequence of length $t = 2p + \dfrac{p-1}{k} - 1$. Then S has an A-weighted zero-sum subsequence of length either p or $2p$.*

Proof For all integers $i = 1, 2, \ldots, t$, we let $w_i = (a_{i1}, a_{i2}, \ldots, a_{i(k+1)}) \in (\mathbb{Z}/p\mathbb{Z})^{k+1}$. We shall consider the following system of equations over $\mathbb{Z}/p\mathbb{Z}$.

$$\sum_{i=1}^{t} a_{i1} x_i^{\frac{p-1}{k}} = 0, \qquad \sum_{i=1}^{t} a_{i2} x_i^{\frac{p-1}{k}} = 0, \ldots, \sum_{i=1}^{t} a_{i(k+1)} x_i^{\frac{p-1}{k}} = 0$$

and

$$\sum_{i=1}^{t} x_i^{p-1} = 0.$$

Note that the sum of the degrees of the polynomials is $(k+1)\frac{p-1}{k} + (p-1) = 2p + \frac{p-1}{k} - 2 < 2p + \frac{p-1}{k} - 1 = t$, the number of variables.

Since the above system has the trivial zero solution, by Theorem 6, there exists a non-zero solution $(y_1, y_2, \ldots, y_t) \in (\mathbb{Z}/p\mathbb{Z})^t$ of the above system.

If we write $I = \{i : y_i \not\equiv 0 \pmod{p}\}$, then from the first $(k+1)$ equations, we get

$$\sum_{i \in I} y_i^{(p-1)/k} (a_{i1}, a_{i2}, \ldots, a_{(k+1)i}) = (0, 0, \ldots, 0)$$

and from the last equation, we get $|I| \equiv 0 \pmod{p}$. Since $y_i \not\equiv 0 \pmod{p}$ for all $i \in I$, we see that $y_i^{(p-1)/k} \in A$. Since $t < 3p$, we get either $|I| = p$ or $|I| = 2p$. Hence, we conclude that the sequence S has an A-weighted zero-sum subsequence of length either p or $2p$. □

Corollary 1 *Let p be an odd prime and let $k \geq 2$ be an even integer which divides $p-1$. Let $A = \{\theta, \theta^2, \ldots, \theta^k = 1\}$ be the subgroup of $(\mathbb{Z}/p\mathbb{Z})^*$ generated by θ which is of order k. Let $S = \prod_{i=1}^{t} w_i \in \mathcal{F}((\mathbb{Z}/p\mathbb{Z})^{k+1})$ be a sequence of length $t = 3p + \dfrac{p-1}{k} - 1$. Then S has an A-weighted zero-sum subsequence of length $2p$.*

Proof Since the given sequence S is of length $t = 3p + \frac{p-1}{k} - 1$ over $(\mathbb{Z}/p\mathbb{Z})^{k+1}$, it has an A-weighted zero-sum subsequence T of length either p or $2p$ by Lemma 5. If T is of length $2p$, then we are done. Otherwise, consider the deleted sequence ST^{-1} which is of length

$$3p + \frac{p-1}{k} - 1 - p = 2p + \frac{p-1}{k} - 1$$

and hence, by Lemma 5, we get ST^{-1} has an A-weighted zero-sum subsequence T_1 of length either p or $2p$. If $|T_1| = 2p$, then we are done. If $|T_1| = p$, then TT_1 is of length $2p$ and it is the required subsequence. □

Corollary 2 *Let p be an odd prime and let $k \geq 2$ be an even integer which divides $p-1$. Let $A = \{\theta, \theta^2, \ldots, \theta^k = 1\}$ be the subgroup of $(\mathbb{Z}/p\mathbb{Z})^*$ generated by θ which is of order k. Let $S = \prod_{i=1}^{t} w_i \in \mathcal{F}((\mathbb{Z}/p\mathbb{Z})^{k+1})$ be a sequence of length $t = 4p + \dfrac{p-1}{k} - 1$. If S has an A-weighted zero-sum subsequence of length p, then it has an A-weighted zero-sum subsequence of length $3p$.*

Proof Since S has an A-weighted zero-sum subsequence T of length p, consider the deleted sequence ST^{-1} which is of length $3p + \frac{p-1}{k} - 1$. Therefore, by Corollary 1, we get an A-weighted zero-sum subsequence T_1 of ST^{-1} of length $2p$. Hence, TT_1 is the required zero-sum subsequence. □

Proof of Theorem 3 For an odd prime p and an even integer $k \geq 2$ such that k divides $p-1$, θ is an element of order k of the multiplicative group $(\mathbb{Z}/p\mathbb{Z})^*$ and A is the subgroup of $(\mathbb{Z}/p\mathbb{Z})^*$ generated by θ. We have to show that

$$\mathsf{s}_{3p,A}((\mathbb{Z}/p\mathbb{Z})^{k+1}) \leq 4p + \frac{p-1}{k} - 1$$

Note that D as defined in Lemma 4 is $A \cup \{0\}$.

Let $S = \prod_{i=1}^{m} w_i \in \mathcal{F}((\mathbb{Z}/p\mathbb{Z})^{k+1})$ be a sequence of length $m = 4p + \frac{p-1}{k} - 1$. For all $i = 1, 2, \ldots, m$, we let $w_i = (a_{i1}, a_{i2}, \ldots, a_{i(k+1)}) \in (\mathbb{Z}/p\mathbb{Z})^{k+1}$. We shall prove that S has an A-weighted zero-sum subsequence of length $3p$.

If possible, suppose that the assertion is false. That is, S has no A-weighted zero-sum subsequence of length $3p$. Therefore, by Corollary 2, S cannot have any A-weighted zero-sum subsequence of length p. Thus, if $T = \prod_{j=1}^{\ell} w_{i_j}$ is a subsequence of S of length $\ell = 3p$ or p, then for any $(z_1, \ldots, z_\ell) \in A^\ell$, we have

$$z_1 w_{i_1} + \cdots + z_\ell w_{i_\ell} \not\equiv (0, 0, \ldots, 0) \pmod{p}. \tag{2}$$

In order to get a contradiction, we need to invoke Lemma 4. For this purpose, we shall introduce some polynomials as follows. Let

$$\sigma(x_1, x_2, \ldots, x_m) = \sum_{\substack{I \subset [1,m], \\ |I| = p}} \prod_{i \in I} x_i^k,$$

be the p-th elementary symmetric polynomial of the variables $x_1^k, x_2^k, \ldots, x_m^k$. We also consider the following polynomials,

$$P_1(x_1, x_2, \ldots, x_m) = \left(\left(\sum_{i=1}^{m} a_{i1}x_i\right)^{p-1} - 1\right)\left(\left(\sum_{i=1}^{m} a_{i2}x_i\right)^{p-1} - 1\right)\ldots\left(\left(\sum_{i=1}^{m} a_{i(k+1)}x_i\right)^{p-1} - 1\right),$$

$$P_2(x_1, x_2, \ldots, x_m) = \left(\left(\sum_{i=1}^{m} x_i^k\right)^{p-1} - 1\right),$$

$$P_3(x_1, x_2, \ldots, x_m) = (\sigma(x_1, x_2, \ldots, x_m) - 2)(\sigma(x_1, x_2, \ldots, x_m) - 4)$$

and

$$P(x_1, x_2, \ldots, x_m) = P_1(x_1, x_2, \ldots, x_m)P_2(x_1, x_2, \ldots, x_m)P_3(x_1, x_2, \ldots, x_m).$$

First, we note that

$$\deg(P) \leq (k+1)(p-1) + k(p-1) + 2kp = 4kp + p - 1 - 2k. \tag{3}$$

Claim $P(\alpha_1, \ldots, \alpha_m) = 0$ *for all* $(\alpha_1, \ldots, \alpha_m) \in D^m \backslash \{(0, 0, \ldots, 0)\}$ *and* $P(0, 0, \ldots, 0) = 8$.

Let $\alpha = (\alpha_1, \ldots, \alpha_m) \in D^m \backslash \{(0, 0, \ldots, 0)\}$ be an arbitrary element.

If the number of non-zero entries of α is not a multiple of p and if we take $I = \{1 \leq i \leq m : \alpha_i \neq 0\}$, then

$$\left(\left(\sum_{i=1}^{m} \alpha_i^k\right)^{p-1} - 1\right) = \left(\left(\sum_{i \in I} \alpha_i^k\right)^{p-1} - 1\right) = 0$$

by Fermat's Little Theorem and hence we get $P_2(\alpha_1, \ldots, \alpha_m) = 0$.

If the number of non-zero entries of α is either p or $3p$, then by (2), we get $P_1(\alpha_1, \ldots, \alpha_m) = 0$.

If the number of non-zero entries of α is $2p$, then $\sigma(\alpha) = \binom{2p}{p} = 2 \in \mathbb{Z}/p\mathbb{Z}$ and if the number of non-zero entries of α is $4p$, then $\sigma(\alpha) = \binom{4p}{p} = 4 \in \mathbb{Z}/p\mathbb{Z}$. Therefore, if the number of non-zero entries of α is either $2p$ or $4p$, then $P_3(\alpha_1, \ldots, \alpha_m) = 0$. Therefore the polynomial $P(x_1, x_2, \ldots, x_m)$ vanishes at all the points of D^m, except at $(0, 0, \ldots, 0)$ and $P(0, 0, \ldots, 0) = 8$, as $(k+1)$ is odd. This proves the claim.

Consider the function $P : D^m \to \mathbb{Z}/p\mathbb{Z}$ in $\mathcal{C}$ given by the polynomial $P(\alpha_1, \ldots, \alpha_m)$.

Now, let $R = 8(1 - x_1^k)(1 - x_2^k)\ldots(1 - x_m^k) \in (\mathbb{Z}/p\mathbb{Z})[x_1, \ldots, x_m]$. Then $R(\alpha_1, \ldots, \alpha_m) = 0$ for all $\alpha = (\alpha_1, \ldots, \alpha_m) \in D^m \backslash \{(0, 0, \ldots, 0)\}$ and $R(0, \ldots, 0) = 8$.

Therefore, $P(x_1, \ldots, x_m)$ and $R(x_1, \ldots, x_m)$ are equal as elements in $\mathcal{C}$.

By Lemma 4, we know that $\mathcal{C}$ has a special basis consisting of monomials of the form $\prod_{1 \leq i \leq m} x_i^{r_i}$, $r_i \in [0, k]$. Now, we write P as a linear combination of these basis elements by replacing each x_i^{tk+r} for some integers $t \geq 1$ and $r \in [1, k]$ by x_i^r and let Q be the polynomial obtained in this way. Also, in this process, the degree of the polynomial Q is not increased. Hence by (3) we get, $\deg Q \leq 4kp + p - 1 - 2k$. Clearly, as elements in $\mathcal{C}$, P and Q are the same. Hence, Q and R are the same as elements in $\mathcal{C}$.

However, $\deg R = mk = 4kp + p - 1 - k > 4kp + p - 1 - 2k \geq \deg Q$.

This will lead to a nontrivial relation among the basis elements consisting of the monomials $\prod_{1 \leq i \leq m} x_i^{r_i}$, which is impossible. □

Acknowledgements The second and the third authors would like to thank Prof. R. Thangadurai for some discussions around the problems.

References

1. S. D. Adhikari, A. A. Ambily and B. Sury, *Zero-sum problems with subgroup weights*, Proc. Indian Acad. Sci. (Math. Sci.), **120 (3)** (2010), 259-266.
2. S. D. Adhikari, R. Balasubramanian, F. Pappalardi and P. Rath, *Some zero-sum constants with weights*, Proc. Indian Acad. Sci. (Math. Sci.), **118 (2)** (2008), 183-188.
3. S.D. Adhikari, Y.G. Chen, *Davenport constant with weights and some related questions II*, J. Combin. Theory A **115**(1) (2008), 178-184.
4. S.D. Adhikari, Y.G. Chen, J.B. Friedlander, S.V. Konyagin, F. Pappalardi, *Contribution to zero-sum problems*, Discrete Math., **306** (2006), 1-10.
5. Sukumar Das Adhikari, David J. Grynkiewicz and Zhi-Wei Sun, On Weighted Zero-Sum Sequences, *Advances in Applied Mathematics* **48** (2012), 506–527.
6. S.D. Adhikari, E. Mazumdar, *Modification of some methods in the study of Zero-sum constants*, Integers **14** (2014), A25.
7. S.D. Adhikari, E. Mazumdar, *The polynomial method in the study of zero-sum theorems*, Int. J. Number Theory **11** (5) (2015), 1451-1461.
8. S. D. Adhikari and P. Rath, *Zero-sum problems in combinatorial number theory*, Ramanujan Math. Soc. Lect. Notes Ser., 2, Ramanujan Math. Soc., Mysore (2006), 1–14.
9. S.D. Adhikari, P. Rath, *Davenport constant with weights and some related questions*, Integers **6** (2006), A30.
10. W. R. Alford, A. Granville and C. Pomerance, *There are infinitely many Carmichael numbers*, Annals of Math., **139** (2) (1994), no. 3, 703-722.
11. N. Alon, M. Dubiner, *A lattice point problem and additive number theory*, Combinatorica, **15** (1995), 301-309.
12. P. van Emde Boas and D. Kruswjik, *A combinatorial problem on finite abelian group III*, Z. W.1969-008 (Math. Centrum, Amsterdam).
13. Y. Caro, *Zero-sum problems, a survey*, Discrete Math. **152** (1996), 93-113.
14. M. N. Chintamani, and B. K. Moriya, *Generalizations of some zero sum theorems*, Proc. Indian Acad. Sci. (Math. Sci.), **122 (1)** (2012), 15–21.
15. Y. Edel, C. Elsholtz, A. Geroldinger, S. Kubertin, L. Rackham, *Zero-sum problems in finite abelian groups and affine caps*, Quart. J. Math. **58** (2007), 159-186.
16. C. Elsholtz, *Lower bounds for multidimensional zero sums*, Combinatorica, **24** (3) (2004), 351-358.
17. P. Erdős, A. Ginzburg, A. Ziv, *Theorem in additive number theory*, Bulletin Research Council Israel **10F** (1961), 41-43.
18. W.D. Gao, A. Geroldinger, *Zero-sum problems and coverings by proper cosets*, European J. Combinatorics **24** (2003), 531-549.

19. W. Gao, A. Geroldinger, *Zero-sum problems in finite abelian groups: a survey*, Expo. Math. **24** (2006), 337-369.
20. W.D. Gao, *On Davenport's constant of finite abelian groups with rank three*, Discrete Math. **222**, no. 1–3, (2000) 111-124.
21. A. Geroldinger and R. Schneider, *On Davenport's constant*, J. Combin. Theory, Ser. A, **61**, no. 1, (1992), 147-152.
22. Simon Griffiths, *The Erdős-Ginzburg-Ziv theorem with units*, Discrete Math., **308**, No. 23, (2008), 5473–5484.
23. D.J. Grynkiewicz, L.E. Marchan, O. Ordaz, *A weighted generalization of two theorems of Gao*. Ramanujan J. **28** (2012), no. 3, 323–340.
24. F. Halter-Koch, *Arithmetical interpretation of weighted Davenport constants*, Arch. Math. (Basel) **103** (2014), no. 2, 125–131.
25. Florian Luca, *A generalization of a classical zero-sum problem*, Discrete Math., 307, No. 13, (2007), 1672-1678.
26. L.E. Marchan, O. Ordaz, I. Santos, W.A. Schmid, *Multi-wise and constrained fully weighted Davenport constants and interactions with coding theory*, J. Combin. Theory Ser. A **135** (2015), 237-267.
27. A. Kemnitz, *On a lattice point problem*, Ars Combin., **16b** (1993), 151-160.
28. W. Narkiewicz, J. Śliwa, *Finite abelian groups and factorization problems - II*, Colloq. Math. **46**, (1982), 115-122.
29. Melvyn B. Nathanson *Additive Number Theory: Inverse Problems and the Geometry of Sumsets*, Springer, 1996.
30. J.E. Olson, *A combinatorial problem in finite abelian groups, I*, J. Number Theory **1**, (1969), 8-10.
31. J.E. Olson, *A combinatorial problem in finite abelian groups, II*, J. Number Theory **1**, (1969), 195-199.
32. C. Reiher, *On Kemnitz' conjecture concerning lattice points in the plane*, Ramanujan J. **13** (2007), 333-337.
33. K. Rogers, *A Combinatorial problem in Abelian groups*, Proc. Cambridge Phil. Soc. **59**, (1963), 559-562.
34. L. Rónyai, *On a conjecture of Kemnitz*, Combinatorica, **20** (2007), 569-573.
35. R. Thangadurai, *A variant of Davenport's constant*, Proc. Indian Acad. Sci. (Math. Sci.), **117**, No. 2, (2007), 147 –158.
36. Xingwu Xia, *Two generalized constants related to zero-sum problems for two special sets*, Integers **7** (2007), A52.
37. Xingwu Xia, Zhigang Li, *Some Davenport constants with weights and Adhikari & Rath's conjecture*, Ars Combin. **88**, (2008), 83-95.
38. P. Yuan, X. Zeng, *Davenport constant with weights*, European Journal of Combinatorics, **31** (2010), 677-680.

Counting Monogenic Cubic Orders

Shabnam Akhtari

Abstract This article is an extension of the author's talk at CANT 2018 conference. In a cubic number field K, we give an absolute upper bound for the number of monogenic orders which have small index compared to the discriminant of O_K, the ring of integers of K. We will also show that a positive proportion of cubic number fields, when ordered by their discriminant, are not monogenic. We will not present any new proofs. We will rather rephrase some of the previous results of the author and collaborators in the language of cubic orders, after giving an overview of the subject. Our main results are stated in Sect. 5.

1 Introduction

A number field K is a finite field extension of $\mathbb{Q}$. In order to generalize the arithmetic of $\mathbb{Q}$ to an algebraic number field K, we might think of a ring $\mathcal{O}$ with the following properties:
(1) K is the quotient field of $\mathcal{O}$.
(2) $\mathcal{O} \cap \mathbb{Q} = \mathbb{Z}$.
(3) The additive group of $\mathcal{O}$ is finitely generated.
A ring in K with these properties is called an order of K.

It is easy to see that for $K \neq \mathbb{Q}$ there are infinitely many orders of K, for instance for every algebraic integer α in K that has degree $[K : \mathbb{Q}]$, the ring $\mathbb{Z}[\alpha]$ is an order. But it is known that there is one maximal order $\mathcal{O}_K$ containing all orders of K. The maximal order $\mathcal{O}_K$ is indeed the familiar ring of integers of K and one of the most important objects in algebraic number theory. The ring of integers of a number field has a number of useful and striking properties.

S. Akhtari (✉)
Department of Mathematics, Fenton Hall, University of Oregon, Eugene, OR 97403-1222, USA
e-mail: akhtari@uoregon.edu

M. B. Nathanson (ed.), *Combinatorial and Additive Number Theory III*,
Springer Proceedings in Mathematics & Statistics 297,
https://doi.org/10.1007/978-3-030-31106-3_2

We note that a significant property of $\mathcal{O}_K$ is that it is integrally closed, however an order $\mathcal{O}$ is not necessarily integrally closed.

A simple way to think of an order is a subring of $\mathcal{O}_K$ which is also a $\mathbb{Z}$-module (a finitely generated additive subgroup of K) of rank $n = [K : Q]$. It is clear from the definition that $\mathbb{Z}$ is the only order in $\mathbb{Q}$. The orders in quadratic number fields are very well understood.

Let K be a quadratic number field. Then there exists a unique square-free $d \in \mathbb{Z}$ such that $K = \mathbb{Q}(\sqrt{d})$. Let

$$\omega = \begin{cases} \frac{1+\sqrt{d}}{2} & \text{if } d \equiv 1 \pmod 4 \\ \sqrt{d} & \text{if } d \equiv 2, 3 \pmod 4. \end{cases}$$

Then $\{1, \omega\}$ is a basis of $\mathcal{O}_K$. It is known that every order in $K = \mathbb{Q}(\sqrt{d})$ has the form $\mathcal{O}_q = \mathbb{Z}[q\omega]$, with a positive rational integer q called the conductor of $\mathcal{O}_q$. As a simple example, consider the quadratic number field $\mathbb{Q}(\sqrt{5})$, with the ring of integer $\mathbb{Z}[\frac{1+\sqrt{5}}{2}]$. We have $\mathbb{Z}[\sqrt{5}] \subset \mathbb{Z}[\frac{1+\sqrt{5}}{2}]$. In [18], the authors consider a similar problem in cubic fields and describe how to find all orders of K with conductor $\mathfrak{q}$, for any cubic number field K and any conductor ideal $\mathfrak{q}$ of K.

For number field K of degree greater than 2 the maximal order $\mathcal{O}_K$ is not always of the form $\mathbb{Z}[\alpha]$. For instance, let $K = \mathbb{Q}(\beta)$ with $\beta^3 + \beta^2 - 2\beta + 8 = 0$. Then $\gamma = \frac{\beta+\beta^2}{2} \in \mathcal{O}_K$ and $1, \beta, \gamma$ is a basis of the module $\mathcal{O}_K$, but there is no $\alpha \in \mathcal{O}_K$ with $\mathcal{O}_K = \mathbb{Z}[\alpha]$ (see, for example, [16] or [23]). This is simply because a quadratic field is uniquely determined by its discriminant, and this is not true for number fields of higher degrees.

Before directing our attention to cubic number fields, let us assume that $\mathcal{O}$ is an order in a number field K of degree n. Since $\mathcal{O}_K$ is also a free $\mathbb{Z}$-module of rank $[K : \mathbb{Q}]$, it follows from the structure theorem for $\mathbb{Z}$-modules that the quotient $\mathcal{O}/\mathcal{O}_K$ is a finite abelian group. The order of this quotient is called the index of the order $\mathcal{O}$ in $\mathcal{O}_K$. Let $\alpha \in \mathcal{O}_K$ be a primitive element of K, that is $K = \mathbb{Q}(\alpha)$. Let $\mathcal{O}_K^+$ and $\mathcal{O}^+$ be the additive groups of the modules $\mathcal{O}_K$ and $\mathcal{O}$, respectively. The index of α is defined by the module index

$$I(\alpha) = (\mathcal{O}_K^+ : \mathbb{Z}[\alpha]^+).$$

We note that $1, \alpha, \ldots, \alpha^{n-1}$ generates an integral basis for $\mathcal{O}_K$ if and only if $I(\alpha) = 1$, and we say α generates a *power integral basis*.

Algebraic integers of index 1 are particularly interesting, as they provide power integral bases for the ring of integers of the number field. In fact, the ring $\mathcal{O}$ is said to be *monogenic* if it is generated by one element as a $\mathbb{Z}$-algebra, i.e., $\mathcal{O} = \mathbb{Z}[\alpha]$ for some $\alpha \in \mathcal{O}$. The element α is then called a *monogenizer*. If α is a monogenizer of $\mathcal{O}$, then so is $\pm\alpha + c$ for any $c \in \mathbb{Z}$. Two monogenizers α and α' are called equivalent if $\alpha' = \pm\alpha + c$ for some $c \in \mathbb{Z}$. It is a deep result of Győry in [15] that any order $\mathcal{O}$ can have at most finitely many monogenizations, which means there are only finitely many equivalence classes of $\alpha \in \mathcal{O}$ such that $\mathcal{O} = \mathbb{Z}[\alpha]$. This naturally raises the question as to how many monogenizations an order $\mathcal{O}$ can have in terms of, say, the degree n of the number field K, or whether such bounds even exist.

In this manuscript, we explore the problem of counting the number of algebraic integers α of a fixed index m in a cubic number field K, which means that the index of the order $\mathbb{Z}[\alpha]$ in $\mathcal{O}_K$ is equal to m. We will relate this problem to counting the number of integral solutions of a family of Diophantine equations, called index form equations. Furthermore, we will see that in a cubic number field K, there is a correspondence between the algebraic integers $\alpha \in \mathcal{O}_K$ with

$$K = \mathbb{Q}(\alpha) \text{ and } I(\alpha) = m,$$

and the number of integral solutions to a Diophantine equation of the shape

$$I(x_2, x_3) = \pm m,$$

where $m \in \mathbb{Z}$ is fixed. Then we will appeal to some known results to give an upper bound for the number of orders of a given index in cubic number fields.

We will show, for every $0 < \epsilon < \frac{1}{8}$, that the ring of integers O_K in a cubic field of discriminant D can have at most $B(\epsilon)$ monogenic subrings of index less than $D^{1/8-\epsilon}$ (see Theorem 2). We will also show that a positive proportion of cubic number fields, when ordered by their discriminant, is not monogenic (see Theorem 5). These statements are consequences of some of previous results of the author and Manjul Bhargava [1, 2]. As a consequence of previous work of Bennett [7] and Okazaki [20], we will see that a cubic order can be monogenized in at most 10 different ways (see Theorems 3 and 4).

The best bounds for the number of monogenizations of an order in a number field of degree greater than 3 can be found in [9]. For a number field K of degree n over $\mathbb{Q}$, let $N_K(B)$ be the number of suborders $\mathcal{O}$ of the ring of integers $\mathcal{O}_K$ of K with $|\text{disc}(\mathcal{O})| \leq B$. In [17], an asymptotic formula is given for $N_K(B)$. In [6] an asymptotic formula is given for the number of cubic orders having bounded discriminant and nontrivial automorphism group. For general treatment of index form equations and their several applications, the reader is referred to [10] and [12].

2 Modules, Lattices and Orders

Modules play an important role in the arithmetic of number fields. By a module $\mathfrak{M}$ in number field K we mean a finitely generated subgroup of K^+, where K^+ is the additive group of the field K. Since K^+ is torsion-free, $\mathfrak{M}$ is a free $\mathbb{Z}$-module of rank $r(\mathfrak{M}) \leq [K : \mathbb{Q}]$. If $r(\mathfrak{M}) = [K : \mathbb{Q}]$, we say $\mathfrak{M}$ is a complete module or a lattice in K. Orders in a number field K arise in a natural way in connection with modules in K. Let $\mathfrak{M}$ be a complete module in K. Then

$$\mathcal{O}(\mathfrak{M}) := \{\alpha \in K : \alpha\mathfrak{M} \subseteq \mathfrak{M}\}$$

is an order. Furthermore, for every order $\mathcal{O}$ in K there exists a complete module $\mathfrak{M}$ with $\mathcal{O} = \mathcal{O}(\mathfrak{M})$; for example, one can take $\mathfrak{M} = \mathcal{O}$.

We note that a lattice in K is an order if it is also a subring (closed under multiplication, and contains 1). For example, for any rational number a, $a\mathbb{Z}$ is a lattice in $\mathbb{Q}$, but the only order in $\mathbb{Q}$ is $\mathbb{Z}$. More generally, a lattice

$$\Gamma = \mathbb{Z}\beta_1 + \cdots + \mathbb{Z}\beta_n$$

is an order if and only if it contains 1, and the rational numbers c_{ij}^l in the expansion

$$\beta_i\beta_j = \sum_{l=1}^{n} c_{ij}^l \beta_l$$

are integers, where $\{\beta_1, \ldots, \beta_n\}$ is a basis for the lattice Γ.

In this manuscript, we consider the problem of counting the number of orders in a cubic number field K with given index. Of course, one can think of an order as a sub-lattice of $\mathcal{O}_K$ and count the number of full sub-lattices. This will provide an upper bound for the number of orders, but there are many sub-lattices that are not closed under multiplication, and therefore we seek a finer way to count the number of orders.

3 The Structure of Index Forms

Let K be an algebraic number field of degree n. The ring of integers $\mathcal{O}_K$ is a finitely-generated $\mathbb{Z}$-module. Indeed, it is a free $\mathbb{Z}$-module, and thus has an integral basis. While the existence and concept of an integral basis is easy enough to understand, computing an integral basis for specific number fields is often very difficult.

First we recall the definition of the discriminant. Let K be a number field of degree n and $\alpha_1, \ldots, \alpha_n$ a linearly independent set of n elements of K. Let $\sigma_1, \ldots, \sigma_n : K \to \mathbb{C}$ be all the embeddings of K. The discriminant of $(\alpha_1, \ldots, \alpha_n)$ is defined as the square of the determinant of an $n \times n$ matrix;

$$D_{K/\mathbb{Q}}(\alpha_1, \ldots, \alpha_n) := \left(\det(\sigma_i\alpha_j)\right)^2,$$

where $i, j \in \{1, \ldots, n\}$.

If $\{\beta_1, \ldots, \beta_n\}$ forms a basis for K, then the discriminant of K is

$$D_K = D_{K/\mathbb{Q}}(\beta_1, \ldots, \beta_n).$$

In [12], one can find a complete account of basic concepts and fundamental facts related to index form equations, as well as elaborated computational methods for several specific number fields. In this section we recall some important statements from [12] that will help us understand the structure of index form equations.

Lemma 1 *Let* $\alpha_1, \ldots, \alpha_n \in \mathcal{O}_K$ *be linearly independent over* $\mathbb{Q}$ *and set*

$$\mathcal{O} = \mathbb{Z}[\alpha_1, \ldots, \alpha_n].$$

then

$$D_{K/\mathbb{Q}}(\alpha_1, \ldots, \alpha_n) = J^2 D_K,$$

where

$$J = (\mathcal{O}_K^+ : \mathcal{O}^+),$$

$\mathcal{O}_K^+$ *and* $\mathcal{O}^+$ *are the additive groups of the modules* $\mathcal{O}_K$ *and* $\mathcal{O}$, *respectively.*

For every $\gamma \in K$, $\gamma^{(i)}$ $(1 \leq i \leq n)$ denote the algebraic conjugates of γ. Let $\{1, \omega_2, \ldots, \omega_n\}$ be an integral basis of K. Let

$$\mathbf{X} = (X_1, \ldots, X_n),$$

and

$$L(\mathbf{X}) = X_1 + \omega_2 X_2 + \cdots + \omega_n X_n, \tag{1}$$

with algebraic conjugates

$$L^{(i)}(\mathbf{X}) = X_1 + \omega_2^{(i)} X_2 + \cdots + \omega_n^{(i)} X_n,$$

$(1 \leq i \leq n)$. Kronecker and Hensel called the form $L(\mathbf{X})$ the *Fundamental form* and

$$D_{K/\mathbb{Q}}(L(\mathbf{X})) = \prod_{1 \leq i < j \leq n} \left(L^{(i)}(\mathbf{X}) - L^{(j)}(\mathbf{X})\right)^2 \tag{2}$$

the *Fundamental discriminante*. The following is Lemma 1.1.2 of [12].

Lemma 2 *We have*

$$D_{K/\mathbb{Q}}(L(\mathbf{X})) = (I(X_1, \ldots, X_n))^2 D_K,$$

where D_K *is the discriminant of field* K, *the linear form* $L(\mathbf{X})$ *and its discriminant are defined in* (1) *and* (2), *and* $I(X_1, \ldots, X_n)$ *is a homogeneous form in* $n-1$ *variables of degree* $\frac{n(n-1)}{2}$ *with integer coefficients.*

The form $I(X_1, \ldots, X_n)$ in the statement of Lemma 2 is called the index form corresponding to the integral basis $\{1, \omega_2, \ldots, \omega_n\}$. A very important property of the index form is that for any primitive algebraic integer (i.e.; $K = \mathbb{Q}(\alpha)$)

$$\alpha = x_1 + x_2\omega_2 + \cdots + x_n\omega_n,$$

we have

$$I(\alpha) = |I(x_2, \ldots, x_n)| .$$

This is a consequence of Lemma 2, as we have

$$\begin{aligned} & (I(x_2, \ldots, x_n))^2 \; D_K \\ &= D_{K/\mathbb{Q}} (L(x_1, \ldots, x_n)) \\ &= D_{K/\mathbb{Q}}(\alpha) \\ &= D_{K/\mathbb{Q}} \left(1, \alpha, \ldots, \alpha^{n-1}\right) \\ &= (I(\alpha))^2 \; D_K . \end{aligned}$$

We conclude that α generates a power integral basis of K if and only if $(x_2, \ldots, x_n) \in \mathbb{Z}^{n-1}$ satisfies the index form equation

$$I(X_2, \ldots, X_n) = \pm 1. \tag{3}$$

Therefore, the problem of determining all power integral bases in K is equivalent to solving the index form Eq. (3). We note that the index form is independent of the variable X_1, for if $\beta = \alpha + a$, where $a \in \mathbb{Z}$, then $I(\alpha) = I(\beta)$.

We will direct our attention to cubic number fields and therefore consider the index form equations of the shape

$$I(x_2, x_3) = \pm m, \tag{4}$$

where $m \in \mathbb{Z}$.

4 The Correspondence of Cubic Forms and Rings

Let $F(x, y) \in \mathbb{Z}[x, y]$ be a binary form and $A = \begin{pmatrix} a & b \\ c & d \end{pmatrix}$, with $a, b, c, d \in \mathbb{Z}$. Define the binary form F_A in the following way.

$$F_A(x, y) := F(ax + by, \; cx + dy).$$

Suppose $A \in \mathrm{GL}_2(\mathbb{Z})$ and that (x_0, y_0) is a solution of $F(x, y) = h$. Then

$$A \begin{pmatrix} x_0 \\ y_0 \end{pmatrix} = \begin{pmatrix} ax_0 + by_0 \\ cx_0 + dy_0 \end{pmatrix}$$

and $(ax_0 + by_0, cx_0 + dy_0)$ is a solution of $F_{A^{-1}}(x, y) = h$. Now assume that $F(x, y)$ factors in $\mathbb{C}$ as

$$F(x, y) = \prod_{i=1}^{n}(\alpha_i x - \beta_i y).$$

The discriminant disc(F) of F is given by

$$\text{disc}(F) = \prod_{i<j}(\alpha_i\beta_j - \alpha_j\beta_i)^2.$$

For any 2×2 matrix A, we have

$$\text{disc}(F_A) = (\det A)^{n(n-1)}\text{disc}(F). \tag{5}$$

If $A \in GL_2(\mathbb{Z})$ then we say that F_A and $-F_A$ are equivalent to F. We denote by $N_{F,m}$ the number of solutions in integers x and y of the equation $F(x, y) = \pm m$. If F_1 and F_2 are equivalent then $N_{F_1,m} = N_{F_2,m}$ and $\text{disc}(F_1) = \text{disc}(F_2)$. This is an equivalence relationship.

The parametrization of cubic rings is due to Levi [19], Delone-Faddeev [8], and in its most general form, Gan-Gross-Savin [14]. The statement is that isomorphism classes of cubic rings are in natural one-to-one-correspondence with classes of integral binary cubic forms. Given

$$F(x, y) = ax^3 + bx^2y + cxy^2 + dy^3 \in \mathbb{Z}[x, y]$$

the associated cubic ring $R(F)$ has $\mathbb{Z}$-basis $1, \omega, \theta$, with multiplication table given by

$$\begin{aligned}
\omega\theta &= -ad \\
\omega^2 &= -ac - b\omega + a\theta \\
\theta^2 &= -bd - d\omega + c\theta.
\end{aligned}$$

Conversely, given a cubic ring R, let $1, \omega, \theta$ be a $\mathbb{Z}$-basis for R. Translating ω and θ by the appropriate elements of $\mathbb{Z}$, we may assume that $\omega\theta \in \mathbb{Z}$. Under this assumption, there exist constants $l, m, n \in \mathbb{Z}$ such that

$$\begin{aligned}
\omega\theta &= n \\
\omega^2 &= m - b\omega + a\theta \\
\theta^2 &= l - d\omega + c\theta.
\end{aligned}$$

We have the associative law

$$\omega^2\theta = \omega\omega\theta \qquad \text{and} \qquad \omega\theta^2 = \omega\theta\theta,$$

which implies that

$$n = -ad, m = -ac, \text{ and } l = -bd,$$

for some $a, b, c, d \in \mathbb{Z}$. This means the cubic ring R can be associated to the binary cubic form

$$F(x, y) = ax^3 + bx^2y + cxy^2 + dy^3,$$

and we have $R = R(F)$.

Not only does this correspondence give a bijection between cubic rings and cubic forms, but it also shows that certain properties and invariants of each type of object translate nicely. For instance, one can easily check that the form $F(x, y)$ and the ring $R(F)$ have the same discriminant. In order to count cubic rings and orders, we can use this parametrization. We remark that in [6] this correspondence is used to establish an asymptotic formula for the number of cubic orders having bounded discriminant. We recommend [6] for more details about this correspondence, and for one of its several striking applications.

5 Thue Equations, and Their Applications

Let $F(x, y)$ be a binary form with integer coefficients, degree $n \geq 3$ and non-zero discriminant. Let m be a non-zero integer and consider the equation $F(x, y) = m$, in integers x and y. It has only finitely many solutions as was first established by Thue [22] in 1909 in the case that F is irreducible over $\mathbb{Q}$. There is an extensive literature dealing with the problem of estimating from above the number of solutions to Thue equations (see e.g., [2–4, 7, 21]).

5.1 *The Number of Elements with Small Index in a Cubic Number Field*

We already observed that an index form equation in a cubic number field K is indeed a Thue equation. Namely, if the ring of integers O_K in the cubic field K has $\mathbb{Z}$-basis $1, \alpha, \beta$, then the product of the pairwise differences of the three algebraic conjugates of $\alpha x + \beta y$, divided by the square root of the discriminant of O_K

$$\frac{\prod_{1 \leq i < j \leq 3} \left[(\alpha^{(i)} - \alpha^{(j)})x + (\beta^{(i)} - \beta^{(j)})y \right]}{\sqrt{Disc(O_K)}},$$

where $\alpha^{(1)} = \alpha$, $\alpha^{(2)}$ and $\alpha^{(3)}$ are the algebraic conjugates of α and $\beta^{(1)} = \beta$, $\beta^{(2)}$ and $\beta^{(3)}$ are the algebraic conjugates of β, is an integral binary cubic form (called the index form of O_K). Conversely, every equivalence class of integral binary cubic form arises uniquely in this way from a unique cubic order. It's called the index form because $F(x, y)$ now has the following interpretation: the monogenic ring $\mathbb{Z}[\alpha x + \beta y]$, for any $x, y \in \mathbb{Z}$, has index $F(x, y)$ in O_K.

Notice that if $F(x_0, y_0) = m$, for some $x_0, y_0 \in \mathbb{Z}$, then $F(-x_0, -y_0) = -m$. Also (x_0, y_0) corresponds to a monogenizer α and $(-x_0, -y_0)$ corresponds to the monogenizer $-\mu$. But we will count μ and $-\mu$ as one monogenizer, as clearly $\mathbb{Z}[\mu] = \mathbb{Z}[-\mu]$. In the following statements about the number of solutions of cubic Thue equations, possible solutions (x_0, y_0) and $(-x_0, -y_0)$ are deemed as one solution.

The following was shown by the author in [2].

Theorem 1 (Akhtari) *Let $F(x, y) \in \mathbb{Z}[x, y]$ be an irreducible cubic binary form of discriminant D. Let m be an integer with*

$$0 < m \leq \frac{|D|^{\frac{1}{8}-\epsilon}}{(3.5)^{3/2}3^{\frac{3}{8}}},$$

where $0 < \epsilon < \frac{1}{8}$. Then the inequality $0 < |F(x, y)| \leq m$ has at most $27 + \frac{3}{4\epsilon}$ solutions in integers x and y with $\gcd(x, y) = 1$.

This result implies that the ring of integers O_K in a cubic field of discriminant D can have at most B monogenic subrings of index less than $D^{1/8-\epsilon}$, where B is a constant depending on ϵ. Let us take $\epsilon = \frac{1}{16}$ in the above theorem. Then we have the following.

Theorem 2 *Let K be a cubic number field of discriminant D. The ring of integers O_K in a cubic field has at most* 39 *monogenic subrings of index less than $\frac{|D|^{\frac{1}{16}}}{(3.5)^{3/2}3^{\frac{3}{8}}}$.*

5.2 *The Number of Monogenizations of a the Ring of Integers of a Cubic Number Field*

The parametrization of cubic orders due to Levi and Delone-Fadveev (see Sect. 4) implies that every order $\mathcal{O}$ in a cubic field is the invariant order of a unique integral binary cubic form up to equivalence. One can also simply observe that any solution to the index form equation

$$I(x_2, x_3) = \pm 1$$

provides a monogenization for the ring of integers $\mathcal{O}_K$. Therefore to count the number of monogenizations of a cubic ring, we have to search for solutions of a cubic Thue equation of the shape

$$F(x, y) = \pm 1,$$

in integers x and y.

The following is the main Theorem in [7].

Theorem 3 (Bennett) *Let $F(x, y)$ be an irreducible cubic form. The Thue equation $F(x, y) = \pm 1$ has at most 10 solutions in integers x, y.*

The above result was improved in [20] for forms of large discriminant (see also the author's work [5]).

Theorem 4 (Okazaki) *Let $F(x, y)$ be an irreducible cubic form. The Thue equation $F(x, y) = \pm 1$ has at most 7 solutions in integers x, y, provided that $|D_F|$ is sufficiently large.*

It follows from Theorems 3 and 4 that an order $\mathcal{O}$ in a cubic field can have at most 10 monogenizations, and if $\mathcal{O}$ has a large discriminant, then it has at most 7 monogenizations.

We should mention that Gaál and Schulte [13] used effective methods for solving Thue equations to determine all power integral bases in totally real and also complex cubic number fields of small discriminants. For degree $n \geq 4$, Evertse and Győry proved in [11] that and order $\mathcal{O}$ in a number field K of degree n can have at most $\left(3 \times 7^{2n!}\right)^{n-2}$ monogenizations. The best known results to date for $n \geq 4$ is due to Evertse [9], who proved that an order $\mathcal{O}$ in a number field K of degree n can have at most $2^{4(n+5)(n-2)}$ monogenizations. As we saw in Sect. 3, when the degree $n \geq 4$, the index form is not a binary form anymore, and therefore the theory of Thue equations cannot be applied directly to count the number of monogenizations of an order. We recommend [9] for a comprehensive review of results on the number of monogenization of orders in higher degree number fields.

5.3 A Positive Proportion of Cubic Orders Is Not Monogenic

It will follow from a work of Bhargava and the author that a positive proportion of cubic orders is not monogenic. To see this, we use Delone-Fadveev correspondence and recall that if a cubic order is monogenic, then the Thue equation $F(x, y) = 1$ must have a solution in integers x, y, where $F(x, y)$ is the binary form corresponding to the order (see Sect. 4).

In [1] we show for every integer $n \geq 3$ that many (indeed, a positive proportion) of binary forms of degree n are not proper subforms, locally represent 1 at every place, but globally fail to represent 1. To state our precise result, first we note that by creating local obstructions, it is possible to construct several cubic forms that do not represent one. As a simple example consider any binary cubic form that is congruent to

$$xy(x + y)$$

modulo 2. Such forms never represent 1 (or any odd number).

Another way to construct forms that do not represent 1 is to work with "subforms" of a given form. For a fixed binary form $F(x, y)$, we can construct subforms $F|_L(x, y)$ in the following way. Let $F(x, y)$ be a binary cubic form and L be any index p sub-lattice of $\mathbb{Z}^2$. There are $p + 1$ such sub-lattices. Let

$$L = \langle (a, b), (c, d) \rangle,$$

with $ad - bc = p$. We define the *subform*

$$F|_L(x, y) := F(ax + by, cx + dy).$$

If the equation $F|_L(x, y) = 1$ has an integral solution, say (x_0, y_0), then $(ax_0 + by_0, cx_0 + dy_0)$ is an integral solution to the equation $F(x, y) = 1$. By Theorem 3, the equation $F(x, y) = 1$ has at most 10 solutions in integers x and y. So we conclude that as L varies over all index p sub-lattices of $\mathbb{Z}^2$, the equations

$$F|_L(x, y) = 1,$$

except at most 10 of them, have no solution in integers x and y. Thus, we have constructed at least $p + 1 - 10$ binary forms that do not represent 1. This way we can construct arbitrarily many sub-forms $G(x, y) = F(ax + by, cx + dy)$ of $F(x, y)$ so that $G(x, y)$ does not represent 1. The intersection of two different sub-lattices of index p has index p^2 and can not represent 1. Therefore a solution to $F(x, y) = 1$ can not correspond to two different sub-lattices of index p.

In [1] we have shown that a positive proportion of cubic binary forms, which are not subforms of other binary forms, does not represent 1 (and in fact, any fixed integer).

Theorem 5 (Akhtari–Bhargava) *When integral binary cubic forms $F(x, y) \in \mathbb{Z}[x, y]$ are ordered by absolute discriminant, a positive proportion of the* $\mathrm{GL}_2(\mathbb{Z})$-*classes of these forms F has the following properties:*

(i) they locally everywhere represent 1 *(i.e., $F(x, y) = 1$ has a solution in $\mathbb{R}^2$ and in $\mathbb{Z}_p^2$ for all p);*
(ii) they globally do not represent 1 *(i.e., $F(x, y) = 1$ has no solution in $\mathbb{Z}^2$); and*
(iii) they are maximal forms (i.e., F is not a proper subform of any other form).

More precisely, let $N_1(3, X)$ denote the number of $\mathrm{GL}_2(\mathbb{Z})$-classes of integral binary cubic forms having absolute discriminant less than X that are maximal, locally represent 1, but do not globally represent 1; and let $N(3, X)$ denote the total number of $\mathrm{GL}_2(\mathbb{Z})$-classes of integral binary cubic forms having absolute discriminant less than X. Then we proved that

$$\liminf_{X \to \infty} \frac{N_1(3, X)}{N(3, X)} > 0. \tag{6}$$

Acknowledgements I would like to thank Professor Manjul Bhargava who originally brought this topic to my attention. I would like to thank Professor Melvyn Nathanson for organizing Combinatorial and Additive Number Theory (CANT 2018) conference, and bringing together a truly diverse group of researchers. I would also like to acknowledge the support from the NSF grant DMS-1601837.

References

1. S. Akhtari and M. Bhargava, A positive proportion of Thue equations fail the integral Hasse principle, to appear in American Journal of Mathematics (2018)
2. S. Akhtari, Representation of small integers by binary forms. Q. J. Math. **66**(4), 1009–1054 (2015)
3. S. Akhtari, Cubic Thue inequalities with positive discriminant. Publ. Math. Debrecen. **83**(4), 727–739 (2013)
4. S. Akhtari, Representation of unity by binary forms. Trans. Amer. Math. Soc. **364**, 2129–2155 (2012)
5. S. Akhtari, Cubic Thue Equations. Publ. Math. Debrecen **75**, 459–483 (2009)
6. M. Bhargava, A. Shnidman, On the number of cubic orders of bounded discriminant having automorphism group C_3, and related problems. Algebra Number Theory **8**(1), 53–88 (2014)
7. M.A. Bennett, On the representation of unity by binary cubic forms. Trans. Amer. Math. Soc. **353**, 1507–1534 (2001)
8. B. N. Delone and D. K. Faddeev, *The Theory of Irrationalities of the Third Degree, AMS Translations of Mathematical Monographs* 10, 1964
9. J.H. Evertse, A survey on monogenic orders. Publ. Math. Debrecen **79**, 411–422 (2011)
10. J.- H. Evertse and K. Gyory, *Discriminant Equations in Diophantine Number Theory*, vol. 32 of New Mathematical Monographs, Cambridge University Press, Cambridge, 2017
11. J.-H. Evertse, K. Győry, On unit equations and decomposable form equations. J. reine angew. Math. **358**, 6–19 (1985)
12. I. Gaál, *Diophantine Equations and Power Integral Bases, New Computational Methods*, Birkhäuser 2002
13. I. Gaál, N. Schulte, Computing all power integral bases of cubic number fields. Math. Comp. **53**, 689–696 (1989)
14. W.T. Gan, B. Gross, G. Savin, Fourier coefficients of modular forms on G_2. Duke Math. J. **115**(1), 105–169 (2002)
15. K. Győry, Sur les polynmes coefficients entiers et de discriminant don III. Publ. Math. Debrecen **23**, 141–165 (1976)
16. H. Hasse, Number Theory, Akademie-Verlag (1979)
17. N. Kaplan, J. Marcinek, R. Takloo- Bighash, Distribution of orders in number fields, Res. Math. Sci. (2015) 2 : 6, 57 pp
18. G. Lettl, C. Prabpayak, Orders in cubic number fields. J. Number Theory. **166**, 415–423 (2016)
19. F. Levi, Kubische Zahlkörper und binäre kubische Formenklassen. Leipz. Ber. **66**, 26–37 (1914)
20. R. Okazaki, Geometry of a cubic Thue equation. Publ. Math. Debrecen **61**, 267–314 (2002)
21. C.L. Stewart, On the number of solutions of polynomial congruences and Thue equations. Journal of American Math. Soc. **4**, 793–838 (1991)
22. A. Thue, Über Annäherungswerte algebraischer Zahlen. J. reine angew. Math. **135**(1909), 284–305 (1909)
23. E. Weiss *Algebraic Number Theory*. New York: Mc Graw Hill (1963)

The Zeckendorf Game

Paul Baird-Smith, Alyssa Epstein, Kristen Flint and Steven J. Miller

Abstract Zeckendorf proved that every positive integer n can be written uniquely as the sum of non-adjacent Fibonacci numbers. We use this to create a two-player game. Given a fixed integer n and an initial decomposition of $n = nF_1$, the two players alternate by using moves related to the recurrence relation $F_{n+1} = F_n + F_{n-1}$, and whoever moves last wins. The game always terminates in the Zeckendorf decomposition, though depending on the choice of moves the length of the game and the winner can vary. We find upper and lower bounds on the number of moves possible. The upper bound is on the order of $n \log n$, and the lower bound is sharp at $n - Z(n)$ moves, where $Z(n)$ is the number of terms in the Zeckendorf decomposition of n. Notably, Player 2 has the winning strategy for all $n > 2$; interestingly, however, the proof is non-constructive.

MSC 2010: 11P99 (primary) · 11K99 (secondary)

This work was supported by NSF Grants DMS1561945 and DMS1659037, Carnegie Mellon and Williams College.
The authors were partially supported by NSF grants DMS1265673 and DMS1561945, the Claire Booth Luce Foundation, and Carnegie Mellon University. We thank the students from the Math 21-499 Spring '16 research class at Carnegie Mellon and the participants from CANT 2016 and 2017 and the 18th Fibonacci Conference, especially Russell Hendel, for many helpful conversations.

P. Baird-Smith (✉) · A. Epstein · K. Flint · S. J. Miller
Department of Computer Science, University of Texas, Austin, TX, USA
e-mail: paul.bairdsmith@gmail.com

A. Epstein
e-mail: wtgalyssa@gmail.com

K. Flint
e-mail: kflint1101@gmail.com

S. J. Miller
e-mail: sjm1@williams.edu

Department of Mathematics and Statistics, Williams College, Williamstown, MA 01267, USA

Department of Mathematics, Carnegie Mellon University, Pittsburgh, PA, USA

M. B. Nathanson (ed.), *Combinatorial and Additive Number Theory III*,
Springer Proceedings in Mathematics & Statistics 297,
https://doi.org/10.1007/978-3-030-31106-3_3

1 Introduction

1.1 History

The Fibonacci numbers are one the most interesting and famous sequences. They appear in many varied settings, from Pascal's triangle to mathematical biology. Among their fascinating properties, the Fibonacci numbers lend themselves to a beautiful theorem of Zeckendorf [7]: each positive integer n can be written uniquely as the sum of distinct, non-adjacent Fibonacci numbers. This is called the *Zeckendorf decomposition* of n and requires that we define the Fibonacci numbers by $F_1 = 1$, $F_2 = 2$, $F_3 = 3$, $F_4 = 5...$ instead of the usual $1, 1, 2, 3, 5...$ to create uniqueness. The Zeckendorf theorem has been generalized many times (see for example [2, 3, 5, 6]), allowing the game explored in this paper potentially to be played similarly on other recurrences. For details on these and other generalizations, as well as references to the literature on generalizations of Zeckendorf's theorem, see the companion papers [1, 4].

1.2 Main Results

We introduce some notation. By $\{1^n\}$ or $\{F_1{}^n\}$ we mean n copies of 1, the first Fibonacci number. If we have 3 copies of F_1, 2 copies of F_2, and 7 copies of F_4, we could write either $\{F_1{}^3 \wedge F_2{}^2 \wedge F_4{}^7\}$ or $\{1^3 \wedge 2^2 \wedge 5^7\}$.

Definition 1.1 (*The Two Player Zeckendorf Game*) At the beginning of the game, there is an unordered list of n 1's. Let $F_1 = 1$, $F_2 = 2$, and $F_{i+1} = F_i + F_{i-1}$; therefore the initial list is $\{F_1{}^n\}$. On each turn, a player can do one of the following moves.

1. If the list contains two consecutive Fibonacci numbers, F_{i-1}, F_i, then a player can change these to F_{i+1}. We denote this move $\{F_{i-1} \wedge F_i \rightarrow F_{i+1}\}$.
2. If the list has two of the same Fibonacci number, F_i, F_i, then
   a. if $i = 1$, a player can change F_1, F_1 to F_2, denoted by $\{F_1 \wedge F_1 \rightarrow F_2\}$,
   b. if $i = 2$, a player can change F_2, F_2 to F_1, F_3, denoted by $\{F_2 \wedge F_2 \rightarrow F_1 \wedge F_3\}$, and
   c. if $i \geq 3$, a player can change F_i, F_i to F_{i-2}, F_{i+1}, denoted by $\{F_i \wedge F_i \rightarrow F_{i-2} \wedge F_{i+1}\}$.

The players alternative moving. The game ends when one player moves to create the Zeckendorf decomposition.

The moves of the game are derived from the recurrence, either combining terms to make the next in the sequence or splitting terms with multiple copies. We first show the game is well-defined, and then provide bounds on its length.

Theorem 1.2 *Every game terminates within a finite number of moves at the Zeckendorf decomposition.*

Now that we know that the Zeckendorf game is playable, we might wonder how long it will take to play.

Theorem 1.3 *The shortest game, achieved by a greedy algorithm, arrives at the Zeckendorf decomposition in* $n - Z(n)$ *moves, where* $Z(n)$ *is the number of terms in the Zeckendorf decomposition of n. The longest game is bounded by* $i * n$*, where i is the index of the largest Fibonacci number less than or equal to n.*

The theoretical upper bound presented here grows on a log-linear scale because the index of the largest Fibonacci number less than or equal to n is less than $\log_\phi(\sqrt{5}F_i + 1/2)$, where ϕ is the golden ratio. This relation comes from Binet's formula. Since there is a wide span between the lower bound and the theoretical bound, we simulated random games and were led to the following conjectures.

Conjecture 1.4 As n goes to infinity, the number of moves in a random game decomposing n into it's Zeckendorf expansion, when all legal moves are equally likely, converges to a Gaussian.

Conjecture 1.5 The longest game on any n is achieved by applying splitting moves whenever possible. Specifically, the longest possible game applies moves in the following order: merging ones, splitting from smallest to largest, and adding consecutives, from smallest to largest.

Conjecture 1.6 The average game is of a length linear with n.

Of course, we are interested not just in how long the game takes, but who wins.

Theorem 1.7 *For all* $n > 2$*, Player 2 has the winning strategy for the Zeckendorf Game.*[1]

Since someone must always make the final move, and the game always terminates, for each n one of the two players must have a winning strategy. In other words, someone must always be able to force their victory. This theorem shows that for all nontrivial games, Player 2 has this strategy. The proof is not constructive: it merely shows the existence of Player 2's winning strategy; we cannot identify how they should move. Though we can give exact winning strategies for small n, we leave the general winning strategy for future research.

[1]If $n = 2$, there is only one move, and then the game is over.

2 The Zeckendorf Game

2.1 The Game Is Playable

In this section, we provide many proofs related to the Zeckendorf Game. We begin with the proof of Theorem 1.2, which shows that the game is well defined and playable, starting with an important lemma.

Lemma 2.1 (Fibonacci Monovariant) *The sum of the square roots of the indices on any given turn is a monovariant.*[2]

Proof Our moves cause the following changes in the proposed monovariant. We observe that we only have to consider the affected terms because the suggested monovariant is a sum, so unaffected terms contribute the same before and after the move. Here, k is the index of F_k, a term in the current decomposition.

- Adding consecutive terms: $-\sqrt{k-2} - \sqrt{k-1} + \sqrt{k}$
- Splitting: $-2\sqrt{k} + \sqrt{k-2} + \sqrt{k+1}$
- Adding 1's: $-2 + \sqrt{2}$
- Splitting 2's: $-2\sqrt{2} + 1 + \sqrt{3}$.

We note that for all positive $k > 2$, in other words all indices not addressed in a special case above, all of these moves cause negative changes. We can see this by the fact that $\sqrt{x}$ is a monotonically increasing, concave function. So this is a monovariant; the sum of the square roots of the indices constantly decrease with each move, so it is strictly decreasing.

With this lemma, we now prove Theorem 1.2.

Proof (*Proof of Theorem* 1.2) At the beginning of the game, we have a sum of the square roots of the indices of our list of numbers equal to $\sqrt{n}$, where n is the number we have chosen for the game. From the monovariant of Lemma 2.1, we know that the listed moves always decrease this sum. Therefore, no two moves can have the same monovariant value, and there will be no repeat turns. Since the game essentially moves among a subset of partitions of n, of which there are a finite number, this implies that the game must always end within a finite number of turns. Moreover, the game always ends at the Zeckendorf decomposition. If it terminated elsewhere, there would either be duplicate terms or the recurrence would apply, by definition. So, there would still be a valid move and the game would not have terminated. This concludes the proof.

[2] For us, monovariant is a quantity which is either non-increasing or non-decreasing.

Now that we know for sure that we can play the Zeckendorf Game, we wonder how long the game will take. First, we address the question of whether the game must always take the same amount of turns. If it does, this game is definitely not fair because it predetermines a victor! Fortunately, this is not the case as long as we choose an n greater than 3.

Lemma 2.2 *Given any positive integer n such that $n > 3$, there are at least two distinct sequences of moves $M = \{m_i\}$ where the application of each set of moves to the initial set, denoted $M(\{F_1\}_n)$, leads to Z_n, the Zeckendorf decomposition of n.*

Proof If we show that there are two distinct sets of moves that arrive at the Zeckendorf decomposition of 4, we have proved the claim because for all $n > 4$: we can follow the two different identified games up to 4, both of which are valid paths to the Zeckendorf decomposition.

The following two sequences of moves result in the Zeckendorf composition of 4:

$$M_1 = \{\{F_1 \wedge F_1 \rightarrow F_2\}, \{F_1 \wedge F_1 \rightarrow F_2\}, \{2F_2 \rightarrow F_1 \wedge F_3\}\}$$
$$M_2 = \{\{F_1 \wedge F_1 \rightarrow F_2\}, \{F_1 \wedge F_2 \rightarrow F_3\}\}$$

Therefore, there are multiple games for any $n > 3$.

Remark 2.3 If $n \leq 3$ there is one unique sequence of moves that arrives at the Zeckendorf decomposition. If $n = 1$, $M = \{\}$. If $n = 2$, $M = \{F_1 \wedge F_1 \rightarrow F_2\}$. If $n = 3$, $M = \{\{F_1 \wedge F_1 \rightarrow F_2\}, \{F_1 \wedge F_2 \rightarrow F_3\}\}$.

Corollary 2.4 *For any positive $n > 3$, there are at least two games with different numbers of moves. Further, there is always a game with an odd number of moves and one with an even number of moves.*

Proof In Lemma 2.2, we showed that two different sets of moves M_1 and M_2 arrive at the Zeckendorf Decomposition of 4. Notice that $|M_1| = 3$ but $|M_2| = 2$. As there are no losing games, for any $n > 4$, we can follow either of these games up to the Zeckendorf decomposition of 4. Regardless of the number or sequence of moves it takes to resolve the rest of the game (call the sequence M_k, with $|M_k| = k$), we have already identified two sets of moves with different orders, $M_1 \wedge M_k$ and $M_2 \wedge M_k$, that describe a complete game. $|M_1 \wedge M_k| = 3 + k$, but $|M_2 \wedge M_k| = 2 + k$. If k is even, $3 + k$ is odd and $2 + k$ is even. If k is odd, $3 + k$ is even and $2 + k$ is even. This proves the claim.

2.2 Bounds on the Lengths of Games

We have now established that this game has variation in both game length and parity. It is natural to ask how much variety there is between short, long, and average games. To this end, we provide a proof of Theorem 1.3. To do so, we first include a lemma about the structure of a game following a greedy algorithm.

Lemma 2.5 *Let $m(n)$ be the number of moves in a deterministic game where the players must always move on the largest valued number. Let $Z(n)$ be the number of terms in the Zeckendorf decomposition of n. Then $m(n) = n - Z(n)$.*

Proof Each player acts on the largest valued summand with an available move. The game on n takes $m(n)$ moves. Looking at the game on $n + 1$, we observe that the list of summands will eventually reach $\{1, a, b, c, \dots\}$ where $\{a, b, c, \dots\}$ is the Zeckendorf decomposition of n. Thus $m(n + 1) = m(n) + k(n + 1)$, where k is a function that is always non-negative.

If the smallest summand in the Zeckendorf decomposition for n is greater than or equal to 3, there are no additional moves that can be made and $k(n + 1) = 0$. However, if the smallest summand is 1 or 2, the smallest summand can be combined with the additional 1. Because an additional move was completed, $k(n + 1) \geq 1$. It then may be possible to now make another move with the decomposition that was just created. For every additional move that can be made, $k(n + 1)$ increases by 1. We also know that for each additional move, the number of terms in the Zeckendorf decomposition decreases by 1, because each move combines two numbers into one. We have

$$\begin{aligned} Z(n+1) &= Z(n) + 1 - k(n+1) \\ m(n+1) &= m(n) + k(n+1). \end{aligned} \tag{1}$$

Define $t(n)$ by

$$t(n) \;:=\; Z(n) + m(n). \tag{2}$$

By adding the equations given by (1) we see that $t(n)$ satisfies a simple recurrence:

$$\begin{aligned} Z(n+1) + m(n+1) &= Z(n) + m(n) + 1 \\ t(n+1) &= t(n) + 1 \\ &= t(n-1) + 2 \\ &= t(n-2) + 3 \\ &\;\;\vdots \\ &= t(1) + n. \end{aligned} \tag{3}$$

Since 1 is a Fibonacci number, the Zeckendorf decomposition of 1 is just 1, and we have $Z(1) = 1$ and $m(1) = 0$. Thus

$$\begin{aligned} t(n+1) &= t(1) + n \\ &= Z(1) + m(1) + n \\ &= 1 + 0 + n \\ t(n+1) &= n + 1. \end{aligned} \tag{4}$$

From this, we see that for any positive integer n, $t(n) = n$ and so, with the definition of $t(n)$, we have that

$$\begin{aligned} t(n) &= Z(n) + m(n) \\ n &= Z(n) + m(n) \\ m(n) &= n - Z(n). \end{aligned} \tag{5}$$

We have shown that for any positive integer n, when starting from a list of length n that contains all 1's, the number of moves it takes to reach the Zeckendorf decomposition for n will be equal to n minus the number of terms in the final Zeckendorf decomposition for n. Thus, we have shown Lemma 2.5.

Proof A quick way to arrive at the Zeckendorf decomposition would be to decrease one term on every move. This would make a short game happen in $n - Z(n)$ moves. No game would be faster, because each possible move decreases the number of terms by at most one. That this game is achievable follows from Lemma 2.5. Since this number of moves is theoretically shortest and is actually possible, it is a sharp lower bound on the number of moves in the Zeckendorf game.

For the longest game, we return to the monovariant established in Lemma 2.1. We observe that the least each move can change the sum is by a splitting move way late into the game. Splitting moves cost at least $2\sqrt{\ell} - \sqrt{\ell - 2} - \sqrt{\ell + 1}$, where ℓ is the index of the largest Fibonacci number less than or equal to n. We notice that $2\sqrt{\ell} - \sqrt{\ell - 2} - \sqrt{\ell + 1} > \sqrt{\ell} - \sqrt{\ell - 1}$ because square root is concave and increasing. Then, we observe that $1 = n - (n - 1) = (\sqrt{n} - \sqrt{n - 1})(\sqrt{n} + \sqrt{n - 1})$, which implies that $\sqrt{n} - \sqrt{n - 1} = \frac{1}{\sqrt{n}+\sqrt{n-1}} > \frac{1}{n}$. So, $2\sqrt{\ell} - \sqrt{\ell - 2} - \sqrt{\ell + 1} > 1/\ell$. This gives that it will take at most $\ell \cdot n$ moves to reach the Zeckendorf decomposition. Since ℓ is a Fibonacci index, we recall Binet's formula to get a bound in terms of n: $F_\ell = \frac{1}{\sqrt{5}}(\phi^\ell - (-\phi)^{-\ell})$. We note that $|\frac{\phi^{-\ell}}{\sqrt{5}}| < \frac{1}{2}$, which implies that $\sqrt{5}F_\ell < \phi^\ell - 1/2$. Taking a base ϕ logarithm of both sides, we get $\log_\phi(\sqrt{5}F_\ell + 1/2) > \ell$. This shows that $\ell \cdot n < \log_\phi(\sqrt{5}n + 1/2)n$.

2.3 Conjectures on Game Lengths

Using Mathematica code (see Appendix 4), we support the conjectures on game length introduced in the introduction with simulation data. We address Conjecture 1.4 first. Observing Fig. 1, the best fit Gaussian seems to align well with the distribution of moves taken over 9,999 simulations of the Zeckendorf Game with $n = 60$. Figure 2 shows the same experiment on $n = 200$ with 9,999 simulations.

To see how Conjecture 1.5 may be true, we provide two pieces of evidence. The first is the move count from simulation of the deterministic algorithm stated in the conjecture. Recall that the order of moves is adding ones, splitting from smallest to largest, then adding consecutives from smallest to largest. Figure 3 shows an array

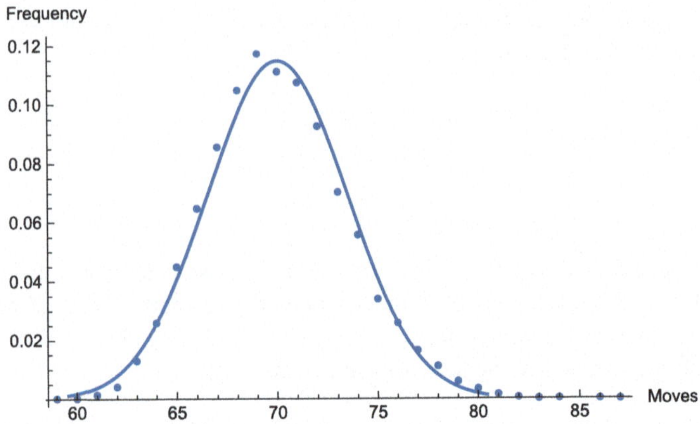

Fig. 1 Frequency graph of the number of moves in 9,999 simulations of the Zeckendorf Game with random moves when $n = 60$ with the best fit Gaussian over the data points

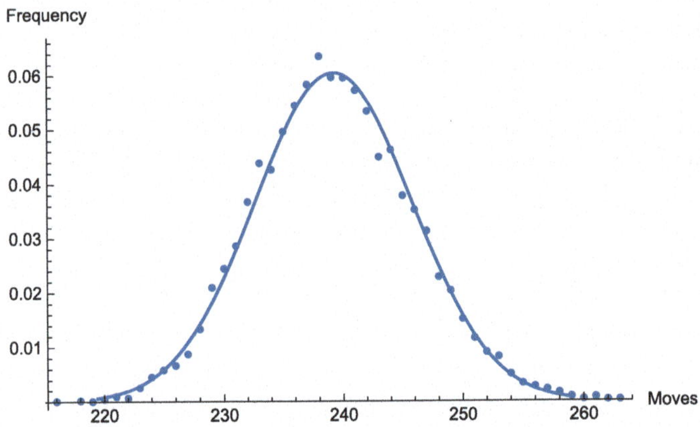

Fig. 2 Frequency graph of the number of moves in 9,999 simulations of the Zeckendorf Game with random moves when $n = 200$ with the best fit Gaussian over the data points

with the x component being n and the y component being the number of moves in the hypothesized deterministic longest game algorithm. The second piece of evidence comes from a Java program, a link to and readme for which is included in Appendix 4. The Java program explores all possible moves in the Zeckendorf game for a given n. The data produced here is the longest possible move length for the n listed. Observe that the two arrays provide identical data. This suggests that the hypothesized longest game algorithm may actually be the theoretically longest game on each n (Fig. 4).

In support of Conjecture 1.6, we offer the graph in Fig. 5. Using data from simulating the Zeckendorf game on varying n, we plot the average number of moves in a game against n. We observe that a best fit line with slope of around 1.2 fits the data points well. Due to computer restraints, we are unable to provide data beyond

```
{{1, 0}, {2, 1}, {3, 2}, {4, 3}, {5, 5}, {6, 6}, {7, 8}, {8, 10},
 {9, 11}, {10, 13}, {11, 15}, {12, 17}, {13, 20}, {14, 21}, {15, 23},
 {16, 25}, {17, 26}, {18, 29}, {19, 31}, {20, 34}, {21, 37},
 {22, 38}, {23, 40}, {24, 42}, {25, 44}, {26, 47}, {27, 48}, {28, 50},
 {29, 53}, {30, 54}, {31, 57}, {32, 60}, {33, 63}, {34, 67}, {35, 68},
 {36, 70}, {37, 72}, {38, 73}, {39, 76}, {40, 78}, {41, 81}, {42, 84},
 {43, 85}, {44, 87}, {45, 89}, {46, 91}, {47, 95}, {48, 96}, {49, 98}}
```

Fig. 3 Data taken from the simulation of the deterministic longest game proposed by the algorithm in Conjecture 1.5

Fig. 4 Computer proven data of the number of moves in the longest route to victory courtesy of the Java code written by Paul Baird-Smith

N	Moves
1	0
2	1
3	2
4	3
5	5
6	6
7	8
8	10
9	11
10	13
11	15
12	17
13	20
14	21
15	23
16	25
17	26
18	29
19	31
20	34
21	37
22	38
23	40
24	42
25	44
26	47
27	48
28	50
29	53
30	54
31	57
32	60
33	63
34	67
35	68
36	70
37	72
38	73
39	76
40	78
41	81
42	84
43	85
44	87
45	89
46	91
47	95
48	96
49	98

$n = 200$ (not pictured in the graph, but included in the data). The average taken on $n = 200$ is 239, very close to $1.2 \cdot 200$.

2.4 *Winning Strategies*

Since someone must always make the final move, and the game always ends at the Zeckendorf decomposition, there are no ties. Therefore one player or the other has a winning strategy on each n. This section is devoted to the proof that Player 2 has

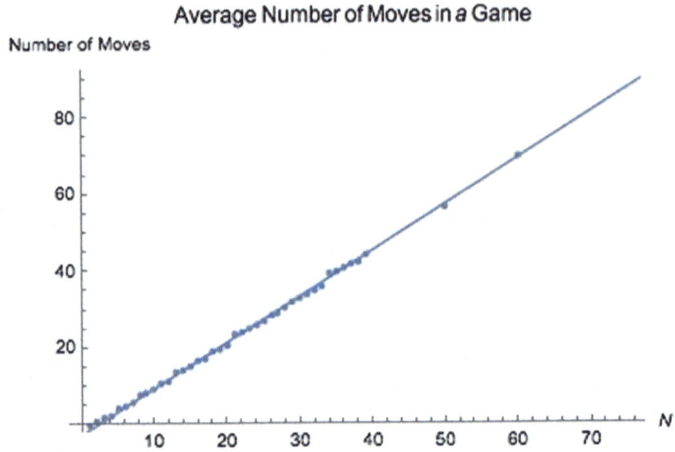

Fig. 5 Graph of the average number of moves in the Zeckendorf game with simulations ranging from 999 to 9,999 for varying n with the best fit line over the data points

the winning strategy for all $n > 2$, the statement of Theorem 1.7. For this proof, we use a visual aid provided in Fig. 6.

Proof Assume that Player 1 wins the game. Therefore, Player 1 must have the winning strategy from the node in the first row with the caption $\{1^{(n)}\}$. We color this node red in Fig. 7. Since this node only has one child, Player 1 must have the winning strategy from $\{1^{(n-2)} \wedge 2\}$ in row two. Player 2 makes the next move, so Player 1 must have the winning strategy from both the nodes in row 3; if not, Player 2 would move to the one from which Player 1 did not have the winning strategy. We focus on the children of the node $\{1^{(n-3)} \wedge 3\}$ in row 3. This node has one descendant only; therefore $\{1^{(n-5)} \wedge 2 \wedge 3\}$ in row 4 must have a winning strategy for Player 1. Player 2 makes the move next, so all three children of $\{1^{(n-5)} \wedge 2 \wedge 3\}$ in row 5 must be a winning strategy for Player 1. Observe that one such child is $\{1^{(n-5)} \wedge 5\}$ in row 5. If Player 1 has the winning strategy from that node in row 5, if that node is on the next layer, in row 6, following the same winning strategy, Player 2 can win from the row 6 node $\{1^{(n-5)} \wedge 5\}$. So we color that node blue on row 6 of Fig. 7 to indicate Player 2 having a winning strategy. Since that node has only one child, $\{1^{(n-7)} \wedge 2 \wedge 5\}$ in row 7, Player 2 must have a winning strategy from that node. This means that any parent of this node must be a winning strategy location for Player 2 because Player 2 could just move to $\{1^{(n-7)} \wedge 2 \wedge 5\}$ in row 7 from those parents. This means that $\{1^{(n-8)} \wedge 2 \wedge 3^{(2)}\}$ in row 6 must have a winning strategy for Player 2; however, since both children in row 6 of $\{1^{(n-6)} \wedge 3^{(2)}\}$ in row 5 have winning strategies for Player 2, this means the row 5 node must be a winning strategy for Player 2, not Player 1 as we had earlier deduced. This leads to a contradiction that proves the claim for n sufficiently large ($n \geq 9$). For the small cases of $2 < n < 9$, computer code such as the one referenced in Appendix 4 can show that Player 2 has the winning strategy by brute force.

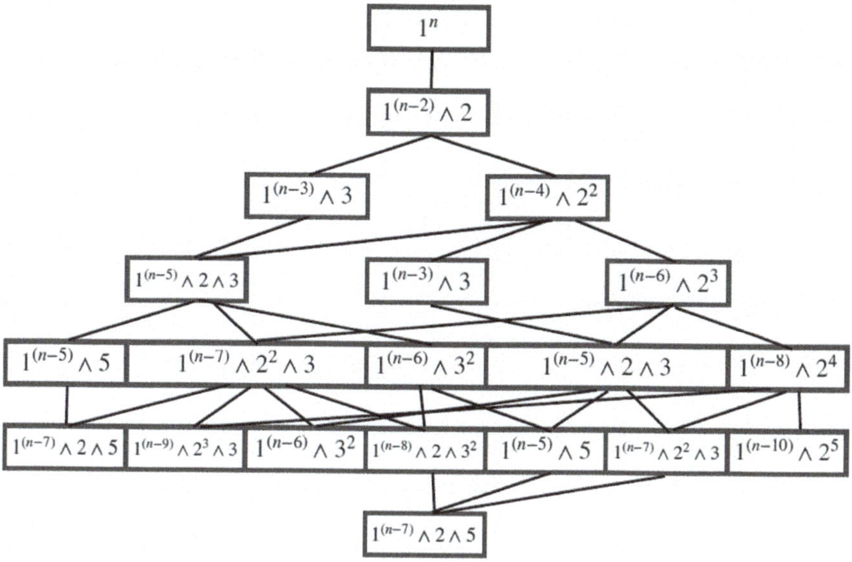

Fig. 6 Tree depicting the general structure of the first several moves of the Zeckendorf game

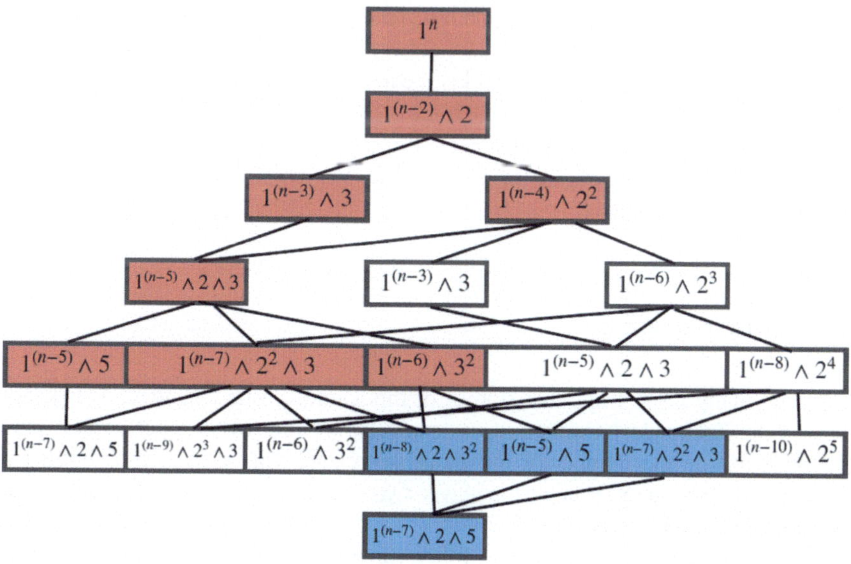

Fig. 7 Tree depicting the proof of Theorem 1.7. Red boxes have a winning strategy for Player 1, and blue boxes indicate a winning strategy for Player 2

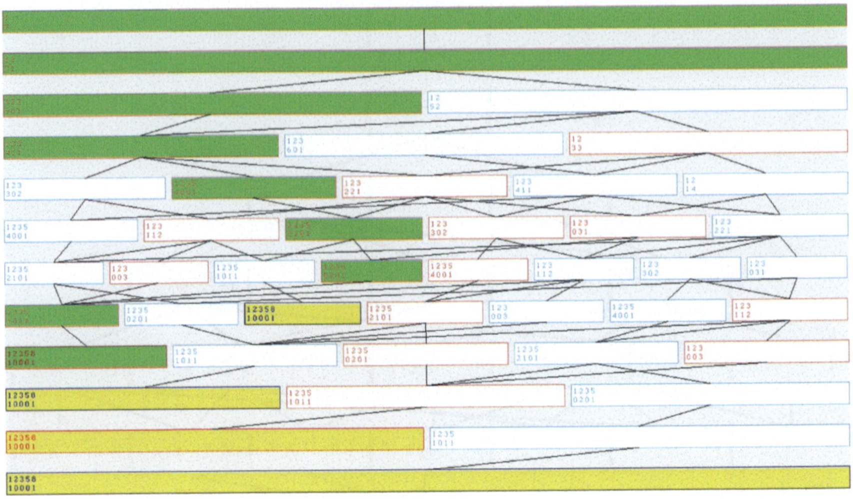

Fig. 8 Game tree for $n = 9$, showing a winning path in green. Image courtesy of the code referenced in Appendix 4

This result is non-trivial and surprising. Game trees for large n have many, many nodes, with no obvious path to victory for either player (see Fig. 8 for $n = 9$ and Fig. 9 for $n = 12$ for an example of how quickly the number of nodes grows). Additionally, this is merely an existence proof, which means we cannot tell how Player 2 should move to achieve his victory. This makes the game less rigged for human players; indeed, random simulations of the games show Players 1 and 2 winning roughly even amounts of the time.

3 Future Work

There are many more ways that studies of this game can be extended. This paper covered the Zeckendorf Game quite extensively, but improved upper bounds may still be found on the number of moves in any game. This work also showed the existence of a winning strategy for player two for all $n > 2$, but it does not show what that strategy is.

The Zeckendorf Game is on the Fibonacci recurrence; however, the fact that Zeckendorf's theorem generalizes means that the game could be played on other recurrences. Finding which classes of recurrences have meaningful games, bounding the moves on those games, and considering winning strategies are all fruitful avenues for further exploration.

Expanding in another direction, the Zeckendorf Game as conceived of by this thesis is a two-player game. What if more players want to join? Who wins in that case, for either the Generalized or regular Zeckendorf Game? The analysis done here

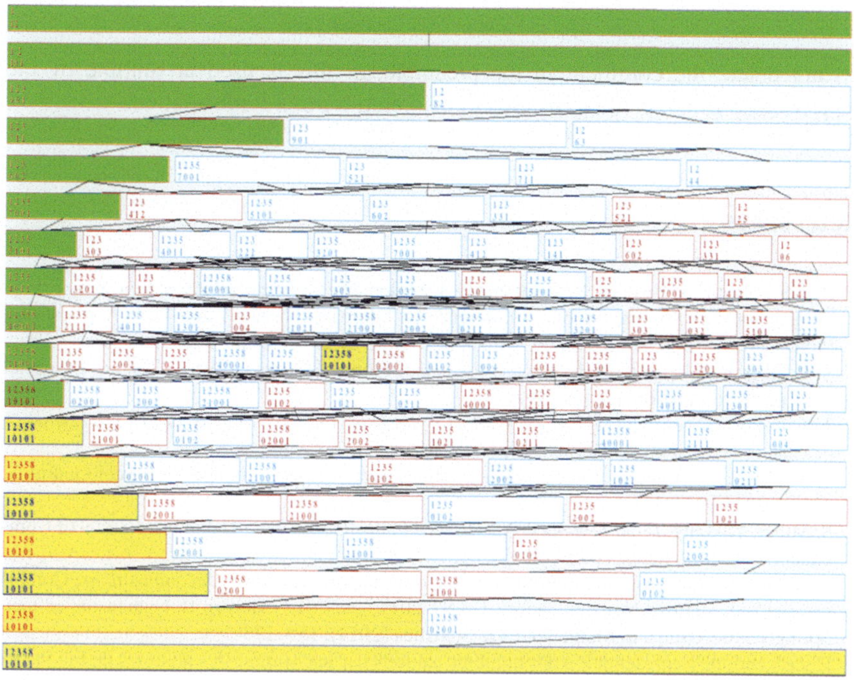

Fig. 9 Game tree for $n = 12$, showing a winning path in green. Image courtesy of the code in Appendix 4

only shows there is a winning strategy that takes an even number of moves for all $n > 2$ for the Zeckendorf Game. It says nothing about the number of moves modulo k, where k is odd and greater than 2!

4 Code

Programs for simulating a random version of the Zeckendorf game, running a deterministic worst game algorithm of the Zeckendorf game, and simulating a random Tribonacci Zeckendorf game is available at

http://github.com/paulbsmith1996/ZeckendorfGame/blob/master/ZeckGameMathematica.nb.

TreeDrawer is used to give a visual representation of the tree structure of the Zeckendorf game. It plays through a specified game, determining all moves that can be made, and draw all possible paths to the end of this game. The ReadMe file can be found at https://github.com/paulbsmith1996/ZeckendorfGame. TreeDrawer can be executed, after compilation, by running the command appletviewer TreeDrawer.java

Do not delete the comment in the preamble, as this is used at runtime by the appletviewer. Email paul.bairdsmith@gmail.com for more information.

References

1. P. Baird-Smith, A. Epstein, K. Flynt and S. J. Miller, *The Generalized Zeckendorf Game*, Proceedings of the 18th International Conference on Fibonacci Numbers and Their Applications, Fibonacci Quarterly **57** (2019), no. 5 (to appear). https://arxiv.org/pdf/1809.04883
2. V. E. Hoggatt, *Generalized Zeckendorf theorem*, Fibonacci Quarterly **10** (1972), no. 1 (special issue on representations), pages 89–93
3. T. J. Keller, *Generalizations of Zeckendorf's theorem*, Fibonacci Quarterly **10** (1972), no. 1 (special issue on representations), pages 95–102
4. S. J. Miller, A. Newlon, *The Fibonacci Quilt Game*, to appear in The Fibonacci Quarterly. arXiv:1909.01938v1
5. S. Miller, Y. Wang, *From Fibonacci Numbers to Central Limit Type Theorems*, Journal of Combinatorial Theory, Series A **119** (2012), no. 7, 1398–1413
6. S. Miller, Y. Wang, *Gaussian Behavior, in Generalized Zeckendorf Decompositions*, Combinatorial and Additive Number Theory, CANT 2011, and 2012 (Melvyn B. Nathanson, editor), Springer Proceedings in Mathematics & Statistics (2014), 159–173
7. E. Zeckendorf, Représentation des nombres naturels par une somme des nombres de Fibonacci ou de nombres de Lucas, Bulletin de la Société Royale des Sciences de Liège **41** (1972), pages 179–182

Iterated Riesel and Iterated Sierpiński Numbers

Holly Paige Chaos and Carrie E. Finch-Smith

Abstract In 1956, Riesel showed that there are infinitely many Riesel numbers—odd positive integers k with the property that $k \cdot 2^n - 1$ is composite for all natural numbers n. A few years later, Sierpiński proved an analagous result using the expression $k \cdot 2^n + 1$ instead. We create iterated Riesel numbers by iterating the process of multiplying by a (fixed) power of 2 and subtracting 1 from the product (or adding 1 to the product, in the case of iterated Sierpiński numbers). In this paper, we show that there are infinitely many iterated Riesel numbers, where we iterate the process 49 times. In addition, we prove an analogous result for iterated Sierpiński numbers.

1 Introduction

In 1956, Hans Riesel showed that 509203 is the smallest number in an infinite sequence of integers in arithmetic progression with a startling property: $509203 \cdot 2^n - 1$ is not prime for any natural number n (see [14]). A few years later, Wacław Sierpiński showed that $15511380746462593381 \cdot 2^n + 1$ is composite for all natural numbers n, and that this is the smallest integer in an infinite arithmetic progression of integers with this property (see [15]). In the years since 1960, numbers with these properties, known as Riesel numbers and Sierpiński numbers in honor of the work of these two mathematicians, have been studied. In particular, the occurrence of Riesel numbers or Sierpiński numbers within other interesting sequences has been a fertile and interesting area of study. For example, there are Riesel numbers and Sierpiński numbers and Riesel-Sierpiński numbers in the sequence of Carmichael numbers (see [4]), in the sequence of Fibonacci numbers (see [12, 13]), in the sequence of Lucas numbers (see [3]), in the sequence of Cullen numbers (see [5]), in the sequence of

H. P. Chaos
Wake Forest University, 1834 Wake Forest Rd, Winston-Salem, NC 27109, USA
e-mail: chaohp18@wfu.edu

C. E. Finch-Smith (✉)
Washington and Lee University, 204 W. Washington St., Lexington, VA 24450, USA
e-mail: finchc@wlu.edu

M. B. Nathanson (ed.), *Combinatorial and Additive Number Theory III*,
Springer Proceedings in Mathematics & Statistics 297,
https://doi.org/10.1007/978-3-030-31106-3_4

perfect powers (see [7, 9, 11]), and in infinitely many sequences of polygonal numbers (see [2]). In addition, there are other nonlinear sequences that contain Riesel numbers, Sierpiński numbers, or Riesel-Sierpiński numbers (see [10]).

The standard method for proving results in this area is to use a covering of the integers, and we also employ this technique in our work. A covering is a set of congruences with the property that every integer satisfies at least one of the congruences. For example, the congruences below form a covering of the integers.

$$\begin{array}{rl} 0 & \pmod{2} \\ 3 & \pmod{4} \\ 2 & \pmod{3} \\ 1 & \pmod{8} \\ 1 & \pmod{12} \\ 21 & \pmod{24} \end{array} \tag{1}$$

In fact, this covering appears for the first time in the 1950 article [8] in which Erdős shows that there are infinitely many odd integers that are *not* the sum of a prime and a power of 2. In the construction of Riesel and Sierpiński numbers, the moduli used in the covering are linked to the prime divisors of $2^M - 1$, where M is the least common multiple of the moduli in the covering. (See Sect. 2 for more details.)

In this paper, we discuss Riesel numbers with the property that all of the following expressions are composite for all natural numbers n; on the right side, we provide an equivalent expression that we prefer to use in computations.

$$\begin{aligned} k \cdot 2^n - 1 &= k \cdot 2^n - 1 \\ (k \cdot 2^n - 1) \cdot 2^n - 1 &= k \cdot 2^{2n} - 2^n - 1 \\ ((k \cdot 2^n - 1) \cdot 2^n - 1) \cdot 2^n - 1 &= k \cdot 2^{3n} - 2^{2n} - 2^n - 1 \\ &\vdots \\ ((\cdots(k \cdot 2^n - 1) \cdot 2^n - 1)\cdots) \cdot 2^n - 1 &= k \cdot 2^{\ell n} - \cdots - 2^{3n} - 2^{2n} - 2^n - 1 \end{aligned}$$

We now establish some notation. Let $\ell \in \mathbb{N}$. If k is an odd positive integer such that all of the expressions above are composite, then we call k an ℓ-iterated Riesel number. When we focus our attention on constructing k so that the expression $k \cdot 2^{\ell n} - \cdots - 2^{3n} - 2^{2n} - 2^n - 1$ is composite for all natural numbers n, we say we are building a covering for level ℓ and that we are constructing a level ℓ Riesel number. Throughout the rest of this article, for an odd prime p, we use order$(2, p)$ to denote the order of 2 modulo p. That is, if $m = \text{order}(2, p)$, then m is the smallest positive integer with the property that p divides $2^m - 1$.

We conclude our introductory remarks with an outline of the remainder of the article. In the next section, we construct an infinite family of 3-iterated Riesel numbers. In Sect. 3, we present a few general theorems that simplify the computations in our work. After this, in Sect. 4, we illustrate how the theorems from Sect. 3 are used to construct a 49-iterated Riesel number. We note that the complete set of data that proves that we attain infinitely many 49-iterated Riesel numbers appears in Sect. 5. We conclude with some conjectures in Sect. 5.

2 Preliminary Results

We begin our discussion by demonstrating a construction of a 3-iterated Riesel number. The implications in the table below, combined with the fact that the congruences for n form a covering of the integers, show that if k is odd and satisfies the given congruences, then k is a Riesel number.

$$\begin{array}{llll}
k \equiv -1 \pmod{3} & \& \; n \equiv 1 \pmod{2} & \Rightarrow k \cdot 2^n - 1 \equiv 0 \pmod{3} \\
k \equiv -1 \pmod{5} & \& \; n \equiv 2 \pmod{4} & \Rightarrow k \cdot 2^n - 1 \equiv 0 \pmod{5} \\
k \equiv -1 \pmod{17} & \& \; n \equiv 4 \pmod{8} & \Rightarrow k \cdot 2^n - 1 \equiv 0 \pmod{17} \\
k \equiv -1 \pmod{257} & \& \; n \equiv 8 \pmod{16} & \Rightarrow k \cdot 2^n - 1 \equiv 0 \pmod{257} \\
k \equiv -1 \pmod{65537} & \& \; n \equiv 16 \pmod{32} & \Rightarrow k \cdot 2^n - 1 \equiv 0 \pmod{65537} \\
k \equiv -1 \pmod{641} & \& \; n \equiv 32 \pmod{64} & \Rightarrow k \cdot 2^n - 1 \equiv 0 \pmod{641} \\
k \equiv 1 \pmod{6700417} & \& \; n \equiv 0 \pmod{64} & \Rightarrow k \cdot 2^n - 1 \equiv 0 \pmod{6700417}
\end{array}$$

Using the Chinese Remainder Theorem, the values of k that satisfy all of the congruences in the table above can be combined into one congruence:

$$k \equiv 2935363327246958234 \pmod{3 \cdot 5 \cdot 17 \cdot 257 \cdot 65537 \cdot 641 \cdot 6700417}. \quad (2)$$

Now, we also include the congruences for k from the following table. Note that the congruences for n in the right column form a covering of the integers. For this table, corresponding congruences for k and n combine to ensure that $k \cdot 2^{2n} - 2^n - 1$ is divisible by the appropriate prime.

$$\begin{array}{ll}
k \equiv -1 \pmod{3} & \& \; n \equiv 0 \pmod{2} \\
k \equiv 0 \pmod{19} & \& \; n \equiv 9 \pmod{18} \\
k \equiv -1 \pmod{7} & \& \; n \equiv 1 \text{ or } 2 \pmod{3} \\
k \equiv -1 \pmod{73} & \& \; n \equiv 3 \text{ or } 6 \pmod{9}
\end{array}$$

Again using the Chinese Remainder Theorem, we combine the congruences for k into a single statement:

$$k \equiv 4598 \pmod{3 \cdot 19 \cdot 7 \cdot 73}. \quad (3)$$

This means that if k is odd and satisfies the congruences in (2) and (3), then k is a 2-iterated Riesel number.

We pause to point out that there is a startling detail apparent in the preceding table. When constructing a standard (non-iterated) Riesel number k, there is exactly one congruence in the covering that is connected to a congruence class for k. However, in our level 2 computations, we see that the choice of residue class for k modulo 7 corresponds to *two* congruences in the covering. This also occurs with the choice of residue class for k modulo 73. This observation lays the groundwork for the theoretical results presented in the next section.

We repeat the process on level 3 to complete our work to construct a 3-iterated Riesel number. We again present a table with congruence requirements for k and a set of covering congruences for n. In this table, the congruences for k and n ensure that $k \cdot 2^{3n} - 2^{2n} - 2^n - 1$ is divisible by the appropriate prime. From this table,

$$
\begin{array}{llr}
k \equiv -1 \pmod{3} & \& \; n \equiv & 1 \pmod{2} \\
k \equiv 0 \pmod{13} & \& \; n \equiv & 4 \text{ or } 8 \pmod{12} \\
k \equiv -1 \pmod{5} & \& \; n \equiv & 2 \pmod{4} \\
k \equiv -1 \pmod{241} & \& \; n \equiv & 6,\ 12,\ \text{or } 18 \pmod{24} \\
k \equiv -1 \pmod{97} & \& \; n \equiv & 24 \pmod{48} \\
k \equiv 3 \pmod{673} & \& \; n \equiv & 0 \pmod{48}
\end{array}
$$

we see that k satisfies the following congruence:

$$k \equiv 450591674 \pmod{3 \cdot 13 \cdot 5 \cdot 241 \cdot 97 \cdot 673}. \tag{4}$$

Again, the congruences for n form a covering of the integers. Moreover, note that if the modulus in a congruence for k already appeared in one of the tables for lower levels, then the information about k is consistent. For example, in all three levels, we have $k \equiv 2 (\text{mod } 3)$, and $k \equiv 4 (\text{mod } 5)$ appears in the first and third tables.

Notice once again in the preceding table that the equivalence classes of k modulo 13 and modulo 241 both yield a family of congruences that are used in the covering. In addition, we note that the congruence class for k in the tables above is frequently either 0 or -1 modulo p. This also foreshadows our work in Sect. 3.

Combining the information about k from the three tables in this section, we see that if

$$
\begin{aligned}
&k \equiv 1 \pmod{2}, \\
&k \equiv 4598 \pmod{3 \cdot 19 \cdot 7 \cdot 73}, \text{ and} \\
&k \equiv 2935363327246958234 \pmod{3 \cdot 5 \cdot 17 \cdot 257 \cdot 65537 \cdot 641 \cdot 6700417},
\end{aligned}
$$

then k is a Riesel number, and all three of the integers

$$k \cdot 2^n - 1, \quad k \cdot 2^{2n} - 2^n - 1, \quad \text{and} \quad k \cdot 2^{3n} - 2^{2n} - 2^n - 1$$

are composite for all natural numbers n. Furthermore, there are infinitely many such k in arithmetic progression with this property. The smallest k in this arithmetic progression is

$$k = 18419953728187341536673053729519,$$

and the common difference of the arithmetic progression is

$$2 \cdot 3 \cdot 5 \cdot 7 \cdot 17 \cdot 19 \cdot 73 \cdot 257 \cdot 65537 \cdot 641 \cdot 6700417.$$

Hence, we have just demonstrated that there are infinitely many 3-iterated Riesel numbers.

We remark now that the expression for level ℓ Sierpiński numbers is

$$k \cdot 2^{\ell n} + 2^{(\ell-1)n} + \cdots + 2^n + 1.$$

Since

$$k \cdot 2^{\ell n} + 2^{(\ell-1)n} + \cdots + 2^n + 1 \equiv 0 \pmod{p}$$

if and only if

$$-k \cdot 2^{\ell n} - 2^{(\ell-1)n} - \cdots - 2^n - 1 \equiv 0 \pmod{p},$$

we see that we can replace k with $-k$ in our work above to obtain infinitely many 3-iterated Sierpiński numbers. That is, if

$$k \equiv 48408050804673756347860890265 91 \pmod{73260758808654717171459142756110},$$

then all three of the integers shown below are composite for all natural numbers n:

$$k \cdot 2^n + 1, \quad k \cdot 2^{2n} + 2^n + 1, \quad \text{and} \quad k \cdot 2^{3n} + 2^{2n} + 2^n + 1.$$

In this section, we have seen that the choice $k \equiv 2 \pmod{3}$ seems to cover either the even or the odd values of n for each level. In fact, an easy computation shows that we have the following result.

Theorem 1 *Let* $k \equiv 2 \pmod{3}$, *and suppose*

$$k \cdot 2^{\ell n} - \cdots - 2^{3n} - 2^{2n} - 2^n - 1 \equiv 0 \pmod{3}.$$

Then either

$$\begin{cases} \ell \equiv 2 \pmod{3} \\ n \equiv 0 \pmod{2} \end{cases} \quad or \quad \begin{cases} \ell \equiv 1 \pmod{2} \\ n \equiv 1 \pmod{2} \end{cases}.$$

We give an analogous result for $k \equiv 4 \pmod{5}$ before moving on to the general setting in the next section.

Theorem 2 *Let* $k \equiv 4 \pmod 5$, *and suppose*

$$k \cdot 2^{\ell n} - \cdots - 2^{3n} - 2^{2n} - 2^{n} - 1 \equiv 0 \pmod 5.$$

Then ℓ *and* n *satisfy one of the pairs of congruences given below.*

$$\begin{cases} \ell \equiv 4 \pmod 5 \\ n \equiv 0 \pmod 4 \end{cases}, \quad \begin{cases} \ell \equiv 3 \pmod 4 \\ n \equiv 1 \pmod 4 \end{cases},$$

$$\begin{cases} \ell \equiv 1 \pmod 2 \\ n \equiv 2 \pmod 4 \end{cases}, \text{ or } \begin{cases} \ell \equiv 3 \pmod 4 \\ n \equiv 3 \pmod 4 \end{cases}.$$

Notice that in both Theorems 1 and 2, there is some overlap in the congruences for ℓ. In particular, we have the following corollaries.

Corollary 1 *If* $k \equiv 2 \pmod 3$ *and* $\ell \equiv 5 \pmod 6$, *then*

$$k \cdot 2^{\ell n} - \cdots - 2^{3n} - 2^{2n} - 2^{n} - 1 \equiv 0 \pmod 3$$

for all natural numbers n.

Corollary 2 *If* $k \equiv 4 \pmod 5$ *and* $\ell \equiv 19 \pmod{20}$, *then*

$$k \cdot 2^{\ell n} - \cdots - 2^{3n} - 2^{2n} - 2^{n} - 1 \equiv 0 \pmod 5$$

for all natural numbers n.

Instead of continuing to pursue results related to specific primes, we move into a more general discussion below.

3 Theoretical Results

Working with congruences to construct a level ℓ Riesel number for smaller values of ℓ obfuscates the connections between the levels, the primes, and the moduli of the congruences used in the coverings. In this section, we present theoretical results illustrating these connections. In particular, these results predict which choice(s) of residue class for k yield multiple congruences in the covering for n.

We approach the construction of a level ℓ Riesel number by first considering how to ensure that $0 \pmod m$ appears in the covering. To this end, suppose p is an odd prime, and let m denote the order of 2 modulo p. Then $n \equiv 0 \pmod m$ gives

$$k \cdot 2^{\ell n} - \cdots - 2^{2n} - 2^{n} - 1 \equiv k - \ell \pmod p.$$

Thus, we have our first general theorem.

Theorem 3 *Let p be an odd prime, and let m=order(2, p). Suppose* $n \equiv 0 \pmod{m}$. *Then*

$$k \cdot 2^{\ell n} - \cdots - 2^{3n} - 2^{2n} - 2^n - 1 \equiv 0 \pmod{p}$$

if and only if $k \equiv \ell \pmod{p}$.

Theorem 3 illustrates that there is a connection between the level ℓ and the prime p. The following theorem begins to shed light on the frequent appearance of $k \equiv 0 \pmod{p}$ and $k \equiv -1 \pmod{p}$ in our work in Sect. 2.

Theorem 4 *Let p be an odd prime, and let* $m = order(2, p)$. *Let* $\ell \in \mathbb{N}$. *Suppose*

$$k \cdot 2^{\ell n} - 2^{(\ell-1)n} - 2^{(\ell-2)n} - \cdots - 2^{2n} - 2^n - 1 \tag{5}$$

is divisible by p for $n \equiv \pm 1 \pmod{m}$. *Then the expression in (5) is divisible for all n with* $1 \le n \le m - 1$. *Moreover, we have either*

$$\begin{cases} k \equiv 0 \pmod{p} \\ \ell \equiv 0 \pmod{m} \end{cases} \quad or \quad \begin{cases} k \equiv -1 \pmod{p} \\ \ell \equiv -1 \pmod{m} \end{cases}.$$

Proof Let p be an odd prime, and let $m = \text{order}(2, p)$. Let $\ell \in \mathbb{N}$. Suppose

$$k \cdot 2^{\ell n} - 2^{(\ell-1)n} - 2^{(\ell-2)n} - \cdots - 2^{2n} - 2^n - 1 \equiv 0 \pmod{p}$$

for $n \equiv \pm 1 \pmod{m}$. When $n \equiv 1 \pmod{m}$, this implies $k \cdot 2^\ell \equiv 2^\ell - 1 \pmod{p}$. This is equivalent to

$$1 \equiv 2^\ell(1-k) \pmod{p},$$

which implies $k \equiv 1 - 2^{-\ell} \pmod{p}$. When $n \equiv -1 \pmod{m}$, we have

$$k \cdot 2^{-\ell} \equiv 1 + 2^{-\ell} + \cdots + 2^{-(\ell-1)} \pmod{p},$$

which implies $k \equiv 2(2^\ell - 1) \pmod{p}$. Equating the two expressions for k, we have $2(2^\ell - 1) \equiv 1 - 2^{-\ell} \pmod{p}$. After rearranging, we see that

$$(2^{\ell+1} - 1)(2^\ell - 1) \equiv 0 \pmod{p}.$$

If p divides $2^{\ell+1} - 1$, then $\ell \equiv -1 \pmod{m}$, and in this case, we have $k \equiv -1 \pmod{p}$. On the other hand, if p divides $2^\ell - 1$, then $\ell \equiv 0 \pmod{m}$, which in turn implies that $k \equiv 0 \pmod{p}$.

Suppose now that $1 < n < m - 1$. Then p does not divide $2^n - 1$.

If $\ell \equiv 0 \pmod{m}$, then p divides $2^{\ell n} - 1$. Since

$$2^{\ell n} - 1 = (2^n - 1)(2^{(\ell-1)n} + 2^{(\ell-2)n} + \cdots + 2^{2n} + 2^n + 1),$$

we see that

$$2^{(\ell-1)n} + 2^{(\ell-2)n} + \cdots + 2^{2n} + 2^n + 1 \equiv 0 \pmod{p}.$$

So if $k \equiv 0 \pmod{p}$, then we have

$$k \cdot 2^{\ell n} - 2^{(\ell-1)n} - 2^{(\ell-2)n} - \cdots - 2^{2n} - 2^n - 1 \equiv 0 \pmod{p}.$$

If $k \equiv -1 \pmod{p}$, then

$$k \cdot 2^{-n}(2^n - 1) \equiv 2^{-n} - 1 \pmod{p}.$$

So if $\ell \equiv -1 \pmod{m}$, we have

$$k \cdot 2^{\ell n}(2^n - 1) \equiv 2^{\ell n} - 1 \pmod{p},$$

and since p does not divide $2^n - 1$, we see that

$$k \cdot 2^{\ell n} \equiv 2^{(\ell-1)n} + 2^{(\ell-2)n} + \cdots + 2^{2n} + 2^n + 1 \pmod{p}.$$

Therefore, if

$$k \cdot 2^{\ell n} - 2^{(\ell-1)n} - 2^{(\ell-2)n} - \cdots - 2^{2n} - 2^n - 1 \equiv 0 \pmod{p}$$

for $n \equiv \pm 1 \pmod{m}$, then we have

$$k \cdot 2^{\ell n} - 2^{(\ell-1)n} - 2^{(\ell-2)n} - \cdots - 2^{2n} - 2^n - 1 \equiv 0 \pmod{p}$$

for all n with $1 \leq n \leq m-1$, provided either $k \equiv 0 \pmod{p}$ and $\ell \equiv 0 \pmod{m}$ or $k \equiv -1 \pmod{p}$ and $\ell \equiv -1 \pmod{m}$. □

The proof of the previous theorem helps us to develop the following partial converse. We note that Theorem 4 and its proof help us to understand how primes, levels, and moduli connected for iterated Riesel numbers, but Theorem 5 is useful in computations related to the construction of level ℓ Riesel numbers.

Theorem 5 *Let p be an odd prime, and let $m = order(2, p)$. Let $\ell \in \mathbb{N}$. If either*

$$\begin{cases} k \equiv 0 \pmod{p} \\ \ell \equiv 0 \pmod{m} \end{cases} \quad or \quad \begin{cases} k \equiv -1 \pmod{p} \\ \ell \equiv -1 \pmod{m} \end{cases},$$

then $k \cdot 2^{\ell n} - 2^{(\ell-1)n} - 2^{(\ell-2)n} - \cdots - 2^{2n} - 2^n - 1$ is divisible by p for all n not divisible by m.

Proof Suppose first that $k \equiv 0 \pmod{p}$ and $\ell \equiv 0 \pmod{m}$. Then p divides $2^\ell - 1$, and hence also $2^{\ell n} - 1$. Moreover, for $n \not\equiv 0 \pmod{m}$, we have $2^n - 1 \not\equiv 0 \pmod{p}$.

This implies that p divides

$$2^{(\ell-1)n} + 2^{(\ell-2)n} + \cdots + 2^{2n} + 2^n + 1.$$

Furthermore, since $k \equiv 0 \pmod{p}$, we see that

$$k \cdot 2^{\ell n} - 2^{(\ell-1)n} - 2^{(\ell-2)n} - \cdots - 2^{2n} - 2^n - 1 \equiv 0 \pmod{p}.$$

Now suppose $k \equiv -1 \pmod{p}$ and $\ell \equiv -1 \pmod{m}$. Then p divides $2^{\ell+1} - 1$, and hence also $2^{(\ell+1)n} - 1$. Thus,

$$2^{\ell n} + 2^{(\ell-1)n} + 2^{(\ell-2)n} + \cdots + 2^{2n} + 2^n + 1 \equiv 0 \pmod{p}.$$

Since $k \equiv -1 \pmod{p}$, we see now that

$$k \cdot 2^{\ell n} - 2^{(\ell-1)n} - 2^{(\ell-2)n} - \cdots - 2^{2n} - 2^n - 1 \equiv 0 \pmod{p}.$$

Putting together Theorems 3 and 5, we have the following corollary. Note that Corollaries 1 and 2 are special cases of Corollary 3.

Corollary 3 *Let $\ell \in \mathbb{N}$. Let p be an odd prime, and let $m = order(2, p)$. If either*

$$\begin{cases} k \equiv 0 \pmod{p} \\ \ell \equiv 0 \pmod{pm} \end{cases} \quad or \quad \begin{cases} k \equiv -1 \pmod{p} \\ \ell \equiv -1 \pmod{pm} \end{cases},$$

then $k \cdot 2^{\ell n} - 2^{(\ell-1)n} - 2^{(\ell-2)n} - \cdots - 2^{2n} - 2^n - 1$ is divisible by p for all natural numbers n.

We remarked in Sect. 2 that some choices of residue class for k correspond to a family of congruences in the covering for n. The next theorem provides insight about these families of congruences.

Theorem 6 *Let p be an odd prime, and let $m = order(2, p)$. Let $\ell \in \mathbb{N}$. Suppose*

$$k \cdot 2^{\ell n} - 2^{(\ell-1)n} - 2^{(\ell-2)n} - \cdots - 2^{2n} - 2^n - 1 \equiv 0 \pmod{p} \tag{6}$$

for $n \equiv \pm a \pmod{m}$ for some a with $1 < a < m - 1$. Then the expression in (6) is divisible by p for all natural numbers n not divisible by m with $n \equiv ad \pmod{m}$ for some integer d.

Proof Let p be an odd prime, and let $m = \text{order}(2, p)$. Let $\ell \in \mathbb{N}$. Suppose

$$k \cdot 2^{\ell n} - 2^{(\ell-1)n} - 2^{(\ell-2)n} - \cdots - 2^{2n} - 2^n - 1 \equiv 0 \pmod{p}$$

for $n \equiv \pm a \pmod{m}$ for some a with $1 < a < m - 1$.

Then we have

$$\begin{aligned}
k\cdot 2^{a\ell} &= 2^{a(\ell-1)}+2^{a(\ell-2)}+\cdots+2^{a}+1 \pmod p\\
(2^a-1)k\cdot 2^{a\ell} &\equiv 2^{a\ell}-1 \pmod p\\
(2^a-1)k &\equiv 1-2^{-a\ell} \pmod p.
\end{aligned}$$

We also have

$$\begin{aligned}
k\cdot 2^{-a\ell} &= 2^{-a(\ell-1)}+2^{-a(\ell-2)}+\cdots+2^{-a}+1 \pmod p\\
k &\equiv 2^{a}+2^{2a}+\cdots+2^{(\ell-1)a}+2^{a\ell} \pmod p\\
k+1 &\equiv k\cdot 2^{a\ell}+2^{a\ell} \pmod p\\
k(1-2^{a\ell}) &\equiv 2^{a\ell}-1 \pmod p\\
(k+1)(2^{a\ell}-1) &\equiv 0 \pmod p.
\end{aligned}$$

Thus, either $k\equiv -1 \pmod p$ or p divides $2^{a\ell}-1$. If $2^{a\ell}-1\equiv 0 \pmod p$, then

$$(2^a-1)k \equiv 1-2^{-a\ell} \pmod p$$

implies k is divisible by p since 2^a-1 is not divisible by p. Moreover, in this case we also have

$$a\ell \equiv 0 \pmod m.$$

If $k\equiv -1 \pmod p$, then we have

$$2^{a\ell}+2^{a(\ell-1)}+2^{a(\ell-2)}+\cdots+2^{a}+1 \equiv 0 \pmod p.$$

This implies $2^{a(\ell+1)}-1$ is divisible by p, whence $a(\ell+1)\equiv 0 \pmod m$. □

Similar to using Theorem 5 to construct level ℓ Riesel numbers, we present the following partial converse of Theorem 6. This result is the one that we use to construct a covering for level ℓ in certain circumstances. We omit the proof due to its similarity with the proof of Theorem 5.

Theorem 7 *Let p be an odd prime, and let $m = order(2, p)$. Let $\ell \in \mathbb{N}$. If either $k\equiv 0 \pmod p$ and $a\ell\equiv 0 \pmod m$ or $k\equiv -1 \pmod p$ and $a(\ell+1)\equiv 0 \pmod m$, then $k\cdot 2^{\ell n}-2^{(\ell-1)n}-2^{(\ell-2)n}-\cdots-2^{2n}-2^{n}-1$ is divisible by p for all n not divisible by m such that $n\equiv ad \pmod m$ with $d\in\mathbb{Z}$.*

4 Construction of 49-Iterated Riesel Numbers

In Sect. 2, we showed the primes and covering congruences used to construct a 3-iterated Riesel number. In this section, we continue our work; the result is a 49-iterated Riesel number. While in Sect. 2 we included the implication for each pair of congruences (information about k modulo p and a congruence for the covering), in this section we simply indicate the level completed along with the congruences.

4.1 Levels Completed Using One Prime

Applying Corollary 3 with the prime $p = 3$ and $k \equiv 2 \pmod 3$, we see that all levels with $\ell \equiv 5 \pmod 6$ are covered using the very simple covering shown below for even and odd integers.

$$\begin{array}{l} 0 \pmod 2 \\ 1 \pmod 2 \end{array}$$

Applying the same corollary with the prime $p = 5$ and $k \equiv 4 \pmod 5$, we have the covering

$$\begin{array}{l} 0 \pmod 4 \\ 1 \pmod 4 \\ 2 \pmod 4 \\ 3 \pmod 4 \end{array}$$

for levels with $\ell \equiv 19 \pmod{20}$.

Similarly, levels with $\ell \equiv 20 \pmod{21}$ are completed using the covering

$$\begin{array}{l} 0 \pmod 3 \\ 1 \pmod 3 \\ 2 \pmod 3 \end{array}$$

by using $k \equiv 6 \pmod 7$ for the prime $p = 7$.

Notice that applying Corollary 3 helps us to show that $k \cdot 2^{\ell n} - 2^{(\ell-1)n} - 2^{(\ell-2)n} - \cdots - 2^{2n} - 2^n - 1$ is composite for all n for 94 of every 420 values of ℓ just using the primes 3, 5, and 7.

4.2 Levels Completed Using Two Primes

Since we have chosen $k \equiv 2 \pmod 3$, we can also see that if ℓ and n are odd, then $k \cdot 2^{\ell n} - 2^{(\ell-1)n} - \cdots - 2^n - 1$ is divisible by 3. Also, using Theorem 3, we see that $k \equiv 4 \pmod 5$ implies that if $\ell \equiv 4 \pmod 5$, then $k \cdot 2^{\ell n} - 2^{(\ell-1)n} - \cdots - 2^n - 1$ is divisible by 5 if n is a multiple of 4. In addition, if $n \equiv 2 \pmod 4$ and ℓ is odd, then $k \cdot 2^{\ell n} - 2^{(\ell-1)n} - \cdots - 2^n - 1$ is divisible by 5. Altogether, these facts means that if $\ell \equiv 9 \pmod{10}$, then $k \cdot 2^{\ell n} - 2^{(\ell-1)n} - \cdots - 2^n - 1$ is divisible by 5 for all natural numbers n. Combining this with the results in the previous subsection, we see that $k \cdot 2^{\ell n} - 2^{(\ell-1)n} - 2^{(\ell-2)n} - \cdots - 2^{2n} - 2^n - 1$ is composite for all n for 74 of every 210 values of ℓ just using the primes 3, 5, and 7.

Our work for level 14 illustrates another approach to finding a covering. Since $k \equiv 2 \pmod 3$ and $\ell = 14$ means $\ell \equiv 2 \pmod 3$, Theorem 3 tells us that $k \cdot 2^{\ell n} - 2^{(\ell-1)n} - 2^{(\ell-2)n} - \cdots - 2^{2n} - 2^n - 1$ is divisible by 3 for all even values of n. In addition, the order of 2 modulo $p = 43$ is $m = 14$. Thus choosing $k \equiv 0 \pmod{43}$, Theorem 5 gives all n other than $n \equiv 0 \pmod{14}$ in the covering since $\ell \equiv 0 \pmod m$.

4.3 Levels Completed Using More Than Two Primes

We illustrate the process used to complete other levels by focusing on a few examples. To start, consider level $\ell = 6$. First, we use $p = 127$ with corresponding $m = 7$. Since $\ell \equiv -1 \pmod{m}$, we use $k \equiv -1 \pmod{p}$ and Theorem 5 to obtain $n \equiv 1a \pmod{7}$ (with $0 < a < 7$) in the covering. This means that we are simply missing values of n that are multiples of 7. Next, we use $p = 337$ with corresponding $m = 21$. Since $7\ell \equiv 0 \pmod{m}$, we use $k \equiv 0 \pmod{p}$ and Theorem 7 to obtain $n \equiv 7a \pmod{21}$ (with $0 < 7a < 21$) in the covering. At this stage, the values of n that are missing in the covering are the multiples of 21. Since there is not another prime with order$(2, p)$ dividing 21 (other than 7, 127, and 337, which have already been used), we expand the least common multiple of the moduli in the covering to 42. The multiples of 21 are now represented by $n \equiv 0 \pmod{42}$ and $n \equiv 21 \pmod{42}$. We handle $n \equiv 0 \pmod{42}$ by choosing $k \equiv \ell \pmod{5419}$ since order$(2, 5419) = 42$. We complete the covering using $p = 43$ with corresponding $m = 14$. Since $7\ell \equiv 0 \pmod{m}$, we use $k \equiv 0 \pmod{p}$ and Theorem 7 to obtain $n \equiv 7a \pmod{14}$ (with $0 < 7a < 14$) in the covering. That is, we now have $n \equiv 7 \pmod{14}$ in the covering, which accounts for $n \equiv 21 \pmod{42}$.

For a final example, consider $\ell = 32$. Since $\ell \equiv 2 \pmod{3}$, we have $n \equiv 0 \pmod{2}$ in the covering by Theorem 1. Next, consider $p = 23$ with corresponding $m = 11$. By Theorem 5, since $\ell \equiv -1 \pmod{m}$, using $k \equiv -1 \pmod{p}$ produces $n \equiv 1a \pmod{11}$ (with $0 < a < 11$) in the covering. Hence, we now only lack odd multiples of 11 in the covering. Now we employ $p = 727$ with corresponding $m = 121$. Since $11(\ell + 1) \equiv 0 \pmod{121}$, using $k \equiv -1 \pmod{p}$ and Theorem 7 yields $n \equiv 11a \pmod{121}$ (with $0 < 11a < 121$) in the covering. Since the least common multiple of 2, 11, and 121 is 242, we see that the only values of n that we need to account for are in the residue class $n \equiv 121 \pmod{242}$. We thus complete the covering using $p = 117371$. Taking $k \equiv 0 \pmod{p}$, since $121\ell \equiv 0 \pmod{242}$, Theorem 7 implies we obtain $n \equiv 121a \pmod{242}$ (with $0 < 121a < 242$) in the covering.

5 Conjectures

The theoretical results in Sect. 3 and experience building coverings for various levels lead us to make the following conjectures.

Conjecture 1 Let $\ell \in \mathbb{N}$. Then there are infinitely many level ℓ Riesel numbers.

While our first conjecture simply posits that we can build a covering for any given level ℓ, the following conjecture makes the stronger claim that we can build coverings for all levels up to and including ℓ (using consistent equivalence class values for k as appropriate).

Conjecture 2 Let $\ell \in \mathbb{N}$. Then there are infinitely many ℓ-iterated Riesel numbers.

Acknowledgements The authors thank Washington and Lee University for support while completing the work in this paper through a Lenfest Summer Research Grant. Moreover, the authors appreciate the opportunity to speak about this work at CANT 2018.

Appendix

Level	Congruences for k	Congruences for n
1	$k \equiv -1 \pmod{3}$	$n \equiv 1 \pmod{2}$
	$k \equiv -1 \pmod{5}$	$n \equiv 2 \pmod{4}$
	$k \equiv -1 \pmod{17}$	$n \equiv 4 \pmod{8}$
	$k \equiv -1 \pmod{257}$	$n \equiv 8 \pmod{16}$
	$k \equiv -1 \pmod{65537}$	$n \equiv 16 \pmod{32}$
	$k \equiv -1 \pmod{641}$	$n \equiv 32 \pmod{64}$
	$k \equiv 1 \pmod{6700417}$	$n \equiv 0 \pmod{64}$
2	$k \equiv -1 \pmod{3}$	$n \equiv 0 \pmod{2}$
	$k \equiv 0 \pmod{19}$	$n \equiv 9 \pmod{18}$
	$k \equiv -1 \pmod{7}$	$n \equiv 1a \pmod{3}$
	$k \equiv -1 \pmod{73}$	$n \equiv 3a \pmod{9}$
3	$k \equiv -1 \pmod{3}$	$n \equiv 1 \pmod{2}$
	$k \equiv 0 \pmod{13}$	$n \equiv 4a \pmod{12}$
	$k \equiv -1 \pmod{5}$	$n \equiv 2 \pmod{4}$
	$k \equiv -1 \pmod{241}$	$n \equiv 6a \pmod{24}$
	$k \equiv -1 \pmod{97}$	$n \equiv 24 \pmod{48}$
	$k \equiv 3 \pmod{673}$	$n \equiv 0 \pmod{48}$
4	$k \equiv 1 \pmod{5}$	$n \equiv 0 \pmod{4}$
	$k \equiv 0 \pmod{41}$	$n \equiv 5a \pmod{20}$
	$k \equiv -1 \pmod{31}$	$n \equiv 1a \pmod{5}$
5	$k \equiv -1 \pmod{3}$	$n \equiv 0, 1 \pmod{2}$
6	$k \equiv -1 \pmod{127}$	$n \equiv 1a \pmod{7}$
	$k \equiv 0 \pmod{43}$	$n \equiv 7 \pmod{14}$
	$k \equiv 0 \pmod{337}$	$n \equiv 7a \pmod{21}$
	$k \equiv 6 \pmod{5419}$	$n \equiv 0 \pmod{42}$
7	$k \equiv -1 \pmod{3}$	$n \equiv 1 \pmod{2}$
	$k \equiv 0 \pmod{43}$	$n \equiv 2a \pmod{14}$
	$k \equiv -1 \pmod{5}$	$n \equiv 2 \pmod{4}$
	$k \equiv -1 \pmod{241}$	$n \equiv 3a \pmod{24}$
	$k \equiv -1 \pmod{97}$	$n \equiv 6a \pmod{48}$
	$k \equiv -1 \pmod{17}$	$n \equiv 1a \pmod{8}$
	$k \equiv 0 \pmod{2017}$	$n \equiv 56a \pmod{336}$
	$k \equiv 7 \pmod{3361}$	$n \equiv 0 \pmod{336}$
8	$k \equiv -1 \pmod{3}$	$n \equiv 0 \pmod{2}$
	$k \equiv -1 \pmod{73}$	$n \equiv 1a \pmod{9}$
	$k \equiv 0 \pmod{19}$	$n \equiv 9 \pmod{18}$
9	$k \equiv -1 \pmod{3}$	$n \equiv 1 \pmod{2}$
	$k \equiv -1 \pmod{5}$	$n \equiv 0 \pmod{2}$

10	$k \equiv -1 \pmod{23}$	$n \equiv 1a \pmod{11}$
	$k \equiv -1 \pmod{727}$	$n \equiv 11a \pmod{121}$
	$k \equiv 10 \pmod{p_1}$	$n \equiv 0 \pmod{121}$
11	$k \equiv -1 \pmod{3}$	$n \equiv 0, 1 \pmod{2}$
12	$k \equiv -1 \pmod{8191}$	$n \equiv 1a \pmod{13}$
	$k \equiv -1 \pmod{4057}$	$n \equiv 13a \pmod{169}$
	$k \equiv 12 \pmod{6740339310641}$	$n \equiv 0 \pmod{169}$
13	$k \equiv -1 \pmod{3}$	$n \equiv 1a \pmod{2}$
	$k \equiv 0 \pmod{2731}$	$n \equiv 2a \pmod{26}$
	$k \equiv 0 \pmod{p_2}$	$n \equiv 26a \pmod{338}$
	$k \equiv 13 \pmod{p_3}$	$n \equiv 0 \pmod{338}$

$p_1 = 1786393878363164227858270210279$
$p_2 = 4929910764223610387$
$p_3 = 1852623864601108673274261404 3$

Level	Congruences for k	Congruences for n
14	$k \equiv -1 \pmod{3}$	$n \equiv 0 \pmod{2}$
	$k \equiv 0 \pmod{43}$	$n \equiv 1a \pmod{14}$
15	$k \equiv -1 \pmod{3}$	$n \equiv 1 \pmod{2}$
	$k \equiv -1 \pmod{5}$	$n \equiv 2 \pmod{4}$
	$k \equiv -1 \pmod{41}$	$n \equiv 2a \pmod{20}$
	$k \equiv -1 \pmod{251}$	$n \equiv 10a \pmod{50}$
	$k \equiv 1 \pmod{101}$	$n \equiv 0 \pmod{100}$
16	$k \equiv -1 \pmod{131071}$	$n \equiv 1a \pmod{17}$
	$k \equiv -1 \pmod{12761663}$	$n \equiv 17a \pmod{289}$
	$k \equiv 12 \pmod{p_4}$	$n \equiv 0 \pmod{289}$
17	$k \equiv -1 \pmod{3}$	$n \equiv 0, 1 \pmod{2}$
18	$k \equiv 0 \pmod{19}$	$n \equiv 1a \pmod{18}$
	$k \equiv 0 \pmod{37}$	$n \equiv 2a \pmod{36}$
	$k \equiv 18 \pmod{109}$	$n \equiv 0 \pmod{36}$
19	$k \equiv -1 \pmod{5}$	$n \equiv 0, 1a \pmod{4}$
20	$k \equiv -1 \pmod{3}$	$n \equiv 0 \pmod{2}$
	$k \equiv 0 \pmod{41}$	$n \equiv 1a \pmod{20}$
21	$k \equiv -1 \pmod{23}$	$n \equiv 1a \pmod{11}$
	$k \equiv 334 \pmod{397}$	$n \equiv 11 \pmod{44}$
	$k \equiv 21 \pmod{683}$	$n \equiv 0 \pmod{22}$
	$k \equiv 2048 \pmod{2113}$	$n \equiv 33 \pmod{44}$
22	$k \equiv -1 \pmod{23}$	$n \equiv 0 \pmod{11}$
	$k \equiv 0 \pmod{89}$	$n \equiv 1a \pmod{11}$
23	$k \equiv -1 \pmod{3}$	$n \equiv 0, 1 \pmod{2}$
24	$k \equiv -1 \pmod{5}$	$n \equiv 0 \pmod{4}$
	$k \equiv 600 \pmod{601}$	$n \equiv 1a \pmod{25}$
	$k \equiv 0 \pmod{41}$	$n \equiv 5a \pmod{20}$
25	$k \equiv -1 \pmod{3}$	$n \equiv 1 \pmod{2}$
	$k \equiv 25 \pmod{1801}$	$n \equiv 0 \pmod{25}$
	$k \equiv 0 \pmod{4051}$	$n \equiv 2a \pmod{50}$
26	$k \equiv -1 \pmod{3}$	$n \equiv 0 \pmod{2}$
	$k \equiv 0 \pmod{2731}$	$n \equiv 1a \pmod{26}$
27	$k \equiv 27 \pmod{113}$	$n \equiv 0 \pmod{28}$
	$k \equiv -1 \pmod{29}$	$n \equiv 1a \pmod{28}$

28	$k \equiv 0 \pmod{43}$	$n \equiv 1a \pmod{14}$
	$k \equiv -1 \pmod{29}$	$n \equiv 0 \pmod{28}$
	$k \equiv 0 \pmod{15790321}$	$n \equiv 14 \pmod{56}$
	$k \equiv 599 \pmod{5153}$	$n \equiv 42 \pmod{112}$
	$k \equiv 45437131183 \pmod{54410972897}$	$n \equiv 98 \pmod{112}$
29	$k \equiv -1 \pmod{3}$	$n \equiv 1 \pmod{2}$
	$k \equiv -1 \pmod{5}$	$n \equiv 0 \pmod{2}$
30	$k \equiv 0 \pmod{11}$	$n \equiv 1a \pmod{10}$
	$k \equiv 0 \pmod{41}$	$n \equiv 10a \pmod{20}$
	$k \equiv 0 \pmod{13}$	$n \equiv 2a \pmod{12}$
	$k \equiv -1 \pmod{31}$	$n \equiv 0 \pmod{5}$
31	$k \equiv -1 \pmod{3}$	$n \equiv 1 \pmod{2}$
	$k \equiv -1 \pmod{5}$	$n \equiv 2 \pmod{4}$
	$k \equiv -1 \pmod{5581}$	$n \equiv 4a \pmod{124}$
	$k \equiv 31 \pmod{8681}$	$n \equiv 0 \pmod{124}$

$p_4 = 17905831260439274251\,1009$

Level	Congruences for k	Congruences for n
32	$k \equiv -1 \pmod{3}$	$n \equiv 0 \pmod{2}$
	$k \equiv -1 \pmod{23}$	$n \equiv 1a \pmod{11}$
	$k \equiv -1 \pmod{727}$	$n \equiv 11a \pmod{121}$
	$k \equiv 0 \pmod{117371}$	$n \equiv 121 \pmod{242}$
33	$k \equiv -1 \pmod{3}$	$n \equiv 1 \pmod{2}$
	$k \equiv -1 \pmod{5}$	$n \equiv 2 \pmod{4}$
	$k \equiv -1 \pmod{17}$	$n \equiv 0 \pmod{4}$
34	$k \equiv -1 \pmod{5}$	$n \equiv 0 \pmod{4}$
	$k \equiv 0 \pmod{41}$	$n \equiv 10 \pmod{20}$
	$k \equiv -1 \pmod{31}$	$n \equiv 1a \pmod{5}$
	$k \equiv 0 \pmod{11}$	$n \equiv 5 \pmod{10}$
35	$k \equiv -1 \pmod{3}$	$n \equiv 0, 1 \pmod{2}$
36	$k \equiv 0 \pmod{37}$	$n \equiv 1a \pmod{36}$
	$k \equiv 0 \pmod{433}$	$n \equiv 2a \pmod{72}$
	$k \equiv 0 \pmod{38737}$	$n \equiv 0 \pmod{72}$
37	$k \equiv 0 \pmod{233}$	$n \equiv 1a \pmod{37}$
	$k \equiv 37 \pmod{616318177}$	$n \equiv 0 \pmod{37}$
38	$k \equiv 0 \pmod{174763}$	$n \equiv 1a \pmod{38}$
	$k \equiv 38 \pmod{524287}$	$n \equiv 0 \pmod{19}$
39	$k \equiv -1 \pmod{5}$	$n \equiv 0, 1a \pmod{4}$
40	$k \equiv 0 \pmod{41}$	$n \equiv 1a \pmod{20}$
	$k \equiv 0 \pmod{11}$	$n \equiv 1a \pmod{10}$
	$k \equiv 0 \pmod{4278255361}$	$n \equiv 2a \pmod{80}$
	$k \equiv 40 \pmod{61681}$	$n \equiv 0 \pmod{40}$
41	$k \equiv -1 \pmod{3}$	$n \equiv 0, 1 \pmod{2}$
42	$k \equiv 0 \pmod{337}$	$n \equiv 1a \pmod{21}$
	$k \equiv 0 \pmod{92737}$	$n \equiv 3a \pmod{63}$
	$k \equiv 42 \pmod{649657}$	$n \equiv 0 \pmod{63}$
43	$k \equiv 0 \pmod{431}$	$n \equiv 1a \pmod{43}$
	$k \equiv 43 \pmod{9719}$	$n \equiv 0 \pmod{43}$
44	$k \equiv -1 \pmod{3}$	$n \equiv 0 \pmod{2}$
	$k \equiv -1 \pmod{73}$	$n \equiv 1a \pmod{9}$
	$k \equiv 0 \pmod{19}$	$n \equiv 9 \pmod{18}$

45	$k \equiv -1 \pmod{3}$	$n \equiv 1 \pmod{2}$
	$k \equiv -1 \pmod{5}$	$n \equiv 2 \pmod{4}$
	$k \equiv 0 \pmod{11}$	$n \equiv 2a \pmod{10}$
	$k \equiv 0 \pmod{251}$	$n \equiv 10a \pmod{50}$
	$k \equiv 0 \pmod{13}$	$n \equiv 4a \pmod{12}$
	$k \equiv -1 \pmod{241}$	$n \equiv 12 \pmod{24}$
	$k \equiv -1 \pmod{97}$	$n \equiv 24 \pmod{48}$
	$k \equiv 45 \pmod{4801}$	$n \equiv 0 \pmod{1200}$
46	$k \equiv 0 \pmod{47}$	$n \equiv 1a \pmod{23}$
	$k \equiv 46 \pmod{178481}$	$n \equiv 0 \pmod{23}$
47	$k \equiv -1 \pmod{3}$	$n \equiv 0, 1 \pmod{2}$
48	$k \equiv 0 \pmod{13}$	$n \equiv 1a \pmod{12}$
	$k \equiv -1 \pmod{7}$	$n \equiv 0 \pmod{3}$
49	$k \equiv -1 \pmod{3}$	$n \equiv 1 \pmod{2}$
	$k \equiv -1 \pmod{5}$	$n \equiv 0 \pmod{2}$

References

1. W. R. Alford, A. Granville, and C. Pomerance, There are infinitely many Carmichael numbers, *Ann. of Math. (2)* **139** (1994), no. 3, 703–722
2. D. Baczkowski, J. Eitner, C. E. Finch, M. Kozek, and B. Suminski, Polygonal, Sierpiński, and Riesel numbers, *J. Integer Seq.* **18** (2015), Article 15.8.1
3. D. Baczkowski, O. Fasoranti, C.E. Finch, Lucas-Sierpiński and Lucas-Riesel numbers. *Fibonacci Quart.* **49** (2011), no (4), 334–339
4. W. Banks, C. Finch, F. Luca, C. Pomerance, P. Stănică, Sierpiński and Carmichael numbers. *Trans. Amer. Math. Soc.* **367** (2015), no (1), 355–376
5. P. Berrizbeita, J. G. Fernandes, M. J. González, F. Luca, V. J. Mejía Huguet, On Cullen numbers which are both Riesel and Sierpiński
6. Y.-G. Chen, On integers of the forms $k^r - 2^n$ and $k^r 2^n + 1$. J. Number Theory **98**, 310–319 (2003)
7. Y.-G. Chen, On integers of the forms $k \pm 2^n$ and $k \cdot 2^n \pm 1$. J. Number Theory **125**, 14–25 (2007)
8. P. Erdős, On integers of the form $2^k + p$ and some related problems. Summa Brasil. Math. **2**, 113–123 (1950)
9. M. Filaseta, C. Finch, M. Kozek, On powers associated with Sierpiński numbers, Riesel numbers and Polignac's conjecture. *J. Number Theory* **128** (2008), no. (7), 1916–1940
10. C. Finch, J. Harrington, L. Jones, Nonlinear Sierpiński and Riesel numbers. J. Number Theory **133**(2), 534–544 (2013)
11. C. Finch, L. Jones, Perfect Power Riesel Numbers. J. Number Theory **150**, 41–46 (2015)
12. D. Ismailescu and P. S. Park, On pairwise intersections of the Fibonacci, Sierpiński, and Riesel sequences. *J. Integer Seq.* **16** (2013), no. 9, Article 13.9.8, 9pp
13. F. Luca and V. J. Mejía Huguet, Fibonacci-Riesel and Fibonacci-Sierpiński numbers, *Fibonacci Quart.* **46/47** (2008/09), no. 3, 216–219
14. H. Riesel, Några stora primtal. Elementa **39**, 258–260 (1956)
15. W. Sierpiński, Sur un problème concernant les nombres $k2^n + 1$. Elem. Math. **15**, 73–74 (1960)

A General Framework for Studying Finite Rainbow Configurations

Mike Desgrottes, Steven Senger, David Soukup and Renjun Zhu

Abstract Given a coloring of a set, classical Ramsey theory looks for various configurations within a color class. Rainbow configurations, also called anti-Ramsey configurations, are configurations that occur across distinct color classes. We present some very general results about the types of colorings that will guarantee various types of rainbow configurations in finite settings, as well as several illustrative corollaries. The main goal of this note is to present a flexible framework for decomposing finite sets while guaranteeing the existence of some desired structure across the decomposition.

1 Introduction

Classical Ramsey problems typically involve partitioning an ambient set up into disjoint subsets called color classes, then looking for conditions under which a given configuration will be present in one of the color classes. One canonical example is Schur's Theorem, which says that if you color the natural numbers with a finite number of colors, then there must be a color class with a triple of the form $(x, y, x+y)$.

Here, we consider so-called anti-Ramsey or *rainbow* problems. Rainbow problems have been studied in different contexts; see [1, 9] and the references therein for some arithmetic rainbow results, and see [2, 10] and the references therein for results in graph theory. We begin with some basic definitions.

Definition 1 A **coloring** of a set X is a function $f : X \to \mathscr{C}$ for some set $\mathscr{C}$ of colors. The preimages $\{f^{-1}(i)\}$, for each $i \in C$, are the **color classes** of the coloring. Notice that the color classes form a partition of X.

Definition 2 A **rainbow configuration** is a k-tuple $(x_1, x_2, \dots, x_k) \in X^k$ such that $x_1, x_2, \dots, x_k$ all belong to distinct color classes.

M. Desgrottes · S. Senger (✉) · D. Soukup · R. Zhu
Missouri State University, Springfield, MO, USA
e-mail: stevensenger@gmail.com

M. B. Nathanson (ed.), *Combinatorial and Additive Number Theory III*,
Springer Proceedings in Mathematics & Statistics 297,
https://doi.org/10.1007/978-3-030-31106-3_5

1.1 Results

In this paper, we prove some very general results on what kinds of colorings will guarantee the existence of various types of rainbow configurations. Our main result is very general, so we first give an illustrative corollary, which can be seen as a rainbow version of Schur's Theorem. We refer the reader to [1, 9] for related results.

Theorem 1 *Suppose we have a coloring of a finite abelian group, G. If no color class has size* $\geq \frac{2}{27}|G|$ *then there must be a rainbow triple of the form* $(x, y, x + y)$.

Theorem 1 is a corollary of our main theorem.

Theorem 2 *Let* X *be a finite set of size* n, *and* $E \subset X^k$ *be a set of* k*-tuples, with the property that there exist* $M_{i,j}$, *such that for any*

$$x = (x_1, x_2, \ldots, x_k) \in E,$$

we have that $|\{y \in E : y_i = x_i, y_j = x_j\}| \leq M_{i,j}$. *Define*

$$M = \sum_{i<j} M_{i,j}.$$

If no color class has size $\geq Cn$, *where* $|E| \geq Dn^2$, *and* $C < \frac{2D}{9M}$, *then there must be a rainbow* k*-tuple in* E.

In this statement, E is the set of k-tuples of elements of X that form whatever configuration we are concerned with (such as $(x, y, x + y)$ in Theorem 1). The quantity M depends on how many of these k-tuples share a pair of coordinates (in Theorem 1, each pair of coordinates uniquely determines a triple). We start with some ambient set, X, that can be decomposed into disjoint subsets, with some control on the size of these subsets. The main idea is that we can guarantee the existence of various types of configurations if we know something about the structure of these configurations (measured by the $M_{i,j}$), as well as how many such configurations there are (measured by D). Note that the structure is quantified without dependence on any particular algebraic structure. This allows Theorem 2 to apply to settings where there is no explicit algebraic structure.

To illustrate the use of Theorem 2, we now list a few other applications. The next result was inspired by the corresponding Ramsey problem: which colorings of various rings admit monochromatic quadruples of the form $(x, y, x + y, xy)$? Note that this result involves both addition and multiplication. We state a similar result in the negative. That is, if there are no rainbow configurations, then we must have a color class that is too large. See [5–7, 11] for background and related Ramsey type problems. In what follows, let $\mathbb{F}_q$ denote the finite field of q elements, where q is an odd prime power.

Corollary 1 *If we color* $\mathbb{F}_q$ *such that each color class has size* $< \frac{(2-o(1))q}{63}$, *then there must be a rainbow quadruple of the form* $(x, y, x + y, xy)$.

Following this, we have a result guaranteeing the existence of long rainbow arithmetic progressions in a wide class of finite abelian groups. See [14] and the references contained therein ([1] in particular), for more on rainbow arithmetic progressions.

Corollary 2 *If we color a finite (additive) abelian group G, with no (nonidentity) element of order $< k$, then if there are no rainbow k-arithmetic progressions, at least one color class has size* $\geq \frac{2}{9\binom{k}{2}}|G|$.

We also include this application of Theorem 2 with an ambient set that is not even a group. Let for integers $a < b$, let $[a \ldots b]$ denote the "interval of integers," $\{a, (a+1), \ldots, (b-1), b\}$.

Corollary 3 *If we color* $[1 \ldots n]$ *such that there are no rainbow triples of the form* $(x, y, x+y)$*, then at least one color class has size* $\geq \frac{1-o(1)}{27} n$.

In this example, we show how Thorem 2 can be used to get quantitative information. Under some slightly stronger conditions, we can estimate how many rainbow configurations must be present.

Corollary 4 *Suppose the conditions of Theorem 2 are satisfied, but are strengthened so that no color class has size* $\geq Cpn$ *for some* $p \in (0, 1]$*. Then there are at least* $(1-p)Dn^2$ *rainbow elements of* E.

We now present a more geometric result, whose proof uses more machinery. In [8], Alex Iosevich and Misha Rudnev used the following definition of distance for use in a vector space over a finite field. For $x, y \in \mathbb{F}_q^2$, we write

$$|x - y| = (x_1 - y_1)^2 + (x_2 - y_2)^2.$$

This notion of distance shares many properties of the Euclidean distance in $\mathbb{R}^2$. The following is a version of the main result in [12], due to the second listed author.

Corollary 5 *If* $\mathbb{F}_q^2$ *is colored so that no color class is of size bigger than* cq^2 *for any positive constant,* c*, and equilateral triangles exist, then there is a rainbow equilateral triangle.*

2 Proof of Theorem 2

The basic idea of this proof will be to assume that we have no rainbow configuration, then find that there must be some large color class. To do this, we will merge color classes together to get a relatively uniform count. Note that merging color classes can destroy but not create rainbow configurations. So if we began without a rainbow configuration, then merging classes will not create one. Once we have a uniform count, we can use the quantities measuring the structure and number of configurations, using M and D respectively, to derive a contradiction on the total number of elements in the ambient set.

Proof Fix $C > 0$ to be determined later. Now, proceed to greedily merge the smallest two color classes pairwise until every class has size between $(1/2)Cn$ and $(3/2)Cn$. That is to say, if it is not the case that every color class is of size between $(1/2)Cn$ and $(3/2)Cn$, then we merge the smallest two color classes and check again. We repeat this merging process until every color class has size between $(1/2)Cn$ and $(3/2)Cn$. Let s denote the number of color classes after this merging.

There are $|E|$ k-tuples in the set E. Fix a color i, and let n_i denote the number of elements from X in color class i. Recall that M bounds how many k-tuples from E share (at least) a pair of coordinates. So the number of k-tuples in E with at least two elements from color i is at most Mn_i^2. Now, if there are no rainbow k-tuples then every k-tuple in E must have at least two coordinates of the same color, so

$$|E| \leq M \sum_{i=1}^{s} n_i^2 \leq M \sum_{i=1}^{s} \left(\frac{3}{2}Cn\right)^2 \leq \frac{9M}{4} sC^2n^2,$$

where we used the assumption that $n_i \leq (3/2)Cn$ for all i. So we have that

$$s \geq \frac{4}{9M}\frac{1}{C^2}\frac{|E|}{n^2} \geq \frac{4D}{9M}\frac{1}{C^2}.$$

But every class has size at least $(1/2)Cn$, which, since they are all disjoint, implies that X has at least

$$\frac{1}{2}Cn \cdot \frac{4D}{9M}C^{-2} = \frac{2D}{9M}\frac{n}{C} > n$$

elements, a contradiction. □

3 Corollaries of Theorem 2

Here, we prove Theorem 1, Corollaries 1, 2, 3, 5, and 4. These illustrate how to use Theorem 2. In particular, they will show how to get a handle the constants $M_{i,j}$ and D in various situations.

3.1 *Proof of Theorem 1*

Proof We will apply Theorem 2 with $X = G$ and $k = 3$. The set $E \subset X^3$ will be the set of triples of G of the form $(x, y, x + y)$. Now, $M_{1,2}$ will be the number of different triples in G that can share the first two coordinates. But any pair of first and second coordinates, x and y, will uniquely determine the third coordinate, $x + y$. There may be other triples in X^3 that share the first two coordinates, but only one of them will be in E, so $M_{1,2} = 1$. Notice that any pair of elements x and $x + y$ will uniquely determine y, so $M_{1,3} = 1$. Similarly, $M_{2,3}$ will be 1, so

$$M = \sum_{i<j} M_{i,j} = M_{1,2} + M_{1,3} + M_{2,3} = 3.$$

All that is left is for us to apply Theorem 2 is to get a value for D. So we need to see how big E is in terms of X. As each pair of elements, x and y, from G generates a distinct triple $(x, y, x + y)$ in E, we see that $|E| = n^2$, so $D = 1$. Plugging everything into Theorem 2 guarantees the existence of a rainbow triple of the form $(x, y, x + y)$ for any coloring where each color class is smaller than Cn, where

$$C < \frac{2D}{9M} = \frac{2}{27}.$$

3.2 *Proof of Corollary 1*

Proof Similar to the proof of Theorem 1, we will set $X = \mathbb{F}_q$, find a set of quadruples, E, and estimate the constants M and D to plug into Theorem 2. Initially, we would start with all quadruples of the form $(x, y, x + y, xy)$ as E, but we drop all quadruples with $xy = 0$, because given $xy = x = 0$ there are still many possible quadruples, and a quadruple with $x = xy$ would necessarily be nonrainbow. So $M_{1,4}$ would be too large to get an effective bound. This leaves the number of possible quadruples that comprise our set E to be

$$q^2 - 2q + 1 = (1 - o(1))q^2.$$

Now, $|\mathbb{F}_q| = q$, so $n = q$. Since $E = (1 - o(1))n^2$, we have that $D = (1 - o(1))$.

Knowing any two of $x, y, x + y$ clearly fixes the rest of the tuple, and knowing xy and either x or y does the same (as $xy \neq 0$). Therefore $M_{i,j} = 1$ for all of the $M_{i,j}$ except $M_{3,4}$. If we know $x + y$ and xy then x and y must be roots of the polynomial $t^2 - (x + y)t + (xy)$, of which there are at most two (so there are at most two quadruples, since we can change the order of x and y). This gives us that $M_{3,4} = 2$, and summing this with the other $M_{i,j}$ gives $M = 7$.

When we put these values into Theorem 2, we get that there must be a rainbow quadruple of the form $(x, y, x + y, xy)$ if all of the color classes are smaller than Cq, for

$$C < \frac{2D}{9M} = \frac{2 - o(1)}{63}.$$

3.3 Proof of Corollary 2

Proof Again, we will find a suitable set E, then compute the corresponding values of M and D. Since our ambient group is G, we have that $n = |G|$. We set E to be the set of all ordered k-tuples whose elements form an ordered k-term arithmetic progression:

$$E := \{(z, z + x, \ldots, z + (k-1)x) : z, x \in G\}.$$

Note that in E, we could have two elements that consist of the same group elements, but in distinct orders. Each of the k-term arithmetic progressions above will be distinct, giving us n^2 distinct elements in E, one for every pair of elements, $(z, x) \in G^2$. So $D = 1$.

To get a handle on M, we need to see how often two k-term arithmetic progressions can have two elements in the same spot (e.g., they have the same fifth element and same ninth element). So suppose that the k-term arithmetic progressions generated in E by (z, x) and (z', x') share the same elements at the slots numbered a and b, for some distinct $a, b \in [0 \ldots (k-1)]$. That is,

$$z + ax = z' + ax' \text{ and } z + bx = z' + bx'.$$

This gives $ax - ax' = z' - z = bx - bx'$. Then $a(x - x') = b(x - x')$, meaning

$$(a - b)(x - x') = 0;$$

so $x - x'$ has order $\leq |a - b|$, meaning $x = x'$ (since $|a - b| < k$), which implies that the two progressions are identical. Thus all the $M_{i,j}$ are 1, and there are $\binom{k}{2}$ of them. So $M = \binom{k}{2}$, $D = 1$ and the result follows. □

3.4 Proof of Corollary 3

Proof This runs essentially the same way as the proof of Theorem 1, with $M = 3$. However, there is a slightly different calculation for D. Note that if we choose $x = c$, there are $n - c$ possible choices for y such that $x + y \in [1 \ldots n]$. Therefore:

$$|E| = \sum_{c=1}^{n} n - c = \binom{n}{2} = \left(\frac{1}{2} - o(1)\right) n^2,$$

giving $D = \frac{1}{2} - o(1)$, and we apply Theorem 2. □

3.5 *Proof of Corollary 4*

Proof Using Theorem 2, we see that there must be a rainbow element of E. Thus we can remove it from E, and we can keep finding rainbow elements as long as $|E| > Dpn^2$. So there are at least $Dn^2 - Dpn^2 = (1-p)Dn^2$ rainbow elements of E. □

3.6 *Proof of Corollary 5*

The proof of this result is a bit more involved, and requires a bit more background than the others.

3.6.1 Existence of Equilateral Triangles

We first have to address what it means that equilateral triangles exist. For some q, there are no triples of points $x, y, z \in \mathbb{F}_q^2$ such that $|x-y| = |y-z| = |x-z|$. See [3], (Lemma 4.1 in particular) by Bennett, Iosevich, and Pakianathan, for more on this point. In this case, it is enough that there is an element $\sigma \in \mathbb{F}_q$ such that $\sigma^2 = 3$.

3.6.2 Geometric Lemmas

Before we dive into the proof, we will need two lemmas, which we will use to get our estimates on M and D. Both can be found in multiple sources, and are stated without proof. The first is pulled from the proof of Theorem 2 of [13] and the second is stated as a special case of Lemma 1.2 from [4].

Lemma 1 *If $E \subset \mathbb{F}_q^2$ and $|E| \gtrsim q^{\frac{3}{2}}$, then*

$$|\{(x, y) \in E \times E : |x-y| = 1\}| \lesssim q^{-1}|E|^2.$$

Lemma 2

$$|\{(x, y) \in \mathbb{F}_q^2 \times \mathbb{F}_q^2 : |x-y| = 1\}| = (1+o(1))q^3.$$

With these results in tow, we are ready to put everything together. Corollary 5 is a corollary of the proof of Theorem 2, as we will need to handle the number of color classes after merging rather carefully.

3.6.3 Finishing the Proof

Proof We will apply Theorem 2 with $X = \mathbb{F}_q^2$, so $n = q^2$. By assumption, we have that no color class is of size proportional to q^2, so after the merging process in the proof of Theorem 2, we will have s different color classes, where s is bigger than any constant[1] with respect to q. Call these color classes $E_1, E_2, \ldots, E_s$, and possibly reorder them so that $|E_j| \geq |E_{j+1}|$ for $j = 1, \ldots, (s-1)$.

Next, we estimate M. Due to the symmetry in an equilateral triangle, all of the $M_{i,j}$ will be equal, so we need only estimate $M_{1,2}$. One of the properties of this notion of distance is that distinct circles can intersect in no more than two points. Fix any color class, E_j. If we take two points, $x, y \in E_j$, that are a unit distance apart, and draw unit circles centered at them, these circles can intersect at most twice. Call those two intersection points z and z'. Then (x, y, z) and (x, y, z') are the only two triples of points from $\mathbb{F}_q^2$ that make unit equilateral triangles with x and y as the first two entries. Now we apply Lemma 1 to each color class and see that

$$M_{1,2} \leq \sum_{j=1}^{s} 2|\{(x, y) \in E_j \times E_j : |x-y| = 1\}| \lesssim q^{-1}s|E_1|^2.$$

Since the $M_{i,j}$ are equal, we have that $M = 3M_{1,2}$. Putting this together with the bound on $M_{1,2}$ gives

$$M \lesssim q^{-1}s|E_1|^2.$$

Now, we apply Lemma 2 to see that there are $(1 + o(1))q^3$ pairs of points $(x, y) \in \mathbb{F}_q^2$ that are a unit distance apart, that is, with $|x - y| = 1$. Note that in any equilateral triangle, there are exactly three pairs of points that are a unit distance apart. Now, as discussed above, each such pair is in exactly two equilateral triangles. Therefore, the total number of equilateral triangles that exist in $\mathbb{F}_q^2$, regardless of color, must be $\frac{2}{3}(1 + o(1))q^3$, giving us that we can set $D = \frac{2}{3}q$.

From here, we finish by putting the estimates on s, M, and D together to see that

$$C < \frac{2D}{9M} \lesssim \frac{q^{-1}s|E_1|^2}{\frac{2}{3}q} \leq \frac{s|E_1|^2}{q^2},$$

This holds as long as $|E_1| \lesssim Cs^{-1}q^2 \leq Cn$, which we know to be true, as s is larger than any constant with respect to q. $\square$

Acknowledgements This work was supported in part by NSF Grant DMS 1559911.

[1] For example, if $s > \log q$, that would suffice, but in fact this works for any slowly growing function in q.

References

1. M. Axenovich and D. Fon-Der-Flaass (2004) On rainbow arithmetic progressions. Elec. J. Comb., 11:R1.
2. L. Babai (1985) An anti-Ramsey theorem. Graphs Combin. 1, 23–28.
3. M. Bennett, A. Iosevich, and J. Pakianathan (2014) Three-point configurations determined by subsets of $\mathbb{F}_q^2$ via the Elekes-Sharir paradigm. Combinatorica 34, no. 6, 689–706.
4. D. Hart, A. Iosevich, D. Koh, S. Senger, I. Uriarte-Tuero (2012) Distance graphs in vector spaces over finite fields. Bilyk, Dmitriy, et al., eds. Recent Advances in Harmonic Analysis and Applications: In Honor of Konstantin Oskolkov. Vol. 25. Springer Science & Business Media.
5. B. Green and T. Sanders (2016) Monochromatic sums and products. Discrete Analysis, pages 1–43, 2016:5.
6. N. Hindman (1979) Partitions and sums and products of integers. Trans. Amer. Math. Soc., 247:227–245.
7. N. Hindman, I. Leader, and D. Strauss (2003) Open problems in partition regularity. Probably. Comput., 12(5–6):571–583. Special issue on Ramsey theory.
8. A. Iosevich and, M. Rudnev (2007) Erdős distance problem in vector spaces over finite fields. Trans. Amer. Math. Soc., 359, 6127–6142.
9. V. Jungić, J. Licht, M. Mahdian, J. Nešetřil, and R. Radoičić (2003) Rainbow Arithmetic Progressions and Anti-Ramsey Results. Combinatorics, Probability and Computing, 12:599–620.
10. H. Lefmann and V. Rödl (1993) On canonical Ramsey numbers for complete graphs versus paths. J. Combin. Theory Ser. B, 58:1–13.
11. J. Moreira (2017) Monochromatic sums and products in $\mathbb{N}$. Ann. Math., 185–3:1069–1090, https://doi.org/10.4007/annals.2017.185.3.10.
12. S. Senger (2017) Rainbow triangles. arXiv:1702.03043.
13. L. A. Vinh (2017) Explicit Ramsey graphs and Erdős distance problem over finite Euclidean and non-Euclidean spaces. Elec. J. Comb. 15(1).
14. M. Young (2016) Rainbow arithmetic progressions in finite abelian groups. arXiv:1603.08153.

Translation Invariant Filters and van der Waerden's Theorem

Mauro Di Nasso

Abstract We present a self-contained proof of a strong version of *van der Waerden's Theorem*. By using translation invariant filters that are maximal with respect to inclusion, a simple inductive argument shows the existence of "piecewise syndetically"-many monochromatic arithmetic progressions of any length k in every finite coloring of the natural numbers. All the presented constructions are constructive in nature, in the sense that the involved maximal filters are defined by recurrence on suitable countable algebras of sets. No use of the axiom of choice or of Zorn's Lemma is needed.

1 Introduction

The importance of maximal objects in mathematics is well-known, starting from the fundamental examples of maximal ideals in algebra, and of ultrafilters in certain areas of topology and of Ramsey theory. In this paper we focus on maximal filters on suitable countable algebras of sets which are stable under translations. By using such maximal objects, along with ultrafilters extending it, we give a proof of a strong version of the following classical result in Ramsey theory:

Theorem ([5]) *In every finite partition* $\mathbb{N} = C_1 \cup \cdots \cup C_r$ *there exists a piece* $C = C_i$ *that contains arbitrarily long arithmetic progressions, that is, for every* k *there exists a progression* $x + y, x + 2y, \ldots, x + ky \in C$.

In fact, we will prove the existence of "piecewise syndetically"-many monochromatic arithmetic progressions of any length k.

Usually, van der Waerden's Theorem is proved either by double induction using elementary, but elaborated, combinatorial arguments in the style of the original proof [5], or by using properties of the smallest ideal $K(\beta\mathbb{N}, \oplus)$ in the algebra of ultrafilters (see [4, Chap.14]; see also [1, 2] for stronger versions). In our proof, for any

M. Di Nasso (✉)
Dipartimento di Matematica, Università di Pisa, Pisa, Italy
e-mail: mauro.di.nasso@unipi.it

M. B. Nathanson (ed.), *Combinatorial and Additive Number Theory III*, Springer Proceedings in Mathematics & Statistics 297,
https://doi.org/10.1007/978-3-030-31106-3_6

given piecewise set, we restrict to a suitable countable algebra of sets, and explicitly construct by recursion a maximal translation invariant filter, and then an ultrafilter extending it. The desired result is finally obtained by a short proof by induction, that is essentially a simplified version of an argument that was used in [3] in the framework of the compact right-topological semigroup $(\beta\mathbb{N}, \oplus)$. It is worth remarking that, contrarily to the usual ultrafilter proof, we make no explicit use of the algebra in the space of ultrafilters; in fact, we make no use of the axiom of choice nor of Zorn's Lemma.

2 Preliminary Notions

$\mathbb{N} = \{1, 2, 3, \ldots\}$ denotes the set of *positive integers*, and $\mathbb{N}_0 = \mathbb{N} \cup \{0\}$ the set of non-negative integers. For $A \subseteq \mathbb{N}$ and $n \in \mathbb{N}_0$, the *leftward shift* of A by n is the set:

$$A - n := \{m \in \mathbb{N} \mid n + m \in A\}$$

Elemental notions in combinatorics of numbers that we will use in this paper are those of thick set, syndetic set, and piecewise syndetic set. For completeness, let us recall them here.

A set $A \subseteq \mathbb{N}$ is *thick* if it includes arbitrarily long intervals. Equivalently, A is thick if every finite set $F = \{n_1, \ldots, n_k\} \subset \mathbb{N}$ has a *rightward shift* included in A, that is, there exists x such that

$$F + x := \{n_1 + x, \ldots, n_k + x\} \subseteq A.$$

Notice that such an x can be picked in A. In terms of intersections, the property of thickness of A can be rephrased by saying that the family $\{A - n \mid n \in \mathbb{N}_0\}$ has the *finite intersection property* (FIP for short), that is, $\bigcap_{i=1}^{k}(A - n_i) \neq \emptyset$ for any $n_1, \ldots, n_k \in \mathbb{N}_0$.

A set $A \subseteq \mathbb{N}$ is *syndetic* if it has "bounded gaps", that is, there exists $k \in \mathbb{N}$ such that A meets every interval of length k. Equivalently, A is syndetic if a finite number of leftward shifts of A covers all the natural numbers, that is, $\mathbb{N} = \bigcup_{i=1}^{k}(A - n_i)$ for suitable $n_1, \ldots, n_k \in \mathbb{N}_0$.

A set is *piecewise syndetic* if it is the intersection of a thick set with a syndetic set. Equivalently, A is piecewise syndetic if a finite number of leftward shifts cover a thick set, that is, $\bigcup_{i=1}^{k}(A - n_i)$ is thick for suitable $n_1, \ldots, n_k \in \mathbb{N}_0$.

Notice that the families of thick, syndetic, and piecewise syndetic sets are all invariant with respect to shifts. A well-known relevant property of piecewise syndetic sets that is satisfied neither by thick sets nor by syndetic sets, is the *Ramsey property* below. For the sake of completeness, we include here a proof.

Proposition 2.1 *In every finite partition $A = C_1 \cup \cdots \cup C_r$ of a piecewise syndetic set A, one of the pieces C_i is piecewise syndetic.*

Proof For simplicity, let us say that an interval I is k-good for the set B if for every sub-interval $J \subseteq I$ of length k one has $J \cap B \neq \emptyset$. By the hypothesis of piecewise syndeticity of A, there exists $k \in \mathbb{N}$ and a sequence of intervals $\langle I_n \mid n \in \mathbb{N}\rangle$ with increasing length such that every I_n is k-good for A. It is enough to consider the case when $A = C_1 \cup C_2$ is partitioned into two pieces, because the general case $A = C_1 \cup \cdots \cup C_r$ where $r \geq 2$ will then follow by induction. We distinguish two cases.

Case # 1: There exists h such that infinitely many intervals I_n are h-good for C_1. In this case C_1 is piecewise syndetic.

Case # 2: For every h, there are only finitely many intervals I_n that are h-good for C_1. So, for every h we can pick an interval I_{n_h} of length $\geq h$ that is not h-good. Let $J_h \subseteq I_{n_h}$ be a sub-interval of length h such that $J_h \cap C_1 = \emptyset$. The sequence of intervals $\langle J_h \mid h \in \mathbb{N}\rangle$ shows that C_2 is piecewise syndetic. Indeed, given h, for every sub-interval $J \subseteq J_h$ of length k we have that $J \cap C_1 \subseteq J_h \cap C_1 = \emptyset$; and so $J \cap C_2 = J \cap A \neq \emptyset$, since $J \subseteq I_{n_h}$ and I_{n_h} is k-good for A. □

3 Maximal Translation Invariant Filters

In the following, by *family* we mean a nonempty collection of subsets of $\mathbb{N}$.

Definition 3.1 A family $\mathcal{G}$ is *translation invariant* if $A \in \mathcal{G} \Rightarrow A - 1 \in \mathcal{G}$ (and hence, $A - n \in \mathcal{G}$ for all $n \in \mathbb{N}_0$).

An *algebra of sets* (on $\mathbb{N}$) is a family that contains $\mathbb{N}$ and is closed under finite unions, finite intersections, and complements. The [translation invariant] algebra *generated* by a family $\mathcal{G}$ is the smallest [translation invariant] algebra of sets that contains $\mathcal{G}$.

Proposition 3.2 *If the family $\mathcal{G}$ is countable, then one can give explicit constructions of both the (countable) algebra generated by $\mathcal{G}$, and the (countable) translation invariant algebra generated by $\mathcal{G}$, in terms of any given enumeration of the sets in $\mathcal{G}$.*

Proof Let $\langle A_n \mid n \in \mathbb{N}\rangle$ be an enumeration of the sets in $\mathcal{G}$, and fix $\langle F_n \mid n \in \mathbb{N}\rangle$ an enumeration of the nonempty finite sets of natural numbers.[1] For $A \subseteq \mathbb{N}$, denote $A^{+1} = A$ and $A^{-1} = A^c$. Then the following family $\mathcal{B}_{\mathcal{G}}$ is the smallest algebra of sets that contains $\mathcal{G}$:

$$\mathcal{B}_{\mathcal{G}} := \left\{ \bigcup_{i=1}^{t} \Big(\bigcap_{k \in F_{n_i}} A_k^{\sigma_i(k)} \Big) \;\Big|\; n_1, \ldots, n_t \in \mathbb{N},\ \sigma_i : F_{n_i} \to \{+1, -1\} \right\}.$$

[1]E.g., if $n = \sum_{k=1}^{\infty} a_{nk} 2^{k-1}$ is written in binary expansion where $a_{nk} \in \{0, 1\}$, then we can let $F_n := \{k \mid a_{nk} = 1\}$.

Notice that if $\mathscr{G}$ is translation invariant, then also $\mathscr{B}_{\mathscr{G}}$ is translation invariant. So, the algebra $\mathscr{B}_{\mathscr{G}'}$ generated by the family of shifts $\mathscr{G}' := \{A - n \mid A \in \mathscr{G}, n \in \mathbb{N}_0\}$ is the smallest translation invariant algebra containing $\mathscr{G}$. □

A *filter* on an algebra of sets $\mathscr{B}$ is a nonempty family $\mathscr{F} \subseteq \mathscr{B}$ such that:

- $\mathscr{F}$ is closed under finite intersections, that is, $A, B \in \mathscr{F} \Rightarrow A \cap B \in \mathscr{F}$;
- $\mathscr{F}$ is closed under supersets, that is, if $B \in \mathscr{B}$ and $B \supseteq A \in \mathscr{F}$ then $B \in \mathscr{F}$.

Every family $\mathscr{G} \subseteq \mathscr{B}$ with the *finite intersection property* (FIP for short) generates a filter $\langle \mathscr{G} \rangle$, namely

$$\langle \mathscr{G} \rangle := \{B \in \mathscr{B} \mid B \supseteq A_1 \cap \cdots \cap A_k \text{ for suitable } A_1, \ldots, A_k \in \mathscr{G}\}.$$

An *ultrafilter* $\mathscr{U}$ on the algebra of sets $\mathscr{B}$ is a filter with the additional property that $A \in \mathscr{U}$ whenever $A \in \mathscr{B}$ and the complement $A^c \notin \mathscr{U}$. It is easily verified that a filter $\mathscr{U}$ is an ultrafilter if and only if the *Ramsey property* holds: If $A_1 \cup \cdots \cup A_k \in \mathscr{U}$ where all sets $A_i \in \mathscr{B}$, then $A_j \in \mathscr{U}$ for some j. Ultrafilters can also be characterized as those filters that are maximal under inclusion and so, by a straight application of *Zorn's Lemma*, it is proved that every filter can be extended to an ultrafilter.

The following objects are the main ingredient in our proof of van der Waerden's Theorem.

Definition 3.3 A *translation invariant filter* (TIF for short) is a filter $\mathscr{F}$ on a translation invariant algebra $\mathscr{B}$ such that $A \in \mathscr{F} \Rightarrow A - 1 \in \mathscr{F}$ (and hence $A - n \in \mathscr{F}$ for all $n \in \mathbb{N}_0$).

Notice that if the algebra $\mathscr{B}$ is translation invariant, and the family $\mathscr{G} \subseteq \mathscr{B}$ is translation invariant, then the generated filter $\langle \mathscr{G} \rangle$ is a TIF.

The notions of TIF and thick set are closely related.

Proposition 3.4 *A set A is thick if and only if it belongs to a TIF $\mathscr{F}$.*

Proof Recall that A is thick if and only if the family $\mathscr{G} = \{A - n \mid n \in \mathbb{N}_0\}$ has the FIP. Since $\mathscr{G}$ is translation invariant, the generated filter $\langle \mathscr{G} \rangle$ on the algebra $\mathscr{B}_{\mathscr{G}}$ is a TIF that contains A.

Conversely, assume that $A \in \mathscr{F}$ for some TIF $\mathscr{F}$. Then trivially the family $\mathscr{G} = \{A - n \mid n \in \mathbb{N}_0\}$ has the FIP because $\mathscr{G} \subseteq \mathscr{F}$. □

Similarly to ultrafilters, by a straightforward application of *Zorn's Lemma* it can be shown that every TIF can be extended to a maximal TIF. However, in the countable case, recursive constructions suffice to produce both ultrafilters and maximal TIFs, which are thus obtained in a constructive manner, without any use of the axiom of choice.

Proposition 3.5 *Let $\mathscr{B} = \{B_n \mid n \in \mathbb{N}\}$ be a countable algebra of sets.*

1. *Given a family $\mathscr{G} \subseteq \mathscr{B}$ with the FIP, inductively define $\mathscr{G}_0 = \mathscr{G}$; $\mathscr{G}_{n+1} = \mathscr{G}_n \cup \{B_n\}$ in case $B_n \cap A \neq \emptyset$ for every $A \in \mathscr{G}_n$; and $\mathscr{G}_{n+1} = \mathscr{G}_n$ otherwise. Then $\mathscr{U} := \bigcup_n \mathscr{G}_n$ is an ultrafilter on $\mathscr{B}$ that extends $\mathscr{G}$.*
2. *Assume that the algebra $\mathscr{B}$ is translation invariant. Given a translation invariant family $\mathscr{G} \subseteq \mathscr{B}$ with the FIP, inductively define $\mathscr{G}_0 = \mathscr{G}$; $\mathscr{G}_{n+1} = \mathscr{G}_n \cup \{B_n - k \mid k \in \mathbb{N}_0\}$ in case that union has the FIP; and $\mathscr{G}_{n+1} = \mathscr{G}_n$ otherwise. Then $\mathscr{M} := \bigcup_n \mathscr{G}_n$ is a maximal TIF that extends $\mathscr{G}$.*

Proof (1). By the definition, it is clear that all families $\mathscr{G}_n$ have the FIP, and so also their increasing union $\mathscr{U}$ has the FIP. Now assume by contradiction that $A \in \mathscr{B}$ is such that both $A, A^c \notin \mathscr{U}$. If $A = B_n$ and $A^c = B_m$ then, by the definition of $\mathscr{U}$, there exist $U \in \mathscr{G}_n$ and $U' \in \mathscr{G}_m$ such that $A \cap U = A^c \cap U' = \emptyset$, and hence $U \cap U' = \emptyset$, against the FIP of $\mathscr{U}$. Finally, if $B \supseteq A$ where $B \in \mathscr{B}$ and $A \in \mathscr{U}$ then $B \in \mathscr{U}$, as otherwise, by what just proved, $B^c \in \mathscr{U}$, and hence $\emptyset = B^c \cap A \in \mathscr{U}$, a contradiction.

(2). By induction, it directly follows from the definition that all families $\mathscr{G}_n$ have the FIP and are translation invariant; so, the same properties hold for $\mathscr{M}$. Now let $B \supseteq A$ where $A \in \mathscr{M}$ and $B \in \mathscr{B}$, say $B = B_n$. Notice that $\mathscr{G}_n \cup \{B - k \mid k \in \mathbb{N}_0\}$ has the FIP because $A - k \subseteq B - k$ for all k and $\mathscr{G}_n \cup \{A - k \mid k \in \mathbb{N}_0\} \subseteq \mathscr{M}$ has the FIP. Then $B \in \mathscr{G}_{n+1} \subseteq \mathscr{M}$, and we can conclude that $\mathscr{M}$ is a TIF. As for the maximality, let $\mathscr{M}' \supseteq \mathscr{M}$ be a TIF. Given $A \in \mathscr{M}'$, pick n with $A = B_n$. The family $\mathscr{G}_n \cup \{A - n \mid n \in \mathbb{N}_0\}$ has the FIP, since it is included in the filter $\mathscr{M}'$, and so $A \in \mathscr{G}_{n+1}$. This shows that $\mathscr{M}' \subseteq \mathscr{M}$, and hence the two TIFs are equal. □

Two properties of maximal TIFs that will be relevant to our purposes are the following.

Proposition 3.6 *Let $\mathscr{B}$ be a translation invariant algebra, and let $\mathscr{U}$ be an ultrafilter on $\mathscr{B}$ that includes a maximal TIF $\mathscr{M}$. Then:*

1. *Every $B \in \mathscr{U}$ is piecewise syndetic.*
2. *For every $B \in \mathscr{U}$, the set $B_{\mathscr{U}} := \{n \in \mathbb{N} \mid B - n \in \mathscr{U}\}$ is syndetic.*[2]

Proof Notice first that for every $B \in \mathscr{U}$ there exist $n_1, \ldots, n_k$ such that the union $\bigcup_{i=1}^k (B - n_i) \in \mathscr{M}$. Indeed, if $\Lambda := \{B^c - n \mid n \in \mathbb{N}_0\}$ then the union $\mathscr{M} \cup \Lambda$ does not have the FIP, as otherwise $\mathscr{M} \cup \Lambda$ would generate a TIF that properly extends $\mathscr{M}$ (since it would contain B^c while $B^c \notin \mathscr{M}$), against the maximality. So, there exist $A \in \mathscr{M}$ and $n_1, \ldots, n_k$ such that $A \cap \bigcap_{i=1}^k (B^c - n_i) = \emptyset$. But then $\bigcup_{i=1}^k (B - n_i) \in \mathscr{M}$, because it is a superset of $A \in \mathscr{M}$.

(1). Pick a finite union of shifts $\bigcup_{i=1}^k (B - n_i) \in \mathscr{M}$. By Proposition 3.4, that union is thick because it is an element of a TIF, and hence B is piecewise syndetic.

(2). As above, pick a finite union of shifts $\bigcup_{i=1}^k (B - n_i) \in \mathscr{M}$. By translation invariance, for every $m \in \mathbb{N}$ one has that $\bigcup_{i=1}^k (B - n_i - m) \in \mathscr{M} \subseteq \mathscr{U}$ and so, by the Ramsey property of ultrafilters, there exists i such that $B - n_i - m \in \mathscr{U}$, that is, $m \in B_{\mathscr{U}} - n_i$. This shows that $\mathbb{N} = \bigcup_{i=1}^k (B_{\mathscr{U}} - n_i)$ is a finite union of shifts of $B_{\mathscr{U}}$, and hence $B_{\mathscr{U}}$ is syndetic. □

[2] We remark that in general the set $B_{\mathscr{U}}$ does not belong to the algebra of sets $\mathscr{B}$.

4 A Strong Version of van der Waerden's Theorem

The following property of piecewise syndetic sets was first proved by exploiting the properties of ultrafilters in the smallest ideal of the right-topological semigroup $(\beta\mathbb{N}, \oplus)$ (see [1, 2]).

Theorem 4.1 *Let A be a piecewise syndetic set. Then for every $k \in \mathbb{N}$, the set* $\mathrm{AP}_k(A) := \{x \in A \mid \exists y \in \mathbb{N} \text{ s.t. } x + iy \in A \text{ for } i = 1, \ldots, k\}$ *is piecewise syndetic.*

Notice that, as a straight consequence, one obtains the following strong version of van der Waerden's Theorem.

Theorem 4.2 *In every finite partition $\mathbb{N} = C_1 \cup \cdots \cup C_r$ there exists a piece $C = C_i$ such that, for every $k \in \mathbb{N}$, the set $\mathrm{AP}_k(C)$ is piecewise syndetic.*

Proof By the Ramsey property of piecewise syndetic sets (see Proposition 2.1), we can pick a color C_i which is piecewise syndetic. □

In this section we will give a new proof of the above theorem which relies on the existence of an ultrafilter $\mathcal{U}$ on an appropriate translation invariant algebra $\mathcal{B}$, which extends a maximal TIF and contains a shift of A.

Proof (of Theorem 4.1) Let $\mathcal{B}$ be the (countable) translation invariant algebra of sets generated by the translation invariant family $\{A - n \mid n \in \mathbb{N}_0\}$. By the property of piecewise syndeticity, a finite union of shifts $T = \bigcup_{j=1}^{m}(A - n_j)$ is thick. Then the translation invariant family $\mathcal{G} := \{T - n \mid n \in \mathbb{N}_0\} \subseteq \mathcal{B}$ has the FIP, and by Proposition 3.5 we can pick a maximal TIF $\mathcal{M}$ on $\mathcal{B}$ with $\mathcal{M} \supseteq \mathcal{G}$, and an ultrafilter $\mathcal{U}$ on $\mathcal{B}$ with $\mathcal{U} \supseteq \mathcal{M}$. The desired result is a consequence of the following general property.

Claim. Let $\mathcal{U}$ be an ultrafilter that extends a maximal TIF. If a shift $B - \ell \in \mathcal{U}$ for some $\ell \in \mathbb{N}_0$, then $B_{\mathcal{U}} - \ell$ contains arbitrarily long arithmetic progressions.

Indeed, let us assume the claim. Since the finite union $T = \bigcup_{j=1}^{m}(A - n_j) \in \mathcal{G} \subseteq \mathcal{U}$, by the Ramsey property of ultrafilters there exists n_j such that $A - n_j \in \mathcal{U}$. Then, for every $k \in \mathbb{N}$ there exist x and y such that $x + iy \in A_{\mathcal{U}} - n_j$ for $i = 0, 1, \ldots, k$. But then $B := \bigcap_{i=0}^{k}(A - n_j - x - iy) \in \mathcal{U}$, and hence also the superset $\mathrm{AP}_k(A) - n_j - x \supseteq B$ belongs to $\mathcal{U}$, as one can easily verify. Now recall that all sets in $\mathcal{U}$ are piecewise syndetic by Proposition 3.6, and so we can conclude that $\mathrm{AP}_k(A)$ is piecewise syndetic because it is a shift of a member of $\mathcal{U}$.

We are left to prove the Claim. We proceed by induction on k, and prove that if $B - \ell \in \mathcal{U}$ for some $\ell \in \mathbb{N}_0$, then $B_{\mathcal{U}} - \ell$ contains a k-term arithmetic progression.[3]

If $B - \ell \in \mathcal{U}$, then the set $(B - \ell)_{\mathcal{U}} = B_{\mathcal{U}} - \ell$ is syndetic by Proposition 3.6. In particular, $B_{\mathcal{U}} - \ell \neq \emptyset$, and this proves the induction base $k = 1$.

Let us turn to the inductive step $k + 1$, and assume that $B - \ell \in \mathcal{U}$. Let $\ell_0 = \ell$. By syndeticity of $B_{\mathcal{U}} - \ell_0$, there exists a finite $F \subset \mathbb{N}_0$ such that for every

[3]This inductive construction uses a simplified version of an argument in [3].

$n \in \mathbb{N}$ there exists $x \in F$ with $\ell_0 + n + x \in B_{\mathcal{U}}$. For convenience, let us assume that $0 \in F$. By the inductive hypothesis, there exist $\ell_1 \in \mathbb{N}_0$ and $y_1 \in \mathbb{N}$ such that $\ell_1 + iy_1 \in B_{\mathcal{U}} - \ell_0$ for $i = 1, \ldots, k$, that is, $\ell_0 + \ell_1 + x_0 + iy_1 \in B_{\mathcal{U}}$ where $x_0 = 0 \in F$. Pick $x_1 \in F$ with $\ell_0 + \ell_1 + x_1 \in B_{\mathcal{U}}$. If $x_1 = x_0$ then we already found a $(k+1)$-term arithmetic progression in $B_{\mathcal{U}} - \ell_0$, as desired. Otherwise, let us consider the intersection

$$B_1 := (B - x_1) \cap \bigcap_{i=1}^{k} (B - x_0 - iy_1).$$

Since $\ell_0 + \ell_1 + x_1 \in B_{\mathcal{U}}$ and $\ell_0 + \ell_1 + x_0 + iy_1 \in B_{\mathcal{U}}$ for all $i = 1, \ldots, k$, the shift $B_1 - \ell_0 - \ell_1 \in \mathcal{U}$ and so, by the inductive hypothesis, there exist $\ell_2 \in \mathbb{N}_0$ and $y_2 \in \mathbb{N}$ such that $\ell_2 + iy_2 \in (B_1)_{\mathcal{U}} - \ell_0 - \ell_1$ for $i = 1, \ldots, k$. In consequence, $\ell_0 + \ell_1 + \ell_2 + x_0 + i(y_1 + y_2) \in B_{\mathcal{U}}$ and $\ell_0 + \ell_1 + \ell_2 + x_1 + iy_2 \in B_{\mathcal{U}}$ for every $i = 1, \ldots, k$. Pick $x_2 \in F$ such that $\ell_0 + \ell_1 + \ell_2 + x_2 \in B_{\mathcal{U}}$. Notice that if $x_2 = x_0$ or $x_2 = x_1$ then we have a $(k+1)$-term arithmetic progression in $B_{\mathcal{U}} - \ell_0$. Otherwise, let us consider the intersection

$$B_2 := (B - x_2) \cap \bigcap_{i-1}^{k} (B - x_1 - iy_2) \cap \bigcap_{i=1}^{k} (B - x_0 - i(y_1 + y_2)).$$

Similarly as above, one can easily verify that $B_2 - \ell_0 - \ell_1 - \ell_2 \in \mathcal{U}$ and so, by the inductive hypothesis, we can pick an arithmetic progression in $B_{\mathcal{U}} - \ell_0 - \ell_1 - \ell_2$ of length k. We iterate the procedure. As the set F is finite, after finitely many steps we will find elements $x_n = x_m$ where $n > m$, and finally obtain the following arithmetic progression of length $k + 1$:

$$\ell_0 + \ell_1 + \cdots + \ell_n + x_n + i(y_{m+1} + \cdots + y_n) \quad i = 0, 1, \ldots, k. \qquad \square$$

5 TIFs and Left Ideals in the Space of Ultrafilters

The usual ultrafilter proof of van der Waerden's Theorem (see [4, Sect. 14.1]) is grounded on the existence of minimal ultrafilters, that is, on ultrafilters that belong to a minimal left ideal of the compact right-topological semigroup $(\beta\mathbb{N}, \oplus)$. In this final section, we show how (maximal) translation invariant filters are in fact related to the closed (minimal) left ideals of $(\beta\mathbb{N}, \oplus)$. Let us recall here the involved notions.

The *space* $\beta\mathbb{N}$ is the topological space of all ultrafilters $\mathcal{U}$ over the full algebra of sets $\mathcal{B} = \mathcal{P}(\mathbb{N})$ where a base of (cl)open sets is given by the family $\{\mathcal{O}_A \mid A \subseteq \mathbb{N}\}$, with $\mathcal{O}_A := \{\mathcal{U} \in \beta\mathbb{N} \mid A \in \mathcal{U}\}$. The space $\beta\mathbb{N}$ is Hausdorff and compact, and coincides with the *Stone-Čech compactification* of the discrete space $\mathbb{N}$.

The *pseudosum* $\mathcal{U} \oplus \mathcal{V}$ of ultrafilters $\mathcal{U}, \mathcal{V} \in \beta\mathbb{N}$ is defined by letting:

$$A \in \mathscr{U} \oplus \mathscr{V} \iff \{n \in \mathbb{N} \mid A - n \in \mathscr{V}\} \in \mathscr{U}.$$

The operation $\oplus$ is associative (but not commutative), and for every $\mathscr{V}$ the map $\mathscr{U} \mapsto \mathscr{U} \oplus \mathscr{V}$ is continuous. This makes $(\beta\mathbb{N}, \oplus)$ a *right-topological semigroup*.

A *left ideal* $L \subseteq \beta\mathbb{N}$ is a nonempty set such that $\mathscr{V} \in L$ implies $\mathscr{U} \oplus \mathscr{V} \in L$ for every $\mathscr{U} \in \beta\mathbb{N}$. The notion of *right ideal* is defined similarly. Left ideals that are minimal with respect to inclusion are particularly relevant objects, as they satisfy special properties. For instance, their union $K(\beta\mathbb{N}, \oplus)$ is shown to be the smallest *bilateral ideal* (i.e., it is both a left and a right ideal). Moreover, all ultrafilters $\mathscr{U}$ in $K(\beta\mathbb{N}, \oplus)$, named *minimal ultrafilters*, have the property that every set $A \in \mathscr{U}$ includes arbitrarily long arithmetic progressions.[4]

It is well-known that there are natural correspondences between families with the finite intersection property on the full algebra $\mathscr{P}(\mathbb{N})$, and closed nonempty subsets of $\beta\mathbb{N}$. Indeed, the following properties are directly verified from the definitions.

- If $\mathscr{G} \subseteq \mathscr{P}(\mathbb{N})$ is a family with the FIP then $\mathfrak{C}(\mathscr{G}) := \{\mathscr{V} \in \beta\mathbb{N} \mid \mathscr{V} \supseteq \mathscr{G}\}$ is a nonempty closed subspace.
- If $X \subseteq \beta\mathbb{N}$ is nonempty then $\mathfrak{F}(X) := \bigcap\{\mathscr{V} \mid \mathscr{V} \in X\}$ is a filter on $\mathscr{P}(\mathbb{N})$.
- $\mathfrak{C}(\mathfrak{F}(X)) = \overline{X}$ (the topological closure of X) for every nonempty $X \subseteq \beta\mathbb{N}$.
- $\mathfrak{F}(\mathfrak{C}(\mathscr{G})) = \langle\mathscr{G}\rangle$ (the filter generated by $\mathscr{G}$) for every family $\mathscr{G} \subseteq \mathscr{P}(\mathbb{N})$ with the FIP.

Proposition 5.1 *If $\mathscr{F}$ is a TIF on $\mathscr{P}(\mathbb{N})$ then $\mathfrak{C}(\mathscr{F})$ is a closed left ideal of $(\beta\mathbb{N}, \oplus)$; and conversely, if L is a left ideal of $(\beta\mathbb{N}, \oplus)$ then $\mathfrak{F}(L)$ is a TIF on $\mathscr{P}(\mathbb{N})$. Moreover, $\mathscr{M}$ is a maximal TIF on $\mathscr{P}(\mathbb{N})$ if and only if $\mathfrak{C}(\mathscr{M})$ is a minimal left ideal of $(\beta\mathbb{N}, \oplus)$; and L is a minimal left ideal of $(\beta\mathbb{N}, \oplus)$ if and only if $\mathfrak{F}(L)$ is a maximal TIF on $\mathscr{P}(\mathbb{N})$.*

Proof Let $\mathscr{V} \in \mathfrak{C}(\mathscr{F})$ and let $\mathscr{U} \in \beta\mathbb{N}$ be any ultrafilter. For every $A \in \mathscr{F}$, by translation invariance we know that $A - n \in \mathscr{F}$ for all n, and so $\{n \mid A - n \in \mathscr{V}\} = \mathbb{N} \in \mathscr{U}$. This shows that $A \in \mathscr{U} \oplus \mathscr{V}$. As this is true for every $A \in \mathscr{F}$, we conclude that $\mathscr{U} \oplus \mathscr{V} \in \mathfrak{C}(\mathscr{F})$, and so $\mathfrak{C}(\mathscr{F})$ is a closed left ideal.

Now let L be a left ideal, and let $A \in \mathfrak{F}(L)$ be in the filter determined by L. For every $\mathscr{V} \in L$, we have that $\mathfrak{U}_1 \oplus \mathscr{V} \in L$, where $\mathfrak{U}_1 := \{B \subseteq \mathbb{N} \mid 1 \in B\}$ is the principal ultrafilter generated by 1. Then $A \in \mathfrak{U}_1 \oplus \mathscr{V}$, which is equivalent to $A - 1 \in \mathscr{V}$. As this holds for every $\mathscr{V} \in L$, we have proved that $A - 1 \in \mathfrak{F}(L)$, and so $\mathfrak{F}(L)$ is a TIF, as desired.

Let $\mathscr{F}$ be a TIF. If the left ideal $\mathfrak{C}(\mathscr{F})$ is not minimal, pick a minimal $L \subsetneq \mathfrak{C}(\mathscr{F})$. Then $\mathscr{F} \subsetneq \mathfrak{F}(L)$, and hence $\mathscr{F}$ is not maximal. Indeed, $L \subseteq \mathfrak{C}(\mathscr{F}) \Rightarrow \mathfrak{F}(L) \supseteq \mathfrak{F}(\mathfrak{C}(\mathscr{F})) = \mathscr{F}$; moreover, $\mathscr{F} \neq \mathfrak{F}(L)$, as otherwise $\mathfrak{C}(\mathscr{F}) = \mathfrak{C}(\mathfrak{F}(L)) = \overline{L} = L$, against our assumptions. (Recall that a minimal left ideal L is necessarily closed because, by minimality, $L = \beta\mathbb{N} \oplus \mathscr{V} := \{\mathscr{U} \oplus \mathscr{V} \mid \mathscr{U} \in \beta\mathbb{N}\}$ for every given $\mathscr{V} \in L$, and $\beta\mathbb{N} \oplus \mathscr{V}$ is closed as it the image of the compact Hausdorff space $\beta\mathbb{N}$ under

[4] For all notions and basic results on the space of ultrafilters $\beta\mathbb{N}$ and on its algebraic structure, including properties of the smallest ideal $K(\beta\mathbb{N}, \oplus)$, we refer the reader to the book [4].

the continuous function $\mathcal{U} \mapsto \mathcal{U} \oplus \mathcal{V}$.) In a similar way, one shows the converse implication: If the TIF $\mathcal{F}$ is not maximal then the left ideal $\mathfrak{C}(\mathcal{F})$ is not minimal. In consequence, $L = \mathfrak{C}(\mathfrak{F}(L))$ is minimal if and only if $\mathfrak{F}(L)$ is maximal, and also the last equivalence is proved. □

As a straight consequence, we obtain the desired characterization.

Proposition 5.2 *An ultrafilter $\mathcal{U}$ on $\mathcal{P}(\mathbb{N})$ includes a maximal TIF if and only if $\mathcal{U}$ belongs to the smallest ideal $K(\beta\mathbb{N}, \oplus)$.*

Proof Recall that $\mathcal{U} \in K(\beta\mathbb{N}, \oplus)$ if and only if $\mathcal{U}$ belongs to some minimal left ideal. Now let $\mathcal{U} \supseteq \mathcal{M}$ where $\mathcal{M}$ is a maximal TIF. Since $\mathcal{M} = \mathfrak{F}(\mathfrak{C}(\mathcal{M}))$, we have that $\mathcal{U} \in \mathfrak{C}(\mathcal{M})$, where $\mathfrak{C}(\mathcal{M})$ is a minimal left ideal. Conversely, let $\mathcal{U} \in L$ where L is a minimal left ideal. Then $\mathfrak{F}(L)$ is a maximal TIF and $\mathcal{U} \supseteq \mathfrak{F}(L)$, since $\mathcal{U} \in L = \mathfrak{C}(\mathfrak{F}(L))$. □

Remark 5.3 One can generalize the contents of this paper from the natural numbers to arbitrary countable semigroups $(S, \cdot)$. Indeed, the notion of translation invariant filter also makes sense in that more general framework.[5] Precisely, for $A \subseteq S$ and $s \in S$, denote by $s^{-1}A := \{t \in S \mid s \cdot t \in A\}$. We say that an algebra $\mathcal{B}$ of subsets of S is *translation invariant* if $B \in \mathcal{B} \Rightarrow s^{-1}B \in \mathcal{B}$ for all $s \in S$. Then one defines a TIF on a translation invariant algebra $\mathcal{B}$ as a filter $\mathcal{F}$ such that $A \in \mathcal{F} \Rightarrow s^{-1}A \in \mathcal{F}$ for all $s \in S$. By the same arguments as the ones used in this paper, one can prove that a reformulation of Theorem 4.1 holds, provided one adopts the appropriate generalization of the notion of piecewise syndetic set.[6]

Acknowledgements I would like to thank the anonymous referee for carefully reading the first version of this paper and for giving comments that were helpful for the final revision.

References

1. V. Bergelson and N. Hindman, Partition regular structures contained in large sets are abundant, *J. Combin. Theory Ser. A* **93** (2001), 18–36.
2. H. Furstenberg and E. Glasner, Subset dynamics and van der Waerden's Theorem, in *Topological Dynamics and Applications* (M.G. Nerurkar, D.P. Dokken, and D.B. Ellis, eds.), Contemp. Math. **215**, Amer. Math. Soc., 1998, 197–203.

[5] Actually, our techniques also apply for uncountable semigroups, but in that case one needs Zorn's Lemma to prove the existence of maximal TIFs and of ultrafilters.

[6] In an arbitrary semigroup $(S, \cdot)$, one defines a subset $T \subseteq S$ to be *thick* if for every finite F there exists $s \in S$ with $F \cdot s := \{x \cdot s \mid x \in F\} \subseteq T$; a set $A \subseteq S$ is *syndetic* if a suitable finite union $\bigcup_{i=1}^{k} s_i^{-1}A = S$ covers the whole semigroup; and finally a set $A \subseteq S$ is *piecewise syndetic* if a suitable finite union $\bigcup_{i=1}^{k} s_i^{-1}A$ is thick (see [4, Sects. 4.4 and 4.5]).

3. N. Hindman, Problems and new results in the algebra of βS and Ramsey Theory, in *Unsolved Problems in Mathematics for the 21st Century* (J. Abe and S. Tanaka, eds.), IOS Press, 2001, 295–305.
4. N. Hindman and D. Strauss, *Algebra in the Stone-Čech Compactification, Theory and Applications* (2nd edition), W. de Gruyter, 2012.
5. B.L. van der Waerden, Beweis einer baudetschen vermutung, *Nieuw. Arch. Wisk.* **15** (1927), 212–216.

Central Values for Clebsch–Gordan Coefficients

Robert W. Donley Jr.

Abstract We develop further properties of the matrices $M(m, n, k)$ defined by the author and W. G. Kim in a previous work. In particular, we continue an alternative approach to the theory of Clebsch–Gordan coefficients in terms of combinatorics and convex geometry. New features include a censorship rule for zeros, a sequence of 36-pointed stars of zeros, and another proof of Dixon's Identity. As a major application, we reinterpret the work of Raynal et al. on vanishing Clebsch–Gordan coefficients as a "middle-out" approach to computing $M(m, n, k)$.

1 Introduction

In the representation theory of $SU(2)$, the Clebsch–Gordan decomposition for tensor products of irreducible representations yields a uniform pattern for highest weights, generally, as an arithmetic progression of integers with difference two, symmetric about zero. Curiously, at the vector level, if one tensors two vectors of weight zero, a similar arithmetic progression of weights occurs, but now with difference four. In the theory of spherical varieties, extensions of this problem consider minimal gap lengths in the weight monoid associated to spherical vector products. An elementary calculation for the case of $SU(2)$ initiates the present work, given in Proposition 15, and we review the open problem of vanishing beyond the weight zero case.

Continuing the work began in [3], this approach to Clebsch–Gordan coefficients substitutes the use of hypergeometric series [12] and the like with elementary combinatorial methods (generating functions, recurrences, finite symmetry groups, Pascal's triangle). Of course, hypergeometric series are fundamental to the theory; a long range goal would be to return this alternative approach, once sufficiently developed, to the hypergeometric context with general parameters. At the practical level, certain computer simulations in nuclear physics and chemistry may require excessively large

R. W. Donley Jr. (✉)
Queensborough Community College, Bayside, NY, USA
e-mail: RDonley@qcc.cuny.edu

M. B. Nathanson (ed.), *Combinatorial and Additive Number Theory III*,
Springer Proceedings in Mathematics & Statistics 297,
https://doi.org/10.1007/978-3-030-31106-3_7

numbers of Clebsch–Gordan coefficients, possibly with large parameters. Methods for an integral theory have immediate scientific applications.

As with [3], this work contains many direct elementary proofs of known, but perhaps lesser known at large, results and gives them a new spatial context. One central item of concern, the domain space, is a set of integer points inside of a five-dimensional cone, equipped with an order 72 automorphism group, which in turn consists of the familiar symmetries for the determinant.

Key observations in this work depend on the fixed points of the symmetry group in the cone, a distinguished subgroup of dihedral type D_{12}, polygonal subsets of the domain invariant under the subgroup action, and the simultaneous use of recurrences with the subgroup's center. One byproduct of the theory is yet another proof of Dixon's Identity and some variants, and, with this identity, our computational viewpoint shifts from the specific "outside-in" algorithm of [3] to a universal "middle-out" approach, adapted from the work in [9].

Section 2 recalls the algorithm of [3] for $M(m, n, k)$, and Sect. 3 develops basic properties of $M(m, n, k)$. Sections 4 and 5 review the theory of the so-called "trivial" zeros, and Sects. 6–9 reconsider the results of [9] as computations near the center of $M(m, n, k)$.

2 Coordinate Vector Matrices for Clebsch–Gordan Sums

As a function in five variables, $c_{m,n,k}(i, j)$ is defined on the integer points of the cone

$$0 \le i \le m, \quad 0 \le j \le n, \quad 0 \le k \le \min(m, n), \quad 0 \le i + j - k \le m + n - 2k.$$

It is convention to extend this domain by zero; here it is enough to do so at least for the matrices $M(m, n, k)$ defined below and for clarity we often omit the corner zeros. In general, the various Clebsch–Gordan coefficients $C_{m,n,k}(i, j)$ and $c_{m,n,k}(i, j)$ differ by a nonzero factor, and our approach to vanishing of Clebsch–Gordan coefficients is through vanishing of the sum $c_{m,n,k}(i, j)$.

From [3], all $c_{m,n,k}(i, j)$ in (3) below may be computed algorithmically as a matrix $M(m, n, k)$. With m, n, k fixed, Proposition 1 and Theorem 1 below produce $M(m, n, k)$, with columns corresponding to coordinate vectors for weight vectors in the subrepresentation $V(m + n - 2k)$ of $V(m) \otimes V(n)$. Consideration of Pascal's identity gives an explicit formula for $c_{m,n,k}(i, j)$ in Proposition 2.

For non-negative integers a, b, c, multinomial coefficients are defined by

$$\binom{a+b}{a} = \frac{(a+b)!}{a!\,b!} \quad \text{and} \quad \binom{a+b+c}{a,\ b,\ c} = \frac{(a+b+c)!}{a!\,b!\,c!}. \tag{1}$$

First the highest weight vectors for $V(m + n - 2k)$ are defined by

Proposition 1 (Leftmost column for $M(m,n,k)$) *With $0 \le i \le k$, define the $(i+1,1)$th entry of $M(m,n,k)$ by*

$$c_{m,n,k}(i,k-i) = (-1)^i \binom{m-i}{k-i}\binom{n-k+i}{i}. \tag{2}$$

Repeated application f to these highest weight vectors produces coordinates for general weight vectors in the corresponding $V(m+n-2k)$, recorded as

Proposition 2 (Entry at coordinate $(i+1, i+j-k+1)$ in $M(m,n,k)$)

$$c_{m,n,k}(i,j) = \sum_{l=0}^{k} (-1)^l \binom{i+j-k}{i-l}\binom{m-l}{k-l}\binom{n-k+l}{l}. \tag{3}$$

We also note the following alternative expression from [3]:

$$c_{m,n,k}(i,j) = \sum_{l=0}^{k} (-1)^l \binom{i+j-k}{i-l}\binom{m-i}{k-l}\binom{n-j}{l}. \tag{4}$$

With this expression, results in later sections may be interpreted by way of Dixon's Identity and its extensions for binomial sums.

Theorem 1 (Definition of Matrix $M(m,n,k)$) *To calculate the coordinate vector matrix $M(m,n,k)$:*

1. *initialize a matrix with $m+1$ rows and $m+n-2k+1$ columns,*
2. *set up coordinates for the highest weight vector in the leftmost column using Proposition 1, and extend the top row value,*
3. *apply Pascal's recurrence rightwards in an uppercase L pattern, extending by zero where necessary,*
4. *for the zero entries in lower-left corner, corresponding entries in the upper-right corner are set to zero, and*
5. *the $(i+1, i+j-k+1)$th entry is $c_{m,n,k}(i,j)$.*

In this work, we often remove corner zeros for visual clarity. Many examples of $M(m,n,k)$ will be given throughout this work.

3 Elementary Rules

A useful parametrization for the domain of $c_{m,n,k}(i,j)$ is given by the set of Regge symbols

$$\left\| \begin{matrix} n-k & m-k & k \\ i & j & m+n-i-j-k \\ m-i & n-j & i+j-k \end{matrix} \right\|. \tag{5}$$

Note that each row or column sums to $J = m + n - k$. That is, the domain space is in one-one correspondence with all 3-by-3 matrices with nonnegative integer entries having this magic square property. Here we follow [12], while [9] switches rows 2 and 3.

In turn, one may define the Regge group of symmetries; each of the 72 determinant symmetries, generated by row exchange, column exchange, and transpose, correspond to a transformation of $c_{m,n,k}(i, j)$.

Of particular interest here is the dihedral subgroup D_{12} of twelve elements generated by column switches and the interchange of rows 2 and 3. Some elements, including a generating set, are given as follows:

Proposition 3 (R23 Symmetry–Weyl Group) *With the asterisks defined by (1),*

$$\binom{m+n-k}{i,\ j,\ *} c_{m,n,k}(m-i, n-j) = (-1)^k \binom{m+n-k}{m-i,\ n-j,\ *} c_{m,n,k}(i, j). \tag{6}$$

Proposition 4 (C12 Symmetry) *When defined,*

$$c_{n,m,k}(j, i) = (-1)^k c_{m,n,k}(i, j). \tag{7}$$

Proposition 5 (C13 Symmetry) *With $i' = i + j - k$ and $m' = m + n - 2k$,*

$$c_{m',n,n-k}(m' - i', j) = (-1)^{n-j} c_{m,n,k}(i, j). \tag{8}$$

Proposition 6 (C123 Symmetry) *With $i' = i + j - k$ and $m' = m + n - 2k$,*

$$c_{m',m,m-k}(m' - i', i) = (-1)^{m-k+i} c_{m,n,k}(i, j). \tag{9}$$

Since the Weyl group symmetry preserves m, n, and k, it transforms $M(m, n, k)$ to itself. In fact, the net effect of this symmetry is to rotate $M(m, n, k)$ by 180°, change signs according to the parity of k, and rescale values by a positive scalar, rational in the five parameters. In particular, the zero locus of $c_{m,n,k}(i, j)$ in $M(m, n, k)$ is preserved under this symmetry.

In the complement of the corner triangles of zeros, the upper, left-most, and lower-left edges in $M(m, n, k)$ have non-vanishing entries by construction. By the Weyl group symmetry, the remaining outer edges also have this property. Thus these edges trace out a polygon, possibly degenerating to a segment, with no zeros on its outer edges.

Definition 1 We refer to the complement of the corner triangles of zeros in $M(m, n, k)$ as the **polygon** of $M(m, n, k)$. A zero in the interior of the polygon of $M(m, n, k)$ is called a **proper zero**. Other terminology for zeros will be noted below.

Opposing edges of this polygon have the same length. The horizontal edges have length $n - k + 1$, the slanted edges have length $m - k + 1$, and the vertical edges have length $k + 1$. When the polygon is a hexagon, one notes that the absolute values of the entries along the top, lower-left, and right-sided edges are constant and equal to, respectively,

$$\binom{m}{k}, \binom{n}{k}, \text{ and } \binom{m+n-2k}{m-k}. \tag{10}$$

These binomial coefficients are composed of parts $m-k$, $n-k$, and k. These values apply to the degenerate cases (parallelogram or line segment) accordingly. The remaining edges link these values through products of binomial coefficients; the indices in the starting coefficient decrease by 1 as the indices in the terminal coefficient increase likewise. See Proposition 1 for the leftmost vertical edge.

We note the decomposition

$$D_{12} \cong S_3 \times C_2, \tag{11}$$

where S_3 represents the permutation subgroup generated by column switches and the Weyl group symmetry generates the two-element group C_2. While column switches do not preserve $M(m, n, k)$ in general, the polygon of $M(m, n, k)$ maps to the polygon of values in another $M(m_2, n_2, k_2)$, differing only by sign changes. Thus there is a well-defined correspondence between the proper zeros of $M(m, n, k)$ and $M(m_2, n_2, k_2)$ under a column switch.

Of the column switches listed, the C12 symmetry changes sign according to k and inverts proper values in each column, and the C13 symmetry changes sign according to $n - j$ and reflects values across the northeasterly diagonal. The column switch C123 rotates values by 120° counter-clockwise and changes sign according to $m - k + i$.

For example, the polygons of $M(3, 4, 2)$, $M(3, 3, 1)$, $M(4, 3, 2)$ below permute among themselves under the D_{12} symmetries:

$$\begin{bmatrix} 3 & 3 & 3 & \\ -6 & -3 & 0 & 3 \\ 6 & 0 & -3 & -3 \\ & 6 & 6 & 3 \end{bmatrix}, \quad \begin{bmatrix} 3 & 3 & 3 & & \\ -3 & 0 & 3 & 6 & \\ & -3 & -3 & 0 & 6 \\ & & -3 & -6 & -6 \end{bmatrix}, \quad \begin{bmatrix} 6 & 6 & & \\ -6 & 0 & 6 & \\ 3 & -3 & -3 & 3 \\ & 3 & 0 & -3 \\ & & 3 & 3 \end{bmatrix}.$$

Next, we note two additional recurrences from [3]; these impose further restrictions on proper zeros within $M(m, n, k)$.

Proposition 7 (Pascal's Recurrence) *When all terms are defined,*

$$c_{m,n,k}(i, j) = c_{m,n,k}(i, j-1) + c_{m,n,k}(i-1, j).$$

Proposition 8 (Reverse Recurrence) *When all terms are defined,*

$$a_1\, c_{m,n,k}(i, j) = a_2\, c_{m,n,k}(i+1, j) + a_3\, c_{m,n,k}(i, j+1)$$

where

1. $a_1 = (i + j - k + 1)(m + n - i - j - k)$,
2. $a_2 = (i + 1)(m - i)$, *and*
3. $a_3 = (j + 1)(n - j)$.

The reverse recurrence is Pascal's recurrence after application of the Weyl group symmetry. It follows a rotated capital-L pattern, only now weighted by positive integers when $0 < i < m$ and $0 < j < n$.

These recurrences immediately yield

Proposition 9 (Censorship Rule) *Suppose $c_{m,n,k}(i, j) = 0$ properly in $M(m, n, k)$. Then adjacent zeros in $M(m, n, k)$ may occur only at the upper-right or lower-left entries. That is, in the following submatrix of $M(m, n, k)$, the bullets must be nonzero:*

$$\begin{bmatrix} \bullet & \bullet & * \\ \bullet & 0 & \bullet \\ * & \bullet & \bullet \end{bmatrix}.$$

Proof First note that, in either recurrence relation, if any two terms in the relation are zero, then so is the third. Now suppose a pair of proper zeros are horizontally adjacent. Then, alternating the two relations, one begins a sequence

$$\begin{bmatrix} * & * & * \\ * & * & * \\ * & 0 & 0 \end{bmatrix} \to \begin{bmatrix} * & * & * \\ * & 0 & * \\ * & 0 & 0 \end{bmatrix} \to \begin{bmatrix} * & * & * \\ 0 & 0 & * \\ * & 0 & 0 \end{bmatrix} \to \begin{bmatrix} 0 & * & * \\ 0 & 0 & * \\ * & 0 & 0 \end{bmatrix} \to \cdots.$$

Eventually this serpentine must place a zero at a nonzero entry on the top row of $M(m, n, k)$, a contradiction. The case of two vertically adjacent zeros follows similarly. Finally, for the diagonal case

$$\begin{bmatrix} 0 & * \\ * & 0 \end{bmatrix},$$

either recurrence rule begins the serpentine. □

As an adjunct to censorship, certain pairs of zeros severely restrict nearby values.

Proposition 10 *With X nonzero, the following allowable pairs of zeros fix adjacent values:*

$$\begin{bmatrix} X & & \\ 0 & X & \\ -X & -X & 0 \end{bmatrix}, \begin{bmatrix} X & & \\ -X & 0 & \\ 0 & -X & -X \end{bmatrix}, \begin{bmatrix} 0 & & \\ X & X & \\ -X & 0 & X \end{bmatrix}.$$

In particular, each triangle implies one of the following three equalities:

$$(i + 1)(m - i) = (i' + 1)(m' - i') = (j + 1)(n - j),$$

where $i' = i + j - k$*,* $m' = m + n - 2k$*, and* $c_{m,n,k}(i, j)$ *corresponds to the middle entry of the leftmost column.*

Proof The restriction on values follows immediately from Pascal's recurrence. The parameter conditions follow from the second recurrence. □

Finally, it will be convenient to note four degenerate cases of $M(m, n, k)$; the first three cases give all conditions for when the polygon is not a hexagon.

1. When $k = 0$ and $n > 0$, the parallelogram of nonzero entries of $M(m, n, 0)$ consists of entries in Pascal's triangle, with columns corresponding to segments of the triangle's rows. When $n = k = 0$, $M(m, 0, 0)$ is an identity matrix of size $m + 1$,
2. When $k = m$ with $n \geq m$, $M(m, n, m)$ contains no zeros; ignoring signs, the upper-right corner corresponds to the peak of Pascal's triangle, with diagonals corresponding to segments of the triangle's rows,
3. When $m = n = k$, $M(m, m, m)$ degenerates to a vertical segment of length $m + 1$, and
4. When $m \geq 4$ even, $n = 2$ and $k = 1$, a central vertical triplet of zeros occurs, but only the central zero is proper.

In fact, with $n > m$, the C23 symmetry carries $M(m, n - m, 0)$ to $M(m, n, m)$ Cases 1–4 are represented below by $M(2, 2, 0)$, $M(2, 4, 2)$, $M(2, 2, 2)$, and $M(4, 2, 1)$, respectively:

$$\begin{bmatrix} 1 & 1 & 1 & 0 & 0 \\ 0 & 1 & 2 & 3 & 0 \\ 0 & 0 & 1 & 3 & 6 \end{bmatrix}, \quad \begin{bmatrix} 1 & 1 & 1 \\ -3 & -2 & -1 \\ 6 & 3 & 1 \end{bmatrix}, \quad \begin{bmatrix} 1 \\ -1 \\ 1 \end{bmatrix}, \quad \begin{bmatrix} 4 & 4 & 0 & 0 & 0 \\ -2 & 2 & 6 & 0 & 0 \\ 0 & -2 & 0 & 6 & 0 \\ 0 & 0 & -2 & -2 & 4 \\ 0 & 0 & 0 & -2 & 4 \end{bmatrix}.$$

4 Diagonal Zeros

Consideration of dihedral reflections leads one to a large family of proper zeros through Propositions 4, 5, and the C23 symmetry. In particular, these zeros occur when a column switch in the Regge symbol fixes an entry of $M(m, n, k)$ and changes parity. Zeros of this type always occur as diagonal subsets in $M(m, n, k)$.

To see this, first observe that when $n = 2k$, $M(m, 2k, k)$ is a square matrix of size $m + 1$, and the C13 symmetry preserves the northeasterly diagonal. In this case, the subgroup of the Regge group generated by C13 and R23, of type $C_2 \times C_2$, preserves both the polygon and the diagonal. Specifically, the top row of the general Regge symbol for these parameters

$$\left\| \begin{matrix} k & m-k & k \\ i & j & m-i-j-k \\ m-i & 2k-j & i+j-k \end{matrix} \right\|. \tag{12}$$

is unchanged by the R23 and C13 symmetries.

With $m \neq 2k$, four edges of the polygon now have length $k+1$, upon which the subgroup acts transitively. When $m = 2k$, the polygon is a regular hexagon preserved by the dihedral subgroup D_{12} of Sect. 3.

Proposition 11 *Suppose $k > 0$ and $m + k$ is odd. Then*

$$c_{m,2k,k}(i, m+k-2i) = 0. \tag{13}$$

Proof Note that coordinates in $M(m, n, k)$ are indexed by $x = i + j - k + 1$ and $y = i + 1$, and the indices for the diagonal in question are solutions to

$$(i + j - k + 1) + (i + 1) = m + 2 \quad \text{or} \quad j = m + k - 2i. \tag{14}$$

Substituting $n = 2k$ into Proposition 5, one obtains

$$c_{m,2k,k}(m + k - i - j, j) = (-1)^j c_{m,2k,k}(i, j). \tag{15}$$

For entries on the diagonal, this equation reduces to

$$c_{m,2k,k}(i, j) = (-1)^{m+k} c_{m,2k,k}(i, j), \tag{16}$$

and $c_{m,2k,k}(i, j)$ vanishes under the given parity condition. □

For example, we have $M(4, 6, 3)$ and $M(5, 4, 2)$, respectively:

$$\begin{bmatrix} 4 & 4 & 4 & 4 & \\ -12 & -8 & -4 & 0 & 4 \\ 20 & 8 & 0 & -4 & -4 \\ -20 & 0 & 8 & 8 & 4 \\ & -20 & -20 & -12 & -4 \end{bmatrix}, \quad \begin{bmatrix} 10 & 10 & 10 & & & \\ -12 & -2 & 8 & 18 & & \\ 6 & -6 & -8 & 0 & 18 & \\ & 6 & 0 & -8 & -8 & 10 \\ & & 6 & 6 & -2 & -10 \\ & & & 6 & 12 & 10 \end{bmatrix}.$$

Next we give a formula for the entries below the diagonal as single term expressions. There are k proper zeros on the diagonal, and we denote the position of such a zero as measured from lower-left to upper-right.

Proposition 12 *The value of the entry directly below the tth proper zero on the diagonal is given as a single term expression by*

$$(-1)^{k+t+1} \binom{2k}{k} \binom{k-1}{t-1} \frac{(2k-2t+1)!}{(2k-1)!} \frac{(\frac{m+k+1}{2})!\,(\frac{m-k-3+2t}{2})!}{(\frac{m-k-1}{2})!\,(\frac{m+k+3-2t}{2})!}. \tag{17}$$

Proof The coordinates for the sth proper zero on the diagonal are

$$(i+1, i+j-k+1) = \left(\frac{m+k-2s+3}{2}, \frac{m-k+2s+1}{2}\right) \tag{18}$$

with

$$i = \frac{m+k-2s+1}{2}, \quad j = 2s-1. \tag{19}$$

The entry directly below the first zero has value

$$c_{m,2k,k}\left(\frac{m+k+1}{2}, 1\right) = (-1)^k \binom{2k}{k}.$$

The second recurrence allows us to compute values below the diagonal by a sequence of factors: at the sth proper zero,

$$\begin{bmatrix} 0 & x \\ y & y \end{bmatrix} \quad \longrightarrow \quad x = -\frac{a_2}{a_3}y = -\frac{(m+k-2s+3)(m-k+2s-1)}{8s(2k-2s+1)}y.$$

Thus the value of the entry below the tth zero on the diagonal is

$$(-1)^{k+t-1}\binom{2k}{k}\prod_{s=1}^{t-1}\frac{(m+k-2s+3)(m-k+2s-1)}{8s(2k-2s+1)}. \tag{20}$$

Since

$$\prod_{s=1}^{t-1}(2N-2s) = \frac{2^{t-1}(N-1)!}{(N-t)!}, \qquad \prod_{s=1}^{t-1}(2N+2s) = \frac{2^{t-1}(N+t-1)!}{N!}, \tag{21}$$

and

$$\prod_{s=1}^{t-1}(2N-2s+1) = \frac{(2N-1)!(N-t)!}{2^{t-1}(2N-2t+1)!(N-1)!}, \tag{22}$$

the result follows by substitution into (20). □

In particular, when m even, k odd and $t = \frac{k+1}{2}$, the entry below the central zero is

$$(-1)^{\frac{k+1}{2}}\frac{2(k+1)^2}{m(m+2)}\begin{pmatrix} & \frac{m+k+1}{2} & \\ \frac{k+1}{2}, & \frac{k+1}{2}, & \frac{m-k-1}{2}\end{pmatrix}. \tag{23}$$

When m is odd, k even and $t = \frac{k}{2}+1$, there is a central square

$$\begin{bmatrix} X & 0 \\ 0 & X \end{bmatrix} \tag{24}$$

with

$$X = (-1)^{\frac{k}{2}} \, \frac{k+2}{m+1} \begin{pmatrix} & \frac{m+k+1}{2} & \\ \frac{k}{2}, & \frac{k+2}{2}, & \frac{m-k-1}{2} \end{pmatrix}. \tag{25}$$

Two other types of zero diagonals may be obtained by applying column switches to (13). Applying the C12 symmetry yields

Proposition 13 *Suppose $k > 0$ and $n + k$ is odd. Then*

$$c_{2k,n,k}(n + k - 2j, j) \;=\; 0. \tag{26}$$

This diagonal of zeros connects the midpoints of the horizontal edges of the polygon. By the censorship rule, no zeros on this diagonal are adjacent, and the diagonal is fixed by the C23 symmetry. Applying the C13 symmetry to the parameters of Proposition 13 gives

Proposition 14 *Suppose $k > 0$ is odd. Then*

$$c_{m,m,k}(i, i) \;=\; 0. \tag{27}$$

This diagonal of zeros connects the midpoints of the vertical edges of the polygon. Again, no zeros on this diagonal are adjacent, and this diagonal is fixed by the C12 symmetry.

From above, $M(4, 6, 3)$ transforms into $M(6, 4, 3)$ and $M(4, 4, 1)$, respectively:

$$\begin{bmatrix} 20 & 20 & & & \\ -20 & 0 & 20 & & \\ 12 & -8 & -8 & 12 & \\ -4 & 8 & 0 & -8 & 4 \\ & -4 & 4 & 4 & -4 \\ & & -4 & 0 & 4 \\ & & & 4 & -4 \end{bmatrix}, \quad \begin{bmatrix} 4 & 4 & 4 & 4 & & & \\ -4 & 0 & 4 & 8 & 12 & & \\ & -4 & -4 & 0 & 8 & 20 & \\ & & -4 & -8 & -8 & 0 & 20 \\ & & & -4 & -12 & -20 & -20 \end{bmatrix}.$$

Next suppose $m = n = 2k$ with k odd; see Fig. 1 below. The polygon in $M(2k, 2k, k)$ is now a regular hexagon with sides of length $k + 1$, and entries on three sides have constant absolute value $\binom{2k}{k}$. The full D_{12} subgroup preserves $M(2k, 2k, k)$, as the general entry has Regge symbol

$$\left\| \begin{matrix} k & k & k \\ i & j & 3k - i - j \\ 2k - i & 2k - j & i + j - k \end{matrix} \right\|. \tag{28}$$

252	252	252	252	252	252					
-756	-504	-252	0	252	504	756				
1176	420	-84	-336	-336	-84	420	1176			
-1176	0	420	336	0	-336	-420	0	1176		
756	-420	-420	0	336	336	0	-420	-420	756	
-252	504	84	-336	-336	0	336	336	-84	-504	252
	-252	252	336	0	-336	-336	0	336	252	-252
		-252	0	336	336	0	-336	-336	0	252
			-252	-252	84	420	420	84	-252	-252
				-252	-504	-420	0	420	504	252
					-252	-756	-1176	-1176	-756	-252

Fig. 1 M(10, 10, 5) with $m = n = 2k$ and $k = 5$ odd

In particular, when $i = j = k$, the central entry is a fixed point under the full Regge group. Since k is odd, the polygon now possesses three diagonals of zeros (six-pointed star) with several interesting consequences; first, when $k \geq 5$ odd, there is an equilateral triangle, centered about the central value and with sides of length 7, with nonzero entries of fixed absolute value:

$$\begin{bmatrix} 0 & & & & & & \\ X & X & & & & & \\ -X & 0 & \mathbf{X} & & & & \\ 0 & -X & -\mathbf{X} & 0 & & & \\ \mathbf{X} & \mathbf{X} & \boxed{0} & -\mathbf{X} & -\mathbf{X} & & \\ -X & 0 & \mathbf{X} & X & 0 & -X & \\ 0 & -X & -\mathbf{X} & 0 & X & X & 0 \end{bmatrix}.$$

By (23), the nonzero entry X is equal to

$$c_{2k,2k,k}(k+1, k-1) = (-1)^{\frac{k+1}{2}} \frac{k+1}{6k} \begin{pmatrix} & \frac{3k+3}{2} & \\ \frac{k+1}{2}, & \frac{k+1}{2}, & \frac{k+1}{2} \end{pmatrix}.$$

Furthermore, the large triangle is an aggregation of the three smaller triangular zero pair patterns in Proposition 10.

Next, as each non-central zero on a diagonal is fixed by an order two subgroup in the Regge group, each orbit under the full Regge group contains 36 zeros. Since each Regge symmetry induces a linear change in indices, we have

Theorem 2 *For each odd $k > 1$, $c_{2k,2k,k}(k, k)$ is a fixed point of the full Regge group and is at the center of both $M(2k, 2k, k)$ and a 36-pointed star of zeros in the five-dimensional Clebsch–Gordan domain space.*

5 Four Cases and Some Special Values

Following [9], we now focus on behavior of $c_{m,n,k}(i, j)$ near the center of $M(m, n, k)$. The shape of the center is determined by parities of m, n, and k; these are determined by the top line of the Regge symbol, and, through the D_{12} symmetries, we can narrow our results to four cases (Table 1):

Suppose both m and n are even. Recall that $M(m, n, k)$ is a matrix of size $m + 1$ by $m + n - 2k + 1$. Thus there is a central entry, $c_{m,n,k}(\frac{m}{2}, \frac{n}{2})$, which we refer to as the **central value** of $M(m, n, k)$, and, with its adjacent entries, we obtain a square submatrix of size 3, which we refer to as the **central square** of $M(m, n, k)$. As seen in the fourth degenerate case, a central square for $k > 0$ can only have non-proper zeros when $n = 2$, $k = 1$, and $m \geq 4$ is even.

Proposition 15 *Suppose m and n are even. The central value* $c_{m,n,k}(\frac{m}{2}, \frac{n}{2}) = 0$ *if and only if k is odd.*

Proof If k is odd then the Weyl group symmetry implies

$$c_{m,n,k}\left(\frac{m}{2}, \frac{n}{2}\right) = -c_{m,n,k}\left(\frac{m}{2}, \frac{n}{2}\right), \tag{29}$$

and the central value vanishes.

In the other direction, suppose the central value vanishes. Consider the central square

$$\begin{bmatrix} -Y & * & * \\ Y & 0 & * \\ * & X & X \end{bmatrix}$$

with X and Y nonzero. From the lower left hook and the second recurrence rule with $i = \frac{m}{2}$ and $j = \frac{n}{2} - 1$, we have

$$a_1 Y = a_2 X,$$

and positivity of a_i implies that X and Y have the same parity. Since $-Y$ and X have opposite parity, the Weyl group symmetry switches signs and k is odd. □

Remark 1 This condition is equivalent to the Regge symbol having matching bottom rows with $J = m + n - k$ odd. This result also corresponds to the linearization for-

Table 1 Central area of $M(m, n, k)$ based on parity

$m - k$	$n - k$	k	Center
Even	Even	Even	Single entry ($\neq 0$)
Odd	Even	Even	Size 2 square
Even	Odd	Odd	Size 2 square
Odd	Odd	Odd	Single entry ($= 0$)

mula for products of Legendre polynomials, as seen, for instance, as Corollary 6.8.3 in [1], and it may be shown directly using the Weyl group and the Casimir operator. In the physics literature, zeros of this type, or their translates under the Regge group, are referred to as "trivial" zeros; all diagonal zeros are of this type. Trivial zeros also correspond to indices outside the polygons and those omitted under raising or lowering operators.

In turn, the proposition may be interpreted as a "gap 4" result when tensoring vectors of weight zero, in contrast with the usual Clebsch–Gordan decomposition, which corresponds to "gap 2." Since a contribution only occurs for k even,

$$f^{m/2}\phi_m \otimes f^{n/2}\phi_n = \sum_{k'=0}^{min(m,n)/2} C_{m,n,2k'}(m/2, n/2) f^{m/2+n/2-2k'}\phi_{m,n,2k'}. \tag{30}$$

Note that the vectors in the sum have nonzero coefficients and correspond to irreducible constituents $V(m+n-4k')$ for $0 \le k' \le \frac{1}{2}\min(m,n)$.

A first step to computing near the center of $M(m,n,k)$ requires knowing either the central value or a near central value. To compute these values inductively, one first notes some values of $c_{m,n,k}(i,j)$ for small k and a four-term recurrence relation. One obtains the following directly from either summation formula:

$$c_{m,n,0}(i,j) = \binom{i+j}{i}, \quad c_{m,n,1}(i,j) = \binom{i+j}{i}\frac{mj-ni}{i+j}, \tag{31}$$

$$c_{m,n,2}(i,j) = \binom{i+j}{i}\frac{jm(j-1)(m-1) - 2ij(m-1)(n-1) + in(i-1)(n-1)}{(i+j)(i+j-1)} \tag{32}$$

Zeros corresponding to $k = 1, 2$ are further classified in [11] and [8], respectively.

Next

Lemma 1 *When all terms are defined,*

$$c_{m+2,n+2,k+2}(i+1, j+1) + c_{m,n,k}(i,j) = c_{m,n+2,k+2}(i, j+1) + c_{m+2,n,k+2}(i+1, j).$$

Proof Using formulas (7.2), (7.3), (7.4), and (7.55) from [3], we have:

$$\begin{aligned}
&c_{m+2,n+2,k+2}(i+1, j+1)\\
&\quad = c_{m+1,n+2,k+2}(i, j+1) + c_{m+2,n+1,k+2}(i+1, j)\\
&\quad = c_{m,n+2,k+2}(i, j+1) + c_{m,n+1,k+1}(i,j)\\
&\qquad\qquad + c_{m+2,n,k+2}(i+1, j) - c_{m+1,n,k+1}(i,j)\\
&\quad = c_{m,n+2,k+2}(i, j+1) + c_{m+2,n,k+2}(i+1, j) - c_{m,n,k}(i,j).
\end{aligned}$$

See [10] for a normalized version of the lemma, along with normalized versions of equations (7.2), (7.3) in [3]. As a special case of the lemma, we obtain another proof of Dixon's Identity, first proven in [5]. Many alternative proofs exist; see, for instance, [4, 6, 13].

Theorem 3 (Dixon's Identity) *When m, n, and k are even, the central value*

$$c_{m,n,k}\left(\frac{m}{2},\frac{n}{2}\right)=\sum_{l=0}^{k}(-1)^l\binom{\frac{m+n-2k}{2}}{\frac{m}{2}-l}\binom{\frac{m}{2}}{k-l}\binom{\frac{n}{2}}{l}=(-1)^{\frac{k}{2}}\binom{\frac{m+n-k}{2}}{\frac{m-k}{2},\frac{n-k}{2},\frac{k}{2}}.$$

Proof We prove the theorem by induction on $N=m+n+k$. For the base case, if $k=0$, the result follows by (31). Now suppose the theorem holds for $N\le m+n+k+4$. Using Lemma 1,

$$\begin{aligned}
&c_{m+2,n+2,k+2}\left(\frac{m+2}{2},\frac{n+2}{2}\right)\\
&=c_{m,n+2,k+2}\left(\frac{m}{2},\frac{n+2}{2}\right)+c_{m+2,n,k+2}\left(\frac{m+2}{2},\frac{n}{2}\right)-c_{m,n,k}\left(\frac{m}{2},\frac{n}{2}\right)\\
&=(-1)^{\frac{k+2}{2}}\left(\binom{\frac{m+n-k}{2}}{\frac{m-k-2}{2},\frac{n-k}{2},\frac{k+2}{2}}+\binom{\frac{m+n-k}{2}}{\frac{m-k}{2},\frac{n-k-2}{2},\frac{k+2}{2}}+\binom{\frac{m+n-k}{2}}{\frac{m-k}{2},\frac{n-k}{2},\frac{k}{2}}\right)\\
&=(-1)^{\frac{k+2}{2}}\frac{m+n-k+2}{k+2}\binom{\frac{m+n-k}{2}}{\frac{m-k}{2},\frac{n-k}{2},\frac{k}{2}}\\
&=(-1)^{\frac{k+2}{2}}\binom{\frac{m+n-k+2}{2}}{\frac{m-k}{2},\frac{n-k}{2},\frac{k+2}{2}}.
\end{aligned}$$

Thus the induction step holds and the theorem is proved. □

With similar proofs, the remaining three cases follow:

Proposition 16 *With m and n even and k odd, the value below the central zero is given by*

$$c_{m,n,k}\left(\frac{m}{2}+1,\frac{n}{2}-1\right)=(-1)^{\frac{k+1}{2}}\frac{2(k+1)(m-k+1)}{m(m+2)}\binom{\frac{m+n-k+1}{2}}{\frac{m-k+1}{2},\frac{n-k-1}{2},\frac{k+1}{2}}.$$

Proposition 17 *With m odd and n and k even, the lower-right central value is given by*

$$c_{m,n,k}\left(\frac{m+1}{2},\frac{n}{2}\right)=(-1)^{\frac{k}{2}}\frac{m-k+1}{m+1}\binom{\frac{m+n-k+1}{2}}{\frac{m-k+1}{2},\frac{n-k}{2},\frac{k}{2}}.$$

Proposition 18 *With m and k odd and n even, the lower-right central value is given by*

$$c_{m,n,k}\left(\frac{m+1}{2},\frac{n}{2}\right)=(-1)^{\frac{k+1}{2}}\,\frac{k+1}{m+1}\begin{pmatrix} & \frac{m+n-k}{2} & \\ \frac{m-k}{2}, & \frac{n-k-1}{2}, & \frac{k+1}{2}\end{pmatrix}.$$

6 Case 1: m, n, and k Even

We now turn our attention to the first of four cases of "non-trivial" zeros. Non-trivial zeros are also called "polynomial" or "structural" in the literature.

For fixed m, n, and k, the use of exponents i and j allow for indexing of $M(m, n, k)$ more or less according to usual matrix notation. When m and n are both even, the polygon rotates or reflects about the central value under the D_{12} symmetries; in the other two cases, the central area may change shape. Positioning the central value as the origin, we develop two infinite lattices of rational functions of m, n, and k, one for each case in the table when m and n are even.

To compute a given $M(m, n, k)$, one extracts the appropriate polygon from the lattice, evaluates the corresponding rational function at m, n, and k, and rescales by the central or near-central value from Sect. 5. An algorithm to construct this lattice follows:

1. obtain a recursive formula to compute down two columns from the center,
2. use Pascal's recurrence to compute down-and-to-the-right from the center, and
3. apply the D_{12} symmetries to extend to the entire lattice.

First we consider the case with k even. The central value X is given by Theorem 3. To begin, we have

Proposition 19 *Suppose m, n, and k are even, and $M(m, n, k)$ has central square*

$$\begin{bmatrix} ZX & * & * \\ YX & \boxed{X} & B_0X \\ * & A_1X & B_1X \end{bmatrix}$$

for nonzero X. Then

$$B_0=\frac{\lambda_{m'}-\lambda_m+\lambda_n}{2\lambda_n},\quad A_1=\frac{\lambda_{m'}-\lambda_m-\lambda_n}{2\lambda_m},\quad B_1=\frac{\lambda_{m'}+\lambda_m-\lambda_n}{2\lambda_m},$$

where $m'=m+n-2k$ and $\lambda_s=s(s+2)$.

Proof Since k is even, X is nonzero. The following four equations follow from Propositions 7 and 8 and the Weyl group symmetry (6):

1. $Z+Y=1$,
2. $1+A_1=B_1$,
3. $\lambda_{m'}Y=\lambda_mA_1+\lambda_n$, and
4. $\lambda_{m'}Z=\lambda_mB_1$.

These equations reduce to a linear system in A_1 and B_1 with the above solutions. The reverse recurrence yields the formula for B_0. □

Consider the following fourth-quadrant submatrix with X in the central position:

$$\begin{bmatrix} \boxed{X} & & & \\ A_1X & B_1X & & \\ A_2X & B_2X & C_2X & \\ A_3X & B_3X & C_3X & D_3X \end{bmatrix}. \tag{33}$$

Alternating between the two main recurrences as in Proposition 19 immediately yields

Theorem 4 *Let X be the central value determined by Theorem 3, and define $\lambda_s = s(s+2)$. The first two columns of matrix (33) are computed recursively by*

$$A_1 = \frac{\lambda_{m'} - \lambda_m - \lambda_n}{2\lambda_m}, \quad B_1 = \frac{\lambda_{m'} + \lambda_m - \lambda_n}{2\lambda_m}, \tag{34}$$

$$A_{s+1} = \frac{(\lambda_{m'} - \lambda_m + \lambda_{2s})A_s + (\lambda_{2s-2} - \lambda_n)B_s}{\lambda_m - \lambda_{2s}}, \tag{35}$$

$$B_{s+1} = \frac{\lambda_{m'}A_s + (\lambda_{2s-2} - \lambda_n)B_s}{\lambda_m - \lambda_{2s}}. \tag{36}$$

In the triangle from the first column to the diagonal, unreduced denominators are equal along rows and increase by a factor of $\lambda_m - \lambda_{2s}$ as we pass from the sth to the $(s+1)$st row. That is, the denominator for index $s+1$ equals

$$d_{s+1} = 2\prod_{l=0}^{s}(\lambda_m - \lambda_{2l}) = 2^{2s+3}\frac{(\frac{m+2s+2}{2})!}{(\frac{m-2s-2}{2})!}. \tag{37}$$

This implies immediately

Corollary 1 *For $s \geq 1$, let $N(A_s)$ and $N(B_s)$ be the numerators in the unreduced expressions of A_s and B_s, respectively. Then $N(A_s)$ and $N(B_s)$ are computed recursively by*

$$N(A_1) = \lambda_{m'} - \lambda_m - \lambda_n, \quad N(B_1) = \lambda_{m'} + \lambda_m - \lambda_n, \tag{38}$$

$$\begin{bmatrix} N(A_{s+1}) \\ N(B_{s+1}) \end{bmatrix} = \begin{bmatrix} \lambda_{m'} - \lambda_m + \lambda_{2s} & \lambda_{2s-2} - \lambda_n \\ \lambda_{m'} & \lambda_{2s-2} - \lambda_n \end{bmatrix} \begin{bmatrix} N(A_s) \\ N(B_s) \end{bmatrix}. \tag{39}$$

To proceed towards the diagonal, for instance, we have for $s \geq 1$,

$$C_{s+1} = B_s + B_{s+1}, \qquad N(C_{s+1}) = (\lambda_m - \lambda_{2s})N(B_s) + N(B_{s+1}). \tag{40}$$

The denominator is non-vanishing for $s < \frac{m}{2}$, and vanishing in a coordinate relative to the the central value reduces to solving the corresponding Diophantine equation, say $N(A_s) = 0$, in m, n, and k.

In [9], three families of zeros corresponding to orders 1, 2, and 3, with 6, 12, and 17 subfamilies, respectively, are classified. Here order is a measure of "distance" using 3-term hypergeometric contiguity relations. It may be computed from the Regge symbol directly (Sect. 4 of [9]). Order 1 subfamilies are indexed I through VI, and, in particular, these zeros admit a full parameterization, as do subfamilies 2.7 and 2.8. Each subfamily of order 2 zeros contains infinitely many zeros. Cardinality in order 3 is an open question, with infinitely many zeros known in types 3.1 and 3.2. We further note that the conjecture by Brudno in footnote 7 of [7] is case I of [9].

For m, n, k even, these subfamilies correspond to positions around the central value as follows:

$$\begin{bmatrix} 3.1 & 3.2 & 3.2 & 3.1 & & & \\ 3.2 & 2.1 & 2.2 & 2.1 & 3.2 & & \\ 3.2 & 2.2 & I & I & 2.2 & 3.2 & \\ 3.1 & 2.1 & I & \boxed{\bullet} & I & 2.1 & 3.1 \\ & 3.2 & 2.2 & I & I & 2.2 & 3.2 \\ & & 3.2 & 2.1 & 2.2 & 2.1 & 3.2 \\ & & & 3.1 & 3.2 & 3.2 & 3.1 \end{bmatrix}$$

with Diophantine equations in the subcentral triangle given by

- I: $N(A_1) = \lambda_{m'} - \lambda_m - \lambda_n = 0$,
- 2.1: $N(A_2) = (\lambda_{m'} - \lambda_m + 8)(\lambda_{m'} - \lambda_m - \lambda_n) - \lambda_n(\lambda_{m'} + \lambda_m - \lambda_n) = 0$,
- 2.2: $N(B_2) = \lambda_{m'}(\lambda_{m'} - \lambda_m - \lambda_n) - \lambda_n(\lambda_{m'} + \lambda_m - \lambda_n) = 0$,
- 3.1: $N(A_3) = (\lambda_{m'} - \lambda_m + 24)N(A_2) - (\lambda_n - 8)N(B_2) = 0$,
- 3.2: $N(B_3) = \lambda_{m'} N(A_2) - (\lambda_n - 8)N(B_2) = 0$.

Under the D_{12} symmetries, the numerators change by permuting m', m, and n and rescaling as in Propositions 3–6.

Now suppose $m = n = 2k$ with k even. The central value reduces to the original Dixon Identity

Corollary 2 (Dixon [2]) *When $m = n = 2k$ and k even, the central value*

$$c_{2k,2k,k}(k,k) = \sum_{l=0}^{k} (-1)^l \binom{k}{l}^3 = (-1)^{\frac{k}{2}} \binom{\frac{3k}{2}}{\frac{k}{2}, \frac{k}{2}, \frac{k}{2}}. \tag{41}$$

Although we no longer have the diagonals of zeros from the odd k case in Sect. 4, there is a central equilateral triangle, with sides of length 4, given by

$$\begin{bmatrix} -X & & & \\ X/2 & -X/2 & & \\ X/2 & \boxed{X} & X/2 & \\ -X & -X/2 & X/2 & X \end{bmatrix}.$$

We note two conjectures, which hold experimentally for $k < 2000$:

1. when k is odd, zeros only occur on one of the three diagonals in the polygon of $M(2k, 2k, k)$, and
2. when k is even, no zeros occur in the polygon of $M(2k, 2k, k)$ unless $k = 8$, in which case there are six doublets forming a hexagon (Fig. 2).

7 Case 2: *m* and *n* Even, *k* Odd

Assume m and n even, and k odd. This section proceeds in a manner similar, but somewhat simpler than the previous section.

Consider the following fourth-quadrant submatrix, with central value 0 and near central value X given by Proposition 16:

$$\begin{bmatrix} \boxed{0} & & & & \\ X & X & & & \\ A_2X & B_2X & C_2X & & \\ A_3X & B_3X & C_3X & D_3X & \\ A_4X & B_4X & C_4X & D_4X & E_4X \end{bmatrix}. \tag{42}$$

Alternating between the two main recurrences as in Proposition 19 immediately yields

Theorem 5 *Let X be the sub-central value determined by Proposition 16, and define $\lambda_s = s(s+2)$. With $s \geq 1$, the first two columns of (42) are computed recursively by*

$$A_1 = 1, \quad B_1 = 1, \tag{43}$$

$$A_{s+1} = \frac{(\lambda_{m'} - \lambda_m + \lambda_{2s})A_s - (\lambda_n - \lambda_{2s-2})B_s}{\lambda_m - \lambda_{2s}}, \tag{44}$$

$$B_{s+1} = \frac{\lambda_{m'}A_s - (\lambda_n - \lambda_{2s-2})B_s}{\lambda_m - \lambda_{2s}}. \tag{45}$$

In the triangle from the first column to the diagonal, unreduced denominators are equal along rows and increase by a factor of $\lambda_m - \lambda_{2s}$ as we pass from the sth to the $(s+1)$st row. That is, with $s \geq 1$, the denominator for index $s+1$ equals

12870	12870	12870	12870	12870	12870	12870	12870	12870								
-57915	-45045	-32175	-19305	-6435	6435	19305	32175	45045	57915							
135135	77220	32175	0	-19305	-25740	-19305	0	32175	77220	135135						
-212355	-77220	0	32175	32175	12870	-12870	-32175	-32175	0	77220	212355					
245025	32670	-44550	-44550	-12375	19800	32670	19800	-12375	-44550	-44550	32670	245025				
-212355	32670	65340	20790	-23760	-36135	-16335	16335	36135	23760	-20790	-65340	-32670	212355			
135135	-77220	-44550	20790	41580	17820	-18315	-34650	-18315	17820	41580	20790	-44550	-77220	135135		
-57915	77220	0	-44550	-23760	17820	35640	17325	-17325	-35640	-17820	23760	44550	0	-77220	57915	
12870	-45045	32175	32175	-12375	-36135	-18315	17325	34650	17325	-18315	-36135	-12375	32175	32175	-45045	12870
	12870	-32175	0	32175	19800	-16335	-34650	-17325	17325	34650	16335	-19800	-32175	0	32175	-12870
		12870	-19305	-19305	12870	32670	16335	-18315	-35640	-18315	16335	32670	12870	-19305	-19305	12870
			12870	-6435	-25740	-12870	19800	36135	17820	-17820	-36135	-19800	12870	25740	6435	-12870
				12870	6435	-19305	-32175	-12375	23760	41580	23760	-12375	-32175	-19305	6435	12870
					12870	19305	0	-32175	-44550	-20790	20790	44550	32175	0	-19305	-12870
						12870	32175	32175	0	-44550	-65340	-44550	0	32175	32175	12870
							12870	45045	77220	77220	32670	-32670	-77220	-77220	-45045	-12870
								12870	57915	135135	212355	245025	212355	135135	57915	12870

Fig. 2 M(16, 16, 8): $m = n = 2k$, k even with proper zeros

$$d_{s+1} = \prod_{l=1}^{s} (\lambda_m - \lambda_{2l}) = \frac{2^{2s+2}}{\lambda_m} \frac{(\frac{m+2s+2}{2})!}{(\frac{m-2s-2}{2})!}. \tag{46}$$

This implies immediately

Corollary 3 *For $s \geq 1$, let $N(A_s)$ and $N(B_s)$ be the numerators in the unreduced expressions of A_s and B_s, respectively. Then $N(A_s)$ and $N(B_s)$ are computed recursively by*

$$N(A_1) = 1, \quad N(B_1) = 1, \tag{47}$$

$$\begin{bmatrix} N(A_{s+1}) \\ N(B_{s+1}) \end{bmatrix} = \begin{bmatrix} \lambda_{m'} - \lambda_m + \lambda_{2s} & \lambda_{2s-2} - \lambda_n \\ \lambda_{m'} & \lambda_{2s-2} - \lambda_n \end{bmatrix} \begin{bmatrix} N(A_s) \\ N(B_s) \end{bmatrix}. \tag{48}$$

To proceed towards the diagonal, for instance, we have for $s \geq 1$,

$$C_{s+1} = B_s + B_{s+1}, \qquad N(C_{s+1}) = (\lambda_m - \lambda_{2s})N(B_s) + N(B_{s+1}). \tag{49}$$

As before, when m is large enough, the denominator is non-vanishing, and vanishing in a coordinate relative to the the central value reduces to solving the corresponding Diophantine equation, say

$$N(A_s) = 0, \tag{50}$$

in m, n, and k. In the classification of [9], diagonal zeros from Sect. 4 closest to the central zero are denoted by R. For m, n even and k odd, these subfamilies correspond to positions around the central value as follows:

$$\begin{bmatrix} 3.15 & 3.16 & 3.17 & 3.16 & 3.15 & & & & \\ 3.16 & 2.11 & 2.12 & 2.12 & 2.11 & 3.16 & & & \\ 3.17 & 2.12 & II & R & II & 2.12 & 3.17 & & \\ 3.16 & 2.12 & R & \bullet & \bullet & R & 2.12 & 3.16 & \\ 3.15 & 2.11 & II & \bullet & \boxed{0} & \bullet & II & 2.11 & 3.15 \\ & 3.16 & 2.12 & R & \bullet & \bullet & R & 2.12 & 3.16 \\ & & 3.17 & 2.12 & II & R & II & 2.12 & 3.17 \\ & & & 3.16 & 2.11 & 2.12 & 2.12 & 2.11 & 3.16 \\ & & & & 3.15 & 3.16 & 3.17 & 3.16 & 3.15 \end{bmatrix}.$$

Diophantine equations in the subcentral triangle are given by

- II: $N(A_2) = \lambda_{m'} - \lambda_m - \lambda_n + 8 = 0$,
- R: $N(B_2) = \lambda_{m'} - \lambda_n = (m - 2k)(m + 2n - 2k + 2) = 0$,
- 2.11: $N(A_3) = (\lambda_{m'} - \lambda_m + 24)N(A_2) - (\lambda_n - 8)N(B_2) = 0$,
- 2.12: $N(B_3) = \lambda_{m'}N(A_2) - (\lambda_n - 8)N(B_2) = 0$,
- 3.15: $N(A_4) = (\lambda_{m'} - \lambda_m + 48)N(A_3) - (\lambda_n - 24)N(B_3) = 0$,
- 3.16: $N(B_4) = \lambda_{m'}N(A_3) - (\lambda_n - 24)N(B_3) = 0$,
- 3.17: $N(C_4) = (\lambda_m - 48)N(B_3) + N(B_4) = 0$.

We have relabeled subfamilies 2.15–2.17 in Table 3 of [9] as 3.15–3.17 here; the groupings naturally correspond to concentric hexagons about the center.

8 Parametrization of Type I and II Zeros

In [9], a full parametrization for zeros of type I–VI are given. For expository purposes, we include an algorithm for generating all (m, n, k) satisfying

$$I : N(A_1) = 0 \quad \text{and} \quad II : N(A_2) = 0.$$

Types III–VI admit similar parameterizations; each case requires solving a Diophantine equation of the form $xy = uv$, where x, y, u, v are linear expressions in m, n, and k.

Proposition 20 *With $k > 0$ and even, all solutions to*

$$m,\ n \text{ even},\quad m' = m + n - 2k,\quad \lambda_{m'} = \lambda_m + \lambda_n \tag{51}$$

are given by

$$m = 2N,\quad n = Q - P - 1,\quad k = N - P \tag{52}$$

for some integers N, P, Q with

1. $N \geq 3$,
2. $PQ = N(N+1)$ *for* $1 \leq P < Q$ *and* $P < N$, *and*
3. P *and* Q *(resp.* P *and* N*) have opposite (resp. same) parity.*

Proof Consider the equation

$$A^2 + 1 = B^2 + C^2 \tag{53}$$

with all $A, B, C > 2$ and odd. Basic algebra yields

$$\frac{A-C}{2}\,\frac{A+C}{2} = \frac{B-1}{2}\,\frac{B+1}{2}. \tag{54}$$

Thus solutions are given precisely when

$$A = P + Q,\quad B = 2N + 1,\quad C = Q - P. \tag{55}$$

Now completing the square in (51) yields

$$(m' + 1)^2 + 1 = (m + 1)^2 + (n + 1)^2, \tag{56}$$

and the proposition follows. □

15	15	15	15	15	15	15	15	15				
-45	-30	-15	0	15	30	45	60	75	90			
45	0	-30	-45	-45	-30	0	45	105	180	270		
	45	45	15	-30	-75	-105	-105	-60	45	225	495	
		45	90	105	75	0	-105	-210	-270	-225	0	495
			45	135	240	315	315	210	0	-270	-495	-495
				45	180	420	735	1050	1260	1260	990	495

Fig. 3 $M(6, 10, 2)$: the smallest example with type I zeros

Table 2 Some basic series of Type I Zeros (Position A1)

P	N	$N+1$
s	$s(2t+1)$	
$2s+1$		$2t(2s+1)$
$2(4s+1)$		$(4s+1)(4t+3)$
$2(4s+3)$		$(4s+3)(4t+1)$
...		...
$2^a(2^{a+1}s+b)$		$(2^{a+1}s+b)(2^{a+1}t+c)$

For example, when $N = 3$, $P = 1$ and $Q = 12$, we have $(m, n, k) = (6, 10, 2)$, see Fig. 3.

Consideration of divisibility properties allows one to directly parametrize some subseries of solutions to (51), as noted in the Table 2. The first line covers all cases where P divides N. The next series gives the general series where the odd part of P divides $N + 1$; in this case, with $0 < b, c < 2^{a+1}$,

$$bc = 1 \ (mod\ 2^a), \quad bc \neq 1 \ (mod\ 2^{a+1}). \tag{57}$$

Noting that N and $N + 1$ have no common factors, we leave it to the reader to generalize to other series.

Associated to $II : N(A_2) = 0$, we have

Proposition 21 *With $k > 1$ and odd, all solutions to*

$$m,\ n\ even, \quad m' = m + n - 2k, \quad \lambda_{m'} + 8 = \lambda_m + \lambda_n \tag{58}$$

are given by

$$m = 2N + 2, \quad n = Q - P - 1, \quad k = N - P + 1 \tag{59}$$

for some integers N, P, Q with

1. $N \geq 3$,
2. $PQ = N(N + 3)$ *for* $1 \leq P < Q$ *and* $P < N$, *and*

3. P and Q (resp. P and N) have opposite (resp. same) parity.

Proof Completing the square in (58) yields

$$(m' + 1)^2 + 9 = (m + 1)^2 + (n + 1)^2, \tag{60}$$

and the proof now follows as in Proposition 20. □

Remark. For example, we have $(m, n, k) = (8, 16, 3)$ when $N = 3$, $P = 1$ and $Q = 18$.

With $l \geq 2$, solutions of (58) of the form $(m, n, k) = (2l, 2, 1)$ correspond to the fourth degenerate case. That is, with respect to $M(m, n, k)$, we obtain a vertical zero triplet with a single proper zero.

9 Cases 3 and 4: *m* Odd, *n* Even

The remaining two cases allow for a simultaneous treatment. In both cases, the center is a square of size 2, and the lower-right entries are given by Propositions 17 and 18.

Proposition 22 *Suppose m is odd and n is even, and $M(m, n, k)$ has central square*

$$\begin{bmatrix} X' & * \\ \boxed{A_0 X} & \boxed{X} \end{bmatrix}$$

for nonzero X. Then

$$A_0 = \frac{m' - m}{m' + 1} \text{ if } k \text{ even}; \quad A_0 = \frac{m' + m + 2}{m' + 1} \text{ if } k \text{ odd}$$

where $m' = m + n - 2k$.

Proof See Proposition 19. In this case, the Weyl group symmetry yields

$$(m + 1)X = (-1)^k (m' + 1)X'. \tag{61}$$

□

Consider the following fourth-quadrant submatrix, where X represents the lower-right entry of the central square:

$$\begin{bmatrix} \boxed{A_0 X} & \boxed{X} & & \\ A_1 X & B_1 X & & \\ A_2 X & B_2 X & C_2 X & \\ A_3 X & B_3 X & C_3 X & D_3 \end{bmatrix}. \tag{62}$$

As before, we have

Theorem 6 *Let X be the lower-right entry determined by Propositions 17 or 18, and define $\lambda_s = s(s+2)$. The first two columns of (62) are computed recursively by*

$$A_0 = \frac{m'-m}{m'+1} \text{ if } k \text{ even}; \quad A_0 = \frac{m'+m+2}{m'+1} \text{ if } k \text{ odd}; \quad B_0 = 1, \tag{63}$$

$$A_{s+1} = \frac{(\lambda_{m'} - \lambda_m + \lambda_{2s+1} + 1)A_s + (\lambda_{2s} - \lambda_n)B_s}{\lambda_m - \lambda_{2s+1}}, \tag{64}$$

$$B_{s+1} = \frac{(\lambda_{m'} + 1)A_s + (\lambda_{2s} - \lambda_n)B_s}{\lambda_m - \lambda_{2s+1}}. \tag{65}$$

In the subcentral triangle, unreduced denominators are equal along rows and increase by a factor of $\lambda_m - \lambda_{2s+1}$ as we pass from the sth to the $(s+1)$st row. That is, with $s \geq 0$, the denominator for index $s+1$ equals

$$d_{s+1} = (m'+1)\prod_{l=0}^{s}(\lambda_m - \lambda_{2l+1}) = 2^{2s+3}\,\frac{m'+1}{m+1}\,\frac{(\frac{m+2s+3}{2})!}{(\frac{m-2s-3}{2})!}. \tag{66}$$

This implies immediately

Corollary 4 *For $s \geq 0$, let $N(A_s)$ and $N(B_s)$ be the numerators in the unreduced expressions of A_s and B_s, respectively. Then $N(A_s)$ and $N(B_s)$ are computed recursively by*

$$N(A_0) = m' - m \text{ if } k \text{ even}; \quad N(A_0) = m' + m + 2 \text{ if } k \text{ odd}; \quad N(B_0) = m' + 1, \tag{67}$$

$$\begin{bmatrix} N(A_{s+1}) \\ N(B_{s+1}) \end{bmatrix} = \begin{bmatrix} \lambda_{m'} - \lambda_m + \lambda_{2s+1} + 1 & \lambda_{2s} - \lambda_n \\ \lambda_{m'} + 1 & \lambda_{2s} - \lambda_n \end{bmatrix} \begin{bmatrix} N(A_s) \\ N(B_s) \end{bmatrix}. \tag{68}$$

As before, numerators correspond to the following positions, up to a $C_2 \times C_2$ symmetry:

m odd, n even, k even

$$\begin{bmatrix} 3.4 & 3.6 & 3.14 & 3.12 & & & & \\ 3.6 & 2.4 & 2.6 & 2.10 & 3.8 & & & \\ 3.14 & 2.6 & IV & VI & 2.8 & 3.10 & & \\ 3.12 & 2.10 & VI & \bullet & R & 2.8 & 3.8 & \\ & 3.8 & 2.8 & \boxed{R} & \boxed{\bullet} & VI & 2.10 & 3.12 \\ & & 3.10 & 2.8 & VI & IV & 2.6 & 3.14 \\ & & & 3.8 & 2.10 & 2.6 & 2.4 & 3.6 \\ & & & & 3.12 & 3.14 & 3.6 & 3.4 \end{bmatrix},$$

- R: $N(A_0) = m' - m = n - 2k = 0$,
- 2.8: $N(A_1) = (m' - m)(\lambda_{m'} - \lambda_m + 4) - \lambda_n(m' + 1) = 0$,
- VI: $N(B_1) = (m' + 1)[(m' - m)(m' + 1) - \lambda_n] = 0$,
- 3.8: $N(A_2) = (\lambda_{m'} - \lambda_m + 16)N(A_1) - (\lambda_n - 8)N(B_1) = 0$,
- 2.10: $N(B_2) = (\lambda_{m'} + 1)N(A_1) - (\lambda_n - 8)N(B_1) = 0$,
- 3.12: $N(B_3) = (\lambda_{m'} + 1)N(A_2) - (\lambda_n - 24)N(B_2) = 0$,
- IV: $N(C_1) = (m' + 1)[\lambda_m - \lambda_n - 3 + (m' - m)(m' + 1)] = 0$,
- 2.6: $N(C_2) = (\lambda_m - 15)N(B_1) + N(B_2) = 0$,
- 3.14: $N(C_3) = (\lambda_m - 35)N(B_2) + N(B_3) = 0$,
- 2.4: $N(D_2) = (\lambda_m - 15)N(C_1) + N(C_2) = 0$,
- 3.6: $N(D_3) = (\lambda_m - 35)N(C_2) + N(C_3) = 0$,
- 3.4: $N(E_3) = (\lambda_m - 35)N(D_2) + N(D_3) = 0$.

$$
m \text{ odd},\ n \text{ even},\ k \text{ odd}
$$

$$
\begin{bmatrix}
3.3 & 3.5 & 3.13 & 3.11 & & & & \\
3.5 & 2.3 & 2.5 & 2.9 & 3.7 & & & \\
3.13 & 2.5 & III & V & 2.7 & 3.9 & & \\
3.11 & 2.9 & V & \bullet & \bullet & 2.7 & 3.7 & \\
 & 3.7 & 2.7 & \boxed{\bullet} & \boxed{\bullet} & V & 2.9 & 3.11 \\
 & & 3.9 & 2.7 & V & III & 2.5 & 3.13 \\
 & & & 3.7 & 2.9 & 2.5 & 2.3 & 3.5 \\
 & & & & 3.11 & 3.13 & 3.5 & 3.3
\end{bmatrix},
$$

- 2.7: $N(A_1) = (m' + m + 2)(\lambda_{m'} - \lambda_m + 4) - \lambda_n(m' + 1) = 0$,
- V: $N(B_1) = (m' + 1)[(m' + m + 2)(m' + 1) - \lambda_n] = 0$,
- 3.7: $N(A_2) = (\lambda_{m'} - \lambda_m + 16)N(A_1) - (\lambda_n - 8)N(B_1) = 0$,
- 2.9: $N(B_2) = (\lambda_{m'} + 1)N(A_1) - (\lambda_n - 8)N(B_1) = 0$,
- 3.11: $N(B_3) = (\lambda_{m'} + 1)N(A_2) - (\lambda_n - 24)N(B_2) = 0$,
- III: $N(C_1) = (m' + 1)[\lambda_m - \lambda_n - 3 + (m' + m + 2)(m' + 1)] = 0$,
- 2.5: $N(C_2) = (\lambda_m - 15)N(B_1) + N(B_2) = 0$,
- 3.13: $N(C_3) = (\lambda_m - 35)N(B_2) + N(B_3) = 0$,
- 2.3: $N(D_2) = (\lambda_m - 15)N(C_1) + N(C_2) = 0$,
- 3.5: $N(D_3) = (\lambda_m - 35)N(C_2) + N(C_3) = 0$,
- 3.3: $N(E_3) = (\lambda_m - 35)N(D_2) + N(D_3) = 0$.

For types 3.9 and 3.10, simultaneously consider the entries Z_1 of this type near A_1. Application of both recurrences yields

$$
N(Z_1) = (\lambda_{m'} + \lambda_m + 2)^2 - 4\lambda_{m'}\lambda_m - 28 - \lambda_n[(m' + 1)N(A_0) + \lambda_m - 3]
$$

and

$$
Z_1 = \frac{A_0 N(Z_1)}{(\lambda_{m'} - 3)}.
$$

10 Computer Implementation

In [9], an analysis is given for certain vanishing $c_{m,n,k}(i, j)$ with $J = m + n - k < 3{,}000$. We leave it to the reader to pursue those details there.

The algorithms in this work require only rudimentary programming expertise, implemented on conventional hardware (2013 MacBook Pro). The figures were constructed using Excel, which was also used to compute all $M(m, n, k)$ above. Other algorithms, such the numerator formulas, were stress-tested using MAPLE.

References

1. Andrews, G.E., Askey, R., Roy, R.: Special functions. Encyclopedia Math. Appl. **71**, Cambridge University Press, Cambridge (1999)
2. Dixon, A.C.: On the sum of cubes of the coefficients in a certain expansion by the binomial theorem. Messenger Math. **20**, 79–80 (1891)
3. Donley, R.W., Jr., Kim, W.G.: A rational theory of Clebsch-Gordan coefficients. In: Representation theory and harmonic analysis on symmetric spaces. American Mathematical Society, Providence. Contemp. Math., Vol. 714, 115–130 (2018)
4. Ekhad, S.: A very short proof of Dixon's theorem. J. Combin. Theory Ser. A **54**, 141–142 (1990)
5. Fjelsted, J.E.: A generalization of Dixon's theorem. Math. Scand. **2**, 46–48 (1954)
6. Gessel, I., Stanton, D.: Short proofs of Saalschütz's and Dixon's theorems. J. Combin. Theory Ser. A. **38**, 87–90 (1985)
7. Heim, T. A., Hinze, J., Rau, A.R.P.: Some classes of 'nontrivial zeroes' of angular momentum addition coefficients. J. Phys. A. **42**, 175203, 11pp (2009)
8. Louck, J. D., Stein, P.R.: Weight-2 zeros of 3j coefficients and the Pell equation. J. Math. Phys. **28**, 2812–2823 (1987)
9. Raynal, J., Van der Jeugt, J., Rao, K.S., Rajeswari, V.: On the zeros of 3j coefficients: Polynomial degree versus recurrence order. J. Phys. A **26**, 2607–2623 (1993)
10. Suresh, R., Rao, K.S.: On the recurrence relations for the 3-j coefficient. Appl. Math. Inf. Sci. **5**, 44–52 (2011)
11. Varshalovich, D.A., Moskalev, A.N., Khersonskii, V.K.: Quantum theory of angular momentum. Translated from the Russian. World Scientific, Teaneck, N.J. (1988)
12. Vilenkin, N.Ja.: Special functions and the theory of group representations. Translated from the Russian by V. N. Singh. Transl. Math. Monogr, Vol. 22, American Mathematical Society, Providence (1968)
13. Ward, J.: 100 years of Dixon's identity. Irish Math. Soc. Bull. **27**, 46–54 (1991)

Numerical Semigroups Generated by Squares and Cubes of Three Consecutive Integers

Leonid G. Fel

Abstract We derive the polynomial representations for minimal relations of the generating set of numerical semigroups $R_n^k = \langle (n-1)^k, n^k, (n+1)^k \rangle, k = 2, 3$. We find also the polynomial representations for degrees of syzygies in the Hilbert series $H\left(z, R_n^k\right)$ of these semigroups, their Frobenius numbers $F\left(R_n^k\right)$ and genera $G\left(R_n^k\right)$. We discuss an extension of polynomial representations for minimal relations on numerical semigroups R_n^k, $k \geq 4$.

2010 Mathematics Subject Classification Primary—20M14 · Secondary—11P81

1 Symmetric and Nonsymmetric Numerical Semigroups $\langle (n-1)^k, n^k, (n+1)^k \rangle$

Numerical semigroups $S_3 = \langle d_1, d_2, d_3 \rangle$, generated by three integers, exhibit a non-trivial example of semigroups with well established relations [2] between degrees of syzygies and values of generators. In this regards, the relations f_k imposed on generators, $f_k(d_1, d_2, d_3) = 0$, may increase a number of semigroups with explicitly computable Hilbert series $H(z, S_3)$, Frobenius numbers $F(S_3)$ and genera $G(S_3)$. Usually the generators of such semigroups are represented as elements of some ordered sets: arithmetic [1], almost arithmetic [11] or geometric [9] progressions, Pythagorean triples [2], Fibonacci [3, 8] or Lucas [3] numbers and others.

Recently, the two 3-generated semigroups were studied [7] to establish an explicit expression for their Frobenius numbers. These are semigroups R_n^2 and R_n^3, generated by squares and cubes of three consecutive positive integers, respectively, where

$$R_n^k = \langle (n-1)^k, n^k, (n+1)^k \rangle, \quad k \in \mathbb{N}. \tag{1}$$

L. G. Fel (✉)
Department of Civil Engineering, Technion–Israel Institute of Technology, 32000 Haifa, Israel
e-mail: lfel@technion.ac.il

M. B. Nathanson (ed.), *Combinatorial and Additive Number Theory III*, Springer Proceedings in Mathematics & Statistics 297,
https://doi.org/10.1007/978-3-030-31106-3_8

Making use of the Euclidean algorithm with negative [12] and positive [10] remainders for computation of the Frobenius number, the authors [7] were able to find polynomial expressions in n for $F\left(R_n^2\right)$ and $F\left(R_n^3\right)$ on residue class of n modulo 4 and 18, respectively,

$$F_j\left(R_n^2\right) = \sum_{i=0}^{3} A_i^j n^i, \quad j = n \pmod 4, \qquad n \neq 3, 4, 5, 6, 9, 13, \quad A_i^j \in \mathbb{Q}, \tag{2}$$

$$F_j\left(R_n^3\right) = \sum_{i=0}^{5} B_i^j n^i, \quad j = n \pmod{18}, \qquad B_i^j \in \mathbb{Q}, \tag{3}$$

$$n \neq 3, 4, 5, 6, 7, 8, 9, 10, 11, 18, 26, 27, 36, 45, 54, 63, 72, 90, 108, 126, 144, 162, 180, 198, 216, 234, 252, 270.$$

A long list of sophisticated formulas for $F_j\left(R_n^2\right)$ and $F_j\left(R_n^3\right)$ in [7], accompanied by 34 exclusions (6 cases for R_n^2 and 28 cases for R_n^3), poses a question to find another representation (Rep) which allows to include all exclusive cases or, at least, to reduce substantially their number. Another reason to discuss this problem again is to find not only the Frobenius numbers, but also the Hilbert series and genera for all semigroups R_n^k, $k = 2, 3$.

Note that the excluding values of n in (2), (3) give rise to the four symmetric semigroups, R_3^2, R_4^2, R_5^2 and R_3^3. The rest of 30 excluding semigroups are nonsymmetric.

Simple considerations (Propositions 1, 2) show that among all semigroups R_n^2 and R_n^3 only the four above mentioned semigroups are symmetric. To prove that we recall necessary conditions when a semigroup $\langle d_1, d_2, d_3\rangle$ becomes symmetric,

$$\begin{aligned} &(a)\quad d_3 \in \langle d_1, d_2\rangle, \quad \gcd(d_1, d_2) = 1, \quad d_j > 3, \qquad \text{or} \\ &(b)\quad \langle d_1, d_2, d_3\rangle, \; d_j > 3, \quad \text{satisfies Watanabe's Lemma [13] adapted to 3 generators,} \end{aligned} \tag{4}$$

A literal quotation of Watanabe's Lemma [13] for arbitrary number of generators reads:

Lemma 1 ([13]) *Let $H_1 = \langle n_1, \ldots, n_k\rangle$ be a semigroup, a and b be positive integers such that:*

- *(i) $a \in H_1$ and $a \neq n_i$, $i = 1, \ldots, k$,*
- *(ii) a and b are relatively prime.*

If we put $H = \langle a, bn_1, \ldots, bn_k\rangle$ (which we will denote by $H = \langle a, bH_1\rangle$) then:

1. *H is a complete intersection if and only if H_1 is a complete intersection.*
2. *H is symmetric if and only if H_1 is symmetric.*

To adapt Watanabe's Lemma for 3-generated semigroups we have to take into account:

- Every symmetric semigroup $\langle d_1, d_2, d_3\rangle$ is always complete intersection and vice-versa (see, e.g., [5].) That makes dichotomy "symmetric" and "complete intersection" not important.
- If $H_1 = \langle n_1, n_2\rangle$, and $a = n_1$, then the 3-generated semigroup $H = \langle n_1, bn_1, bn_2\rangle$ can be reduced up to $H = \langle n_1, bn_2\rangle$. But the 2-generated semigroup is always symmetric. That is why the claim in Watanabe's Lemma about $a \neq n_i$, $i = 1, 2$, i.e., in the case of the 3-generated semigroup H, may be omitted.

Thus, we arrive at the following version of Watanabe's Lemma [13] for 3-generated semigroups.

Lemma 2 *Let a numerical semigroup $\langle d_1, d_2, d_3\rangle$ be given such that $d_1 = a\delta_1$, $d_2 = a\delta_2$, $\delta_j \geq 2$, and $\gcd(\delta_1, \delta_2) = \gcd(a, d_3) = 1$. Then $\langle d_1, d_2, d_3\rangle$ is symmetric iff $d_3 \in \langle \delta_1, \delta_2\rangle$.*

For semigroups, satisfying Lemma 2, the following formula for the Frobenius number holds [5, 6],

$$F(\langle d_1, d_2, d_3\rangle) = aF(\langle \delta_1, \delta_2\rangle) + (a-1)d_3, \qquad F(\langle \delta_1, \delta_2\rangle) = \delta_1\delta_2 - \delta_1 - \delta_2. \tag{5}$$

1.1 Symmetric Numerical Semigroups R_n^2 and R_n^3

Prove an exclusive property of semigroups R_3^2, R_4^2, R_5^2.

Proposition 1 *There exist only three symmetric numerical semigroups R_n^2, $n = 3, 4, 5$.*

Proof Note that semigroups R_n^2, $n = 3, 4$, are symmetric due to the condition (4a). Find more n which satisfy (4a),

$$(n+1)^2 = a_1(n-1)^2 + a_2 n^2, \quad a_1, a_2 \in \mathbb{N}, \quad n > 4. \tag{6}$$

Simplifying the last equality we obtain the Diophantine equation

$$(a_1 + a_2 - 1)(n-4)^2 + 2(3a_1 + 4a_2 - 5)(n-4) + 9a_1 + 16a_2 - 25 = 0,$$

with constraints (6) on three variables a_1, a_2, n which has no solutions.

Consider another way to symmetrize R_n^2 by providing condition (4b) according to Lemma 2. This may occur only when $n = 2p + 1$ and results in the Diophantine equation in b_1, b_2, p,

$$(2p+1)^2 = b_1 p^2 + b_2(p+1)^2, \quad b_1, b_2 \in \mathbb{N}, \quad p \geq 2. \tag{7}$$

Solving (7) as a quadratic equation, we obtain

$$p=\frac{2-b_2\pm\Theta}{b_1+b_2-4},\qquad \Theta^2=b_1+b_2-b_1b_2. \tag{8}$$

Combining the last expressions with the constraint $p\geq 2$ in (7) we obtain

$$4b_1^2+9b_2^2+13b_1b_2-41b_1-61b_2+100\leq 0. \tag{9}$$

Find a canonical representation of a quadratic form in (9) by substitution $\mathcal{T}$,

$$\mathcal{T}:\ b_1=q_1+11/5,\quad b_2=q_2+9/5,\qquad\rightarrow\qquad 4q_1^2+9q_2^2+13q_1q_2\leq 0. \tag{10}$$

An inequality (10) is satisfied when $-1\leq q_2/q_1\leq -4/9$. Apply an inverse substitution $\mathcal{T}^{-1}$ to the double inequality and keep in mind constraints (7) and obtain four linear inequalities,

$$b_1+b_2\geq 4,\qquad 4b_1+9b_2\leq 25,\qquad b_1,b_2\geq 1,$$

which have one solution, $b_1=4$, $b_2=1$. By (8) it leads to $p=2$ and gives rise to R_5^2. □

The next Propositions deal with symmetric semigroups R_n^3.

Proposition 2 *There exists only one symmetric numerical semigroup* R_n^3, $n=3$.

Its proof is similar to proof of Proposition 1 but more cumbersome and therefore is given in Appendix 5.

1.2 Nonsymmetric Numerical Semigroups Generated by Three Integers

According to Propositions 1 and 2 there are exactly four symmetric semigroups among the whole set of R_n^2, R_n^3, $3\leq n<\infty$, and calculation of their Frobenius numbers, genera and Hilbert series does not need general formulas. The situation changes drastically if we turn to nonsymmetric semigroups. In the present paper we calculate the Hilbert series for nonsymmetric semigroups R_n^k, $k=2,3$, making use of an approach of minimal relations for three generators d_1,d_2,d_3. Recall this approach following [2].

A nonsymmetric numerical semigroup $S_3=\langle d_1,d_2,d_3\rangle$,

$$\mathsf{S}_3=\left\{s\in\mathbb{N}\cup\{0\}\mid s=\sum_{i=1}^{3}x_id_i,\ x_i,d_j\in\mathbb{N}\cup\{0\}\right\}, \tag{11}$$

$$\gcd(d_1,d_2,d_3)=1,\quad d_j\geq 3,$$

is said to be generated by *a minimal set* of three integers d_j, if none of them is linearly representable by the rest of them. By [2, 6] there exists a matrix A_3 of minimal relations,

$$\mathsf{A}_3\begin{pmatrix} d_1\\ d_2\\ d_3\end{pmatrix}=\begin{pmatrix}0\\0\\0\end{pmatrix},\ \mathsf{A}_3=\begin{pmatrix} a_{11} & -a_{12} & -a_{13}\\ -a_{21} & a_{22} & -a_{23}\\ -a_{31} & -a_{32} & a_{33}\end{pmatrix},\ \begin{cases}\gcd(a_{11},a_{12},a_{13})=1\\ \gcd(a_{21},a_{22},a_{23})=1\\ \gcd(a_{31},a_{32},a_{33})=1\end{cases}, \tag{12}$$

$$\begin{aligned} a_{11} &= \min\{v_{11}\mid v_{11}\geq 2,\ v_{11}d_1=v_{12}d_2+v_{13}d_3,\ v_{12},v_{13}\in\mathbb{N}\cup\{0\}\},\\ a_{22} &= \min\{v_{22}\mid v_{22}\geq 2,\ v_{22}d_2=v_{21}d_1+v_{23}d_3,\ v_{21},v_{23}\in\mathbb{N}\cup\{0\}\},\\ a_{33} &= \min\{v_{33}\mid v_{33}\geq 2,\ v_{33}d_3=v_{31}d_1+v_{32}d_2,\ v_{31},v_{32}\in\mathbb{N}\cup\{0\}\}.\end{aligned} \tag{13}$$

such that the nine matrix elements a_{ij} satisfy the six Diophantine equations,

$$a_{11}=a_{21}+a_{31},\quad a_{22}=a_{12}+a_{32},\quad a_{33}=a_{13}+a_{23},\quad a_{jj}\geq 2, \tag{14}$$
$$a_{ij}\geq 1,\ i\neq j,$$
$$d_1=a_{22}a_{33}-a_{23}a_{32},\qquad d_2=a_{33}a_{11}-a_{31}a_{13},\qquad d_3=a_{11}a_{22}-a_{12}a_{21}. \tag{15}$$

The generating function $H(z,S_3)$ of numerical semigroups S_3,

$$H(z,S_3)=\sum_{s\,\in\,S_3} z^s,\qquad F(S_3)=\max\{\mathbb{N}\setminus S_3\},\quad G(S_3)=\#\{\mathbb{N}\setminus S_3\},$$

is referred to as *the Hilbert series* of S_3. The rational Rep of $H(z,S_3)$, the degrees e_i and t_i of the 1st and 2nd syzygies, the Frobenius number $F(S_3)$ and genus $G(S_3)$ read [2],

$$H(z,S_3)=\left(1-z^{e_1}-z^{e_2}-z^{e_3}+z^{t_1}+z^{t_2}\right)\prod_{i=1}^{3}\left(1-z^{d_i}\right)^{-1},\quad e_i=a_{ii}d_i, \tag{16}$$
$$\begin{aligned} &t_1=D_0+D_1,\quad t_2=D_0+D_2,\quad e_1+e_2+e_3=t_1+t_2,\quad F_1=t_1-D_3,\\ &F(S_3)=\max\{F_1,F_2\},\quad 2G(S_3)=1+D_0+D_1+D_2-D_3,\quad F_2=t_2-D_3,\\ &D_0=a_{11}a_{22}a_{33},\quad D_1=a_{12}a_{23}a_{31},\quad D_2=a_{13}a_{32}a_{21},\quad D_3=d_1+d_2+d_3,\end{aligned}$$

Prove Lemma on uniqueness of matrix A_3.

Lemma 3 *Let a nonsymmetric numerical semigroup S_3 be given. Then there exists a unique matrix A_3 of minimal relations (12), (13) which satisfies (14).*

Proof If a numerical semigroup S_3 be given, then its Hilbert series $H(z,S_3)$ is defined in (16) by unique values of syzygies degrees e_j, t_j and generators d_j due to uniqueness of the product of polynomial $\prod_{i=1}^{3}\left(1-z^{d_i}\right)$ and infinite series, and both of them are uniquely defined,

$$\prod_{i=1}^{3}\left(1-z^{d_i}\right)\sum_{s\in S_3} z^s = 1 - z^{e_1} - z^{e_2} - z^{e_3} + z^{t_1} + z^{t_2}.$$

A standard proof of uniqueness of matrix A_3 by contradiction is to suppose that there are two different matrices (a_{ij}) and (b_{ij}), which provide similar equalities, and to prove that these equalities have identically the same solutions.

First, equating degrees e_j of the 1st syzygy according to (16) we arrive at $a_{ii}d_i = b_{ii}d_i$, that results in equalities, $a_{ii} = b_{ii}$, $i = 1, 2, 3$.

Next, to explore the uniqueness of degrees t_j for two matrices, let us make use of the other expressions for t_j, which may be found if we combine (15) and (16), see also formulas (117) in [2],

$$t_1 = a_{33}d_3 + a_{12}d_2 = b_{33}d_3 + b_{12}d_2, \qquad t_2 = a_{22}d_2 + a_{13}d_3 = b_{22}d_2 + b_{13}d_3.$$

Keeping in mind a coincidence of diagonal elements $a_{ii} = b_{ii}$ and uniqueness of generators d_j, we obtain $a_{12} = b_{12}$, $a_{13} = b_{13}$, that lead due to (15) to complete coincidence of two matrices $a_{ij} = b_{ij}$, $i, j = 1, 2, 3$. □

Corollary 1 *Let a numerical semigroup S_3 be generated by a minimal set $\{d_1, d_2, d_3\}$ according to (11)–(14). There exists only one way to partition $a_{jj}d_j$, $j = 1, 2, 3$, $a_{jj} \geq 2$, in a sum of d_i and d_k, $i, k \neq j$, with integer coefficients $a_{ik} \geq 1$.*

Lemma 3 has one more Corollary which is important in a view of a subject of the present paper, namely, a polynomial Rep (PRep) of the matrix A_3. Let all matrix elements of A_3 for semigroup R_n^k, have the PRep in n on residue class of n modulo T_k, i.e.,

$$\text{if} \quad n = T_k m + q \quad \text{and} \quad 0 \leq q < T_k, \quad \text{then} \quad a_{ij} = A_{ij}(m; q) \quad \text{is a polynomial in} \quad m. \tag{17}$$

Corollary 2 *Let a numerical semigroup R_n^k be generated by a minimal set $\{(n-1)^k, n^k, (n+1)^k\}$ and matrix elements a_{ij} of A_3 have the PRep in n according to (17). Then such PRep is unique.*

Proof By (17), for every fixed k and given n the matrix A_3 has PRep, $a_{ij} = A_{ij}(m; q)$. Let, by way of contradiction, there exists another PRep $a_{ij} = B_{ij}(m; q)$, which differs from $A_{ij}(m; q)$. On the other hand, by Lemma 3 the valuations of $A_{ij}(m; q)$ and $B_{ij}(m; q)$ coincide for all integers m, q. Prove that the both polynomials are coincided identically, i.e., $A_{ij}(m; q) \equiv B_{ij}(m; q)$.

Let $A_{ij}(m; q)$ and $B_{ij}(m; q)$ are polynomials in m with the following PReps,

$$A_{ij}(m; q) = \sum_{k=0}^{\alpha_{ij}} A_{ij,k}(q)m^k, \quad \alpha_{ij} = \deg A_{ij}(m; q),$$

$$B_{ij}(m; q) = \sum_{k=0}^{\beta_{ij}} B_{ij,k}(q)m^k, \quad \beta_{ij} = \deg B_{ij}(m; q).$$

Consider a polynomial difference $C_{ij}(m;q)$

$$C_{ij}(m;q) = A_{ij}(m;q) - B_{ij}(m;q) = \sum_{k=0}^{\gamma_{ij}} \mathcal{C}_{ij,k}(q) m^k,$$

$$\mathcal{C}_{ij,k}(q) = \mathcal{A}_{ij,k}(q) - \mathcal{B}_{ij,k}(q), \qquad \gamma_{ij} = \deg C_{ij}(m;q) = \max\{\alpha_{ij}, \beta_{ij}\} < \infty.$$

The polynomial $C_{ij}(m;q)$ has a finite degree and infinite number of zeroes $m \in \mathbb{N}$, since the values of $A_{ij}(m;q)$ and $B_{ij}(m;q)$ are coincided for all integers m. The only polynomial, which satisfies the above requirements, is the zero polynomial, i.e., $A_{ij}(m;q) \equiv B_{ij}(m;q)$. □

In Sects. 2 and 3 we derive the PRep for matrix elements of minimal relations A_3 in numerical semigroups R_n^2 and R_n^3, respectively, and reduce a large number of exclusive semigroups [7] with $F(R_n^2)$ and $F(R_n^3)$, which differ from polynomials (2), (3), by making use of expression (16). These exclusions were happen when in different ranges of n a difference $D_1(n) - D_2(n)$ has changed its sign. In other words, the both sequences F_1 and F_2 contribute to the PRep of the Frobeniius numbers. In Sect. 4 we discuss briefly the results of numerical calculations [4] of the PRep for minimal relations A_3 in semigroups R_n^4, and pose Conjecture 1 about the PRep for numerical semigroups R_n^k with arbitrary degrees k.

1.3 Polynomial Representations for Numerical Semigroups R_n^k

Consider a semigroup R_n^k, generated by integer degrees of consecutive integers, and write a matrix equation (12), (13) supplemented by relations (15),

$$\begin{pmatrix} a_{11} & -a_{12} & -a_{13} \\ -a_{21} & a_{22} & -a_{23} \\ -a_{31} & -a_{32} & a_{33} \end{pmatrix} \begin{pmatrix} (n-1)^k \\ n^k \\ (n+1)^k \end{pmatrix} = \begin{pmatrix} 0 \\ 0 \\ 0 \end{pmatrix}, \quad \begin{cases} a_{22}a_{33} - a_{23}a_{32} = (n-1)^k, \\ a_{33}a_{11} - a_{31}a_{13} = n^k, \\ a_{11}a_{22} - a_{12}a_{21} = (n+1)^k, \end{cases} \tag{18}$$

where a_{ij} are natural numbers satisfying three equalities (14) and strict inequalities,

$$a_{ij} \in \mathbb{N}, \qquad a_{11} > a_{12},\ a_{13}, \qquad a_{22} > a_{23}. \tag{19}$$

which follow by comparison of terms in the l.h.s. and r.h.s. of matrix equation (18). Keeping in mind three identities in (18), let us distinguish two cases of even and odd degrees $k > 1$ and assume that the PRep for matrix elements $a_{ij}(n)$ may be chosen as follows:

1. If $k = 2p$ then all nine elements $a_{ij}(n)$ have PRep of degree p. (20)
2. If $k = 2p - 1$ then only six elements $a_{1j}(n)$, $a_{2j}(n)$ have PRep of degree p while the other three elements $a_{3j}(n)$ – of degree $p - 1$.

We justify these assumptions in the next sections for the cases $k = 2, 3$. We give a detailed derivation of the PRep for matrix elements of minimal relations by solving the Diophantine equations for semigroups R_n^2 and R_n^3 in Sects. 2 and 3, respectively. We find their Frobenius numbers and genera in all residue classes of n modulo 4 for R_n^2 and modulo 18 for R_n^3.

By Lemma 3 and Corollary 2 of uniqueness, the obtained PRep for matrix elements of A_3, Frobenius numbers and genera do not depend on the way of their derivation. However, in the general case of semigroups R_n^k with arbitrary degrees k the problem of PRep is left still open.

2 Numerical Semigroups R_n^2, $n \geq 6$

Consider a semigroup R_n^2 with a linear Rep of the matrix elements $a_{ij}(n) = \Gamma_{ij} + \Upsilon_{ij} n$, where $\Gamma_{ij} \in \mathbb{Z}$, $\Upsilon_{ij} \in \mathbb{N} \cup \{0\}$, however $a_{ij}(n) \in \mathbb{N}$, and substitute it into the matrix equation (18), and balance degrees of n in the l.h.s. and r.h.s. of quadratic equalities in n. That gives a reduction of 18 indeterminates Γ_{ij}, Υ_{ij} up to six independent rational variables $\xi_i \in \mathbb{Q}$, $1 \leq j \leq 6$.

$$\begin{aligned}
a_{11} &= \xi_1 + \left(2\xi_1 + \tfrac{\xi_2}{4}\right) n, & a_{12} &= -\xi_2 + 4\xi_1 n, & a_{13} &= \xi_1 + \left(-2\xi_1 + \tfrac{\xi_2}{4}\right) n, \\
a_{21} &= \xi_3 + \left(2\xi_3 + \tfrac{\xi_4}{4}\right) n, & a_{22} &= -\xi_4 + 4\xi_3 n, & a_{23} &= -\xi_3 + \left(2\xi_3 - \tfrac{\xi_4}{4}\right) n, \\
a_{31} &= -\xi_5 + \left(-2\xi_5 + \tfrac{\xi_6}{4}\right) n, & a_{32} &= \xi_6 + 4\xi_5 n, & a_{33} &= -\xi_5 + \left(2\xi_5 + \tfrac{\xi_6}{4}\right) n.
\end{aligned} \tag{21}$$

Due to relations (14), (15) the variables ξ_i satisfy the following equalities,

$$\xi_1 = \xi_3 - \xi_5, \quad \xi_2 = \xi_4 + \xi_6, \quad \xi_1\xi_6 + \xi_2\xi_5 = \xi_3\xi_6 + \xi_4\xi_5 = \xi_2\xi_3 - \xi_1\xi_4 = 1, \tag{22}$$

where the last three equalities are equivalent. Require a non-negativeness of the linear (in n) part (ℓp_n) of $a_{ij}(n)$ in (21), that together with (19) gives

$$\begin{aligned}
&\ell p_n(a_{13}): \ \xi_2 \geq 8\xi_1, \quad \ell p_n(a_{23}): \ 8\xi_3 \geq \xi_4, \quad \ell p_n(a_{31}): \ \xi_6 \geq 8\xi_5, \\
&\ell p_n(a_{12}): \ \xi_1 \geq 0, \quad \ell p_n(a_{22}): \ \xi_3 \geq 0, \quad \ell p_n(a_{32}): \ \xi_5 \geq 0, \\
&a_{11} > a_{13}: \ \xi_1 > 0, \quad a_{22} > a_{23}: \ (2n+1)\xi_3 + \left(\frac{n}{4} + 1\right)\xi_4 > 0.
\end{aligned} \tag{23}$$

Consider a matrix element $a_{33} \in \mathbb{N}$ in (21) and minimize it over $\xi_5, \xi_6 \geq 0$ satisfying (23) and preserving $a_{3i} \in \mathbb{N}$, $i = 1, 2$, for the other two elements in (22). That results in $\xi_5 = 0$, $\xi_6 \in \mathbb{N}$. Summarizing the last result and relations (22), (23), we obtain

$$\xi_1 > 0, \ \xi_2 \geq 8\xi_1, \ \xi_3 = \xi_1, \ -\infty \leq \xi_4 \leq 8\xi_1, \ \xi_5 = 0, \ \xi_6 > 0 \ \rightarrow \ a_{31} = a_{33}. \tag{24}$$

To provide all entries in A_3 to be integers, we consider four different cases, i.e., $T_2 = 4$. Substitute successively $n = 4m + j$, $0 \leq j \leq 3$, into (21), and minimize

$a_{11}, a_{22}, a_{33} \in \mathbb{N}$ over $\xi_1, \xi_2, \xi_4, \xi_6$, which satisfy equalities (22) and inequalities (24). E.g., if $n = 4m$, then

$$a_{11} = \xi_1 + (8\xi_1 + \xi_2)m, \qquad a_{22} = -\xi_4 + 16\xi_1 m, \qquad a_{33} = \xi_6 m.$$

Minimizing a_{11} over ξ_1, ξ_2 we get due to the two first inequalities in (24): $\xi_1 = 1$, $\xi_2 = 8$. In the similar way (minimizing a_{33} over ξ_6 and keeping in mind the last inequality in (24)) we get $\xi_6 = 1$. Finally, due to the second identity in (22), we obtain $\xi_4 = 7$.

1. $n = 4m$.

$$\begin{array}{lll} a_{11} = \xi_1 + (8\xi_1 + \xi_2)m, & a_{12} = -\xi_2 + 16\xi_1 m, & a_{13} = \xi_1 + (-8\xi_1 + \xi_2)m, \\ a_{21} = \xi_1 + (8\xi_1 + \xi_4)m, & a_{22} = -\xi_4 + 16\xi_1 m, & a_{23} = -\xi_1 + (8\xi_1 - \xi_4)m, \\ a_{31} = \xi_6 m, & a_{32} = \xi_6, & a_{33} = \xi_6 m. \end{array}$$

$$\xi_1 = 1, \quad \xi_2 = 8, \quad \xi_4 = 7, \quad \xi_6 = 1.$$

$$\begin{pmatrix} 16m+1 & -8(2m-1) & -1 \\ -(15m+1) & 16m-7 & -(m-1) \\ -m & -1 & m \end{pmatrix}, \quad \begin{array}{l} G = 4m(34m^2 - 21m + 2), \\ F = 20, \ \text{if } m = 1, \\ F = 272m^3 - 168m^2 + m - 2, \ \text{if } m \geq 2. \end{array}$$

2. $n = 4m + 2$.

$$\begin{array}{lll} a_{11} = \frac{\xi_2}{2} + 5\xi_1 + (8\xi_1 + \xi_2)m, & a_{12} = 8\xi_1 - \xi_2 + 16\xi_1 m, & a_{13} = \frac{\xi_2}{2} - 3\xi_1 + (\xi_2 - 8\xi_1)m, \\ a_{21} = \frac{\xi_4}{2} + 5\xi_1 + (8\xi_1 + \xi_4)m, & a_{22} = 8\xi_1 - \xi_4 + 16\xi_1 m, & a_{23} = -\frac{\xi_4}{2} - 3\xi_1 + (8\xi_1 - \xi_4)m, \\ a_{31} = \frac{\xi_6}{2} + \xi_6 m, & a_{32} = \xi_6, & a_{33} = \frac{\xi_6}{2} + \xi_6 m. \end{array}$$

$$\xi_1 = 1/2, \quad \xi_2 = 5, \quad \xi_4 = 3, \quad \xi_6 = 2.$$

$$\begin{pmatrix} 9m+5 & -(8m-1) & -(m+1) \\ -(7m+4) & 8m+1 & -m \\ -(2m+1) & -2 & 2m+1 \end{pmatrix}, \quad \begin{array}{l} G = m(80m^2 + 71m + 16), \\ F = 312, \ \text{if } m = 1, \\ F = 160m^3 + 128m^2 + 10m - 9, \ \text{if } m \geq 2. \end{array}$$

3. $n = 4m + 1$.

$$\begin{array}{lll} a_{11} = \frac{\xi_2}{4} + 3\xi_1 + (\xi_2 + 8\xi_1)m, & a_{12} = 4\xi_1 - \xi_2 + 16\xi_1 m, & a_{13} = \frac{\xi_2}{4} - \xi_1 + (\xi_2 - 8\xi_1)m, \\ a_{21} = \frac{\xi_4}{4} + 3\xi_1 + (\xi_4 + 8\xi_1)m, & a_{22} = 4\xi_1 - \xi_4 + 16\xi_1 m, & a_{23} = \xi_1 - \frac{\xi_4}{4} + (8\xi_1 - \xi_4)m, \\ a_{31} = \frac{\xi_6}{4} + \xi_6 m, & a_{32} = \xi_6, & a_{33} = \frac{\xi_6}{4} + \xi_6 m. \end{array}$$

$$\xi_1 = 1/4, \quad \xi_2 = 5, \quad \xi_4 = 1, \quad \xi_6 = 4.$$

$$\begin{pmatrix} 7m+2 & -4(m-1) & -(3m+1) \\ -(3m+1) & 4m & -m \\ -(4m+1) & -4 & 4m+1 \end{pmatrix}, \quad \begin{array}{l} G = 2m(32m^2+9m+1), \\ F = 112m^3+48m^2+8m-1, \ \text{if} \ m \le 3, \\ F = 128m^3-20m-5, \ \text{if} \ m \ge 4. \end{array}$$

4. $n = 4m + 3$.

$$\begin{array}{lll} a_{11} = \frac{3\xi_2}{4} + 7\xi_1 + (\xi_2 + 8\xi_1)m, & a_{12} = 12\xi_1 - \xi_2 + 16\xi_1 m, & a_{13} = \frac{3\xi_2}{4} - 5\xi_1 + (\xi_2 - 8\xi_1)m, \\ a_{21} = \frac{3\xi_4}{4} + 7\xi_1 + (\xi_4 + 8\xi_1)m, & a_{22} = 12\xi_1 - \xi_4 + 16\xi_1 m, & a_{23} = 5\xi_1 - \frac{3\xi_4}{4} + (8\xi_1 - \xi_4)m, \\ a_{31} = \frac{3\xi_6}{4} + \xi_6 m, & a_{32} = \xi_6, & a_{33} = \frac{3\xi_6}{4} + \xi_6 m. \end{array}$$

$$\xi_1 = 1/4, \quad \xi_2 = 3, \quad \xi_4 = -1, \quad \xi_6 = 4.$$

$$\begin{pmatrix} 5m+4 & -4m & -(m+1) \\ -(m+1) & 4(m+1) & -(3m+2) \\ -(4m+3) & -4 & 4m+3 \end{pmatrix}, \quad \begin{array}{l} G = 2(32m^3+57m^2+33m+6), \\ F = 128m^3+2242+124m+19, \ \text{if} \ m \ge 1, \\ F = 23, \ G = 12, \ \text{if} \ m = 0. \end{array}$$

A coincidence of formulas for $F(R_n^2)$ with those obtained in [7] when $t_2 > t_1$ occur in the cases

$$n = 4m, \ m \ge 2, \quad n = 4m+1, \ m \ge 4, \quad n = 4m+2, \ m \ge 2, \quad n = 4m+3, \ m \ge 1.$$

Otherwise ($t_2 < t_1$) we obtain the other six formulas related to the six excluded cases (2),

$$n = 4m, \ m = 1, \quad n = 4m+1, \ 1 \le m \le 3, \quad n = 4m+2, \ m = 1, \quad n = 4m+3, \ m = 0.$$

3 Numerical Semigroups R_n^3, $n \ge 4$

Consider a semigroup R_n^3 and, instead of a straightforward construction of PRep as in Sect. 2, start with a simple algebraic observation. Write the third minimal relation in (18) as follows,

$$a_{32}n^3 = (a_{33} - a_{31})\left(n^3 + 3n\right) + (a_{33} + a_{31})(3n^2 + 1), \tag{25}$$

and, according to assumption (20.2), simplify it by choosing

$$a_{31} = pn + q, \qquad a_{33} = pn - q, \quad p, q \in \mathbb{Q}. \tag{26}$$

A choice of the linear Rep (26) for a_{31}, a_{33} will be justified later, according to Lemma 3, by uniqueness of PRep of all matrix elements in A_3 found in Sects. 3.1, 3.2 and satisfied six equalities (14), (15). Substitute (26) into Eq. (25) and obtain,

$$a_{32}n^2 = 2p(3n^2+1) - 2q(n^2+3), \quad \text{or} \quad a_{32} = 6p - 2q + 2\,\frac{p-3q}{n^2}.$$

To eliminate the dependence of a_{32} on n^{-2} in the last relation we put

$$p = 3q \quad \rightarrow \quad a_{31} = q(3n+1), \quad a_{32} = 16q, \quad a_{33} = q(3n-1).$$

To satisfy $\gcd(a_{31}, a_{32}, a_{33}) = 1$ in (12), we have to distinguish two different cases: $q = 1$ if $n = 2N$ and $q = 1/2$ if $n = 2N+1$.

3.1 Numerical Semigroups R_n^3, $n = 0 \pmod 2$

The matrix A_3 of minimal relations (12), (13) and three equalities (15) read,

$$\begin{pmatrix} a_{21}+6N+1 & 16-a_{22} & a_{23}-(6N-1) \\ -a_{21} & a_{22} & -a_{23} \\ -(6N+1) & -16 & (6N-1) \end{pmatrix},$$

$$\begin{aligned} &(2N-1)^3 = a_{22}(6N-1) - 16a_{23}, \quad 8N^3 = a_{21}(6N-1) + a_{23}(6N+1), \\ &(2N+1)^3 = a_{22}(6N+1) + 16a_{21}. \end{aligned} \tag{27}$$

According to (20.2), choose elements of the 1st and 2nd rows in A_3 as quadratic polynomials in N on residue class of N modulo τ_3 which will be found later, i.e., $N = \tau_3 m + j, 0 \le j < \tau_3$,

$$\begin{aligned} &a_{21} = r_2m^2 + r_1m + r_0, \quad a_{22} = k_2m^2 + k_1m + k_0, && (28)\\ &a_{23} = l_2m^2 + l_1m + l_0, \\ &a_{11} = r_2m^2 + (6\tau_3 + r_1)m + r_0 + 6j + 1, \quad a_{12} = k_2m^2 + k_1m + 16 + k_0, && (29)\\ &a_{13} = -l_2m^2 + (6\tau_3 - l_1)m + 6j - 1 - l_0. \end{aligned}$$

By (19), require $a_{2i}, a_{1i} > 0$ and $a_{11} > a_{12}$ for the large $m \ge 1$ and $a_{11}, a_{22} > 0$ when $m = 0$,

$$\begin{aligned} &(a) \quad l_2 = 0, \quad 0 < l_1 < 6\tau_3, \\ &(b) \quad r_2 > k_2 > 0, \quad \text{or if } r_2 = k_2 > 0 \quad \text{then} \quad 6\tau_3 + r_1 > k_1, \\ &(c) \quad r_0 + 6j + 1 > 0, \quad k_0 > 0. \end{aligned} \tag{30}$$

Substitute (28), (29) into (27) and balance degrees of m in the l.h.s. and r.h.s. of cubic equalities in m. Nine variables $r_i, k_i, l_i \in \mathbb{Z}$, $i = 0, 1, 2$, (including $l_2 = 0$) satisfy twelve linear equations, but not all of them are independent. The system may be reduced substantially,

$$k_2 = r_2 = \frac{4\tau_3^2}{3}, \quad l_2 = 0, \quad k_1 = \frac{8\tau_3}{9}(3j-2), \tag{31}$$

$$3\tau_3 k_0 - 8l_1 = \tau_3\left(4j^2 - \frac{16}{3}j + \frac{19}{9}\right), \quad (6j-1)k_0 - 16l_0 = (2j-1)^3, \tag{32}$$

$$3\tau_3 k_0 + 8r_1 = \tau_3\left(4j^2 + 16j + \frac{35}{9}\right), \quad (6j+1)k_0 + 16r_0 = (2j+1)^3,$$

and the variables satisfy constraints (30). The minimal value of τ_3, providing $k_1 \in \mathbb{Z}$ in (31), is $\tau_3 = 9$, so that $T_3 = 18$. The five variables, r_0, r_1, k_0, l_0, l_1, depend on j and satisfy four Diophantine equations (32), supplemented by constraints (30), and may be solved in nine different cases, $0 \le j \le 8$. We present here, as an example, the Diophantine equations and their detailed numerical solutions, satisfying constraints (30), for two first cases, (a) $j = 0$ and (b) $j = 1$, and skip the other,

$$\begin{aligned}
(a)\quad & 27k_0 - 8l_1 = 19, \quad 27k_0 + 8r_1 = 35, \quad k_0 + 16l_0 = 1, \quad k_0 + 16r_0 = 1\ : \\
& k_0 = r_1 = l_1 = 1, \quad l_0 = r_0 = 0, \quad k_2 = r_2 = 108, \quad k_1 = -16, \\
(b)\quad & 27k_0 - 8l_1 = 7, \quad 27k_0 + 8r_1 = 215, \quad 5k_0 - 16l_0 = 1, \quad 7k_0 + 16r_0 = 27\ : \\
& k_0 = 13, \quad r_1 = -17, \quad l_1 = 43, \quad l_0 = 4, \quad r_0 = -4, \\
& k_2 = r_2 = 108, \quad k_1 = 8.
\end{aligned}$$

1. $n = 18m,\ m \ge 1$,

$$\begin{pmatrix} 108m^2 + 55m + 1 & -(108m^2 - 16m - 15) & -(53m - 1) \\ -m(108m + 1) & 108m^2 - 16m + 1 & -m \\ -(54m + 1) & -16 & 54m - 1 \end{pmatrix},$$

$$\begin{aligned}
G &= 314928m^5 + 110808m^4 + 16632m^3 + 532m^2 - 62m, \\
F &= 629856m^5 + 215784m^4 + 34020m^3 + 1890m^2 - 109m - 1, \quad \text{if } m \le 15, \\
F &= 629856m^5 + 221616m^4 - 58320m^3 + 1944m^2 - 108m - 1, \quad \text{if } m \ge 16,
\end{aligned}$$

2. $n = 18m + 2,\ m \ge 1; \quad n \ne 2$

$$\begin{pmatrix} 108m^2 + 37m + 3 & -(108m^2 + 8m - 3) & -(11m + 1) \\ -(108m^2 - 17m - 4) & 108m^2 + 8m + 13 & -(43m + 4) \\ -(54m + 7) & -16 & 54m + 5 \end{pmatrix},$$

$$\begin{aligned}
G &= 314928m^5 + 285768m^4 + 104760m^3 + 15244m^2 + 786m + 6, \\
F &= 629856m^5 + 571536m^4 + 190512m^3 + 31752m^2 + 2548m + 75,
\end{aligned}$$

3. $n = 18m + 4,\ m \ge 1; \quad n \ne 4.$

$$\begin{pmatrix} 108m^2+55m+7 & -(108m^2+32m+1) & -(5m+1) \\ -(108m^2+m-6) & 108m^2+32m+17 & -(49m+10) \\ -(54m+13) & -16 & 54m+11 \end{pmatrix},$$

$$G = 314928m^5 + 460728m^4 + 270648m^3 + 77476m^2 + 10674m + 564,$$
$$F = 629856m^5 + 921456m^4 + 532656m^3 + 153144m^2 + 21812m + 1223,$$

4. $n = 18m + 6,\ m \geq 1; \quad n \neq 6.$

$$\begin{pmatrix} 108m^2+109m+25 & -(108m^2+56m-3) & -(35m+11) \\ -(108m^2+55m+6) & 108m^2+56m+13 & -(19m+6) \\ -(54m+19) & -16 & 54m+17 \end{pmatrix},$$

$$G = 314928m^5 + 635688m^4 + 514296m^3 + 202780m^2 + 38434m + 2778,$$
$$F = 629856m^5 + 1271376m^4 + 968112m^3 + 355752m^2 + 63828m + 4499,$$

5. $n = 18m + 8,\ m \geq 0,$

$$\begin{pmatrix} 108m^2+145m+44 & -(108m^2+80m+1) & -(47m+20) \\ -(108m^2+91m+19) & 108m^2+80m+17 & -(7m+3) \\ -(54m+25) & -16 & 54m+23 \end{pmatrix},$$

$$G = 314928m^5 + 810648m^4 + 835704m^3 + 428308m^2 + 108658m + 10888,$$
$$F = 629856m^5 + 1580472m^4 + 1604772m^3 + 821178m^2 + 210979m + 21700,$$
$$\text{if } m - 0, 1,$$
$$F = 629856m^5 + 1621296m^4 + 1590192m^3 + 753624m^2 + 173908m + 15695,$$
$$\text{if } m \geq 2.$$

6. $n = 18m + 10,\ m \geq 0,$

$$\begin{pmatrix} 108m^2+163m+58 & -(108m^2+104m+13) & -(41m+22) \\ -(108m^2+109m+27) & 108m^2+104m+29 & -(13m+7) \\ -(54m+31) & -16 & 54m+29 \end{pmatrix},$$

$$G = 314928m^5 + 985608m^4 + 1234872m^3 + 769612m^2 + 237666m + 29022,$$
$$F = 55222, \quad \text{if } m = 0,$$
$$F = 629856m^5 + 1971216m^4 + 2398896m^3 + 1429704m^2 + 419252m + 48539,$$
$$\text{if } m \geq 1,$$

7. $n = 18m + 12,\ m \geq 0,$

$$\begin{pmatrix} 108m^2+163m+61 & -(108m^2+128m+33) & -(17m+11) \\ -(108m^2+109m+24) & 108m^2+128m+49 & -(37m+24) \\ -(54m+37) & -16 & 54m+35 \end{pmatrix},$$

$$G = 314928m^5 + 1160568m^4 + 1711800m^3 + 1257796m^2 + 459058m + 66444,$$
$$F = 629856m^5 + 2321136m^4 + 3394224m^3 + 2466936m^2 + 892404m + 128663,$$

8. $n = 18m + 14,\ m \geq 0,$

$$\begin{pmatrix} 108m^2+199m+90 & -(108m^2+152m+45) & -(29m+22) \\ -(108m^2+145m+47) & 108m^2+152m+61 & -(25m+19) \\ -(54m+43) & -16 & 54m+41 \end{pmatrix},$$

$$G = 314928m^5 + 1335528m^4 + 2266488m^3 + 1917916m^2 + 807378m + 135042,$$
$$F = 629856m^5 + 2671056m^4 + 4482864m^3 + 3730536m^2 + 1541908m + 253539,$$

9. $n = 18m + 16,\ m \geq 0,$

$$\begin{pmatrix} 108m^2+217m+108 & -(108m^2+176m+65) & -(23m+20) \\ -(108m^2+163m+59) & 108m^2+176m+81 & -(31m+27) \\ -(54m+49) & -16 & 54m+47 \end{pmatrix},$$

$$G = 314928m^5 + 1510488m^4 + 2898936m^3 + 2776756m^2 + 1325266m + 251824,$$
$$F = 629856m^5 + 3020976m^4 + 5758128m^3 + 5458968m^2 + 2576660m + 484767.$$

3.2 Numerical Semigroups R_n^3, $n = 1 \pmod 2$

The matrix A_3 of minimal relations (12), (13) and three equalities (15) read,

$$\begin{pmatrix} a_{21}+(3N+2) & 8-a_{22} & a_{23}-(3N+1) \\ -a_{21} & a_{22} & -a_{23} \\ -(3N+2) & -8 & 3N+1 \end{pmatrix},$$

$$(2N)^3 = a_{22}(3N+1) - 8a_{23}, \quad (2N+1)^3 = a_{21}(3N+1) + a_{23}(3N+2),$$
$$(2N+2)^3 = a_{22}(3N+2) + 8a_{21}. \tag{33}$$

We skip the intermediate calculations repeating the procedure performed in Sect. 3.1.
1. $n = 18m + 1,\ m \geq 1; \quad n \neq 1.$

$$\begin{pmatrix} 216m^2+29m+1 & -(216m^2-8m) & -m \\ -(216m^2+2m-1) & 216m^2-8m+8 & -(26m+1) \\ -(27m+2) & -8 & 27m+1 \end{pmatrix},$$

$$G = 629856m^5 + 160380m^4 + 23220m^3 + 2330m^2 + 73m,$$
$$F = 1259712m^5 + 320760m^4 + 44712m^3 + 4644m^2 + 154m - 1,$$

2. $n = 18m+3,\ m \geq 1; \qquad n \neq 3.$

$$\begin{pmatrix} 216m^2+83m+8 & -(216m^2+40m) & -(7m+1) \\ -(216m^2+56m+3) & 216m^2+40m+8 & -(20m+3) \\ -(27m+5) & -8 & 27m+4 \end{pmatrix},$$

$$G = 629856m^5 + 510300m^4 + 172260m^3 + 30134m^2 + 2657m + 91,$$
$$F = 181, \text{ if } m = 0,$$
$$F = 1259712m^5 + 1020600m^4 + 332424m^3 + 55404m^2 + 4698m + 157, \text{ if } m \geq 1.$$

3. $n = 18m+5,\ m \geq 0,$

$$\begin{pmatrix} 216m^2+128m+19 & -(216m^2+88m+8) & -(4m+1) \\ -(216m^2+101m+11) & 216m^2+88m+16 & -(23m+6) \\ -(27m+8) & -8 & 27m+7 \end{pmatrix},$$

$$G = 629856m^5 + 860220m^4 + 476820m^3 + 134186m^2 + 18993m + 1066,$$
$$F = 1259712m^5 + 1720440m^4 + 946728m^3 + 263412m^2 + 37082m + 2107,$$

4. $n = 18m+7,\ m \geq 0,$

$$\begin{pmatrix} 216m^2+191m+42 & -(216m^2+136m+16) & -(19m+7) \\ -(216m^2+164m+31) & 216m^2+136m+24 & -(8m+3) \\ -(27m+11) & -8 & 27m+10 \end{pmatrix},$$

$$G = 629856m^5 + 1210140m^4 + 936900m^3 + 365246m^2 + 71609m + 5637,$$
$$F = 10745, \text{ if } m = 0,$$
$$F = 1259712m^5 + 2420280m^4 + 1840968m^3 + 693468m^2 + 129322m + 9537, \\ \text{if } m \geq 1,$$

5. $n = 18m+9,\ m \geq 0,$

$$\begin{pmatrix} 216m^2+245m+69 & -(216m^2+184m+32) & -(25m+12) \\ -(216m^2+218m+55) & 216m^2+184m+40 & -(2m+1) \\ -(27m+14) & -8 & 27m+13 \end{pmatrix},$$

$$G = 629856m^5 + 1560060m^4 + 1552500m^3 + 776234m^2 + 195001m + 19684,$$
$$F = 1259712m^5 + 3108456m^4 + 3083184m^3 + 1537596m^2 + 385666m + 38919, \quad \text{if } m \le 3,$$
$$F = 1259712m^5 + 3120120m^4 + 3061800m^3 + 1488132m^2 + 358074m + 34087, \quad \text{if } m \ge 4,$$

6. $n = 18m + 11$, $m \ge 0$,

$$\begin{pmatrix} 216m^2 + 290m + 97 & -(216m^2 + 232m + 56) & -(22m + 13) \\ -(216m^2 + 263m + 80) & 216m^2 + 232m + 64 & -(5m + 3) \\ -(27m + 17) & -8 & 27m + 16 \end{pmatrix},$$

$$G = 629856m^5 + 1909980m^4 + 2323620m^3 + 1417694m^2 + 433745m + 53223,$$
$$F = 103589, \quad \text{if } m = 0,$$
$$F = 1259712m^5 + 3819960m^4 + 4609224m^3 + 2766636m^2 + 826058m + 98125, \quad \text{if } m \ge 1,$$

7. $n = 18m + 13$, $m \ge 0$,

$$\begin{pmatrix} 216m^2 + 326m + 123 & -(216m^2 + 280m + 90) & -(10m + 7) \\ -(216m^2 + 299m + 103) & 216m^2 + 280m + 96 & -(17m + 12) \\ -(27m + 20) & -8 & 27m + 19 \end{pmatrix},$$

$$G = 629856m^5 + 2259900m^4 + 3250260m^3 + 2342114m^2 + 845465m + 122286,$$
$$F = 1259712m^5 + 3120120m^4 + 3061800m^3 + 1488132m^2 + 358074m + 34087.$$

8. $n = 18m + 15$, $m \ge 0$,

$$\begin{pmatrix} 216m^2 + 380m + 167 & -(216m^2 + 328m + 120) & -(16m + 13) \\ -(216m^2 + 353m + 144) & 216m^2 + 328m + 128 & -(11m + 9) \\ -(27m + 23) & -8 & 27m + 22 \end{pmatrix},$$

$$G = 629856m^5 + 2609820m^4 + 4332420m^3 + 3601550m^2 + 1499153m + 249937,$$
$$F = 1259712m^5 + 5219640m^4 + 8637192m^3 + 7135452m^2 + 2943162m + 484897.$$

9. $n = 18m + 17$, $m \ge 0$,

$$\begin{pmatrix} 216m^2 + 425m + 209 & -(216m^2 + 376m + 160) & -(13m + 12) \\ -(216m^2 + 398m + 183) & 216m^2 + 376m + 168 & -(14m + 13) \\ -(27m + 26) & -8 & 27m + 25 \end{pmatrix},$$

$$G = 629856m^5 + 2959740m^4 + 5570100m^3 + 5247626m^2 + 2474713m + 467304,$$
$$F = 1259712m^5 + 5919480m^4 + 11117736m^3 + 10433124m^2 + 4892186m + 917039.$$

Formulas for $F(R_n^3)$ in the cases

$$\begin{array}{lll} n = 18m,\ m \geq 16, & n = 18m+2,\ m \neq 0, & n = 18m+4,\ m \neq 0, \\ n = 18m+6,\ m \neq 0, & n = 18m+8,\ m \geq 2, & n = 18m+10,\ m \geq 1, \\ n = 18m+12,\ m \geq 0, & n = 18m+14,\ m \geq 0, & n = 18m+16,\ m \geq 0, \\ n = 18m+1,\ m \geq 0, & n = 18m+3,\ m \geq 1, & n = 18m+5,\ m \geq 0, \\ n = 18m+7,\ m \geq 1, & n = 18m+9,\ m \geq 4, & n = 18m+11,\ m \geq 1, \\ n = 18m+13,\ m \geq 0, & n = 18m+15,\ m \geq 0, & n = 18m+17,\ m \geq 0, \end{array}$$

coincide with those obtained in [7] when $t_2 > t_1$. In the opposite case ($t_2 < t_1$)

$$\begin{array}{lll} n = 18m,\ 1 \leq m \leq 15, & n = 18m+2,\ m = 0, & n = 18m+4,\ m = 0, \\ n = 18m+6,\ m = 0, & n = 18m+8,\ m = 0, 1, & n = 18m+10,\ m = 0, \\ n = 18m+3,\ m = 0, & n = 18m+7,\ m = 0, & n = 18m+9,\ 0 \leq m \leq 3, \\ n = 18m+11,\ m = 0. & & \end{array}$$

we get the other 28 formulas. These are exactly the 28 excluded cases (3).

3.3 Exceptional Nonsymmetric Semigroups R_n^3, $n = 4, 6$

$$R_4^3 : \begin{pmatrix} 7 & -1 & -1 \\ -1 & 18 & -9 \\ -6 & -17 & 10 \end{pmatrix}, \quad \begin{array}{l} G = 558 \\ F = 1098 \end{array}, \qquad R_6^3 : \begin{pmatrix} 31 & -10 & -5 \\ -6 & 13 & -6 \\ -25 & -3 & 11 \end{pmatrix}, \quad \begin{array}{l} G = 2670 \\ F = 5249 \end{array}.$$

4 Extension on Numerical Semigroups R_n^k, $k \geq 4$

It is quite natural to ask about PRep of minimal relations for semigroups R_n^k with arbitrary k. Albeit nothing rigorous is known for $k \geq 4$, the case $k = 4$ is worth to discuss briefly.

Making use of the Euclidean numerical algorithm with negative [12] and positive [10] remainders for computation of the Frobenius numbers, in the article [7], besides a calculation of the PRep for $F(R_n^2)$, $F(R_n^3)$, there were put forward weak arguments to predict a value of modulus T_4 of residue class. Namely, $F(R_n^4)$ is given by polynomial expressions in n on residue class of n modulo 88 *'whereas experimental tests make us believe that we need 40 formulas'*.

Motivated by the last observation [7] and making use of assumption (20.1), we have undertaken an attempt [4] to verify numerically an assertion $T_4 = 40$ and, if so, to calculate the PRep of minimal relations for semigroups R_n^4. If such PRep does exist and is computable then by Lemma 3 and Corollary 2 it is unique and does not depend on the way of its derivation.

To perform calculations, we define $n = T_4 m + q$ and choose a qudratic in m Rep of all matrix elements $a_{ij}(n) = \mathcal{L}_{ij}(q)m^2 + \mathcal{M}_{ij}(q)m + \mathcal{N}_{ij}(q)$. Due to (14), among 27 indeterminates $\mathcal{L}_{ij}(q), \mathcal{M}_{ij}(q), \mathcal{N}_{ij}(q)$ with fixed q there are only 18 linearly independent variables. In fact, their number may be decreased much stronger if we impose three (quartic in m) equalities in (18). However, instead of treating non-linear equalities, we choose another way to solve three linear Diophantine equations (18) for $k = 4$ and fixed q. For this purpose, choose three different m and find $\mathcal{L}_{ij}(q), \mathcal{M}_{ij}(q), \mathcal{N}_{ij}(q)$ according to minimality requirement (13) and Corollary 1. Finally, verify a validity of matrix A_3 of minimal relations by non-linear equalities (18).

Below we list the main results of this study [4].

- There exist only three symmetric numerical semigroups R_n^4, $n = 3, 5, 7$.
- If $n = 0 \pmod 2$ then all matrix elements of A_3 for semigroup R_n^4, have the quadratic Rep in n on residue class of n modulo $T_4 = 40$.
- If $n = 1 \pmod 2$ then all matrix elements of A_3 for semigroup R_n^4, have the quadratic Rep in n on residue class of n modulo $T_4 = 20$.
- There exist 15 exceptional nonsymmetric semigroups R_n^4, where $n = 6, 9, 10, 13, 14, 17, 20, 26, 27, 30, 33, 40, 60, 80, 120$.

The whole list of quadratic Rep of minimal relations for semigroups R_n^4 comprise 30 matrix formulas for which we refer the readers to preprint [4].

In spite of numerical results, discussed in this section, the problem of PRep of minimal relations for semigroups R_n^4 cannot be considered solved finally so far we have not a pure proof on the value of T_4. The last question is still awaiting its answer.

4.1 Conjecture and Question

In this section we state a conjecture and put a question, which concerned with numerical semigroups R_n^k, $n > 3$, $k > 4$, where an appearance of symmetric semigroups seems very rare.

Indeed, besides semigroups $R_3^k, 5 \le k < \infty$, numerical calculations give only two symmetric numerical semigroups according to Lemma 2,

$$R_5^{11} \;:\; 5^{11} = 1093 \cdot 2^{11} + 263 \cdot 3^{11}, \qquad R_5^{13} \;:\; 5^{13} = 51118 \cdot 2^{13} + 503 \cdot 3^{13},$$

among others numerical semigroups R_{2p+1}^k, $2 \le p \le 50$, $5 \le k \le 10^3$.

Conjecture 1 *Let a numerical semigroup R_n^k, $n = T_k m + j$, be given by their matrix* A_3 *of minimal relations on residue class of n modulo T_k,*

$$\mathsf{A}_3\left(R^k_{T_k m+j}\right) = \begin{pmatrix} K_{11}^{(k)}(m,j) & -K_{12}^{(k)}(m,j) & -K_{13}^{(k)}(m,j) \\ -K_{21}^{(k)}(m,j) & K_{22}^{(k)}(m,j) & -K_{23}^{(k)}(m,j) \\ -K_{31}^{(k)}(m,j) & -K_{32}^{(k)}(m,j) & K_{33}^{(k)}(m,j) \end{pmatrix}, \quad 0 \le j < T_k. \tag{34}$$

Then polynomial expressions in m for $K_{ir}^{(k)}(m,j)$ read in two different cases.

If $k = 2q$, then

$$K_{ir}^{(2q)}(m,j) = A_{ir}(j)m^q + B_{ir}(j)m^{q-1} + \cdots + C_{ir}(j)m + D_{ir}(j), \quad 1 \le i, r \le 3, \tag{35}$$

and the Frobenius number and genus have the asymptotic behavior: $F(n), G(n) = \mathcal{O}\left(n^{3q}\right)$.

If $k = 2q+1$, then the matrix elements with $(i,r) = (1,1), (1,2), (2,1), (2,2)$ are given by

$$K_{ir}^{(2q+1)}(m,j) = E_{ir}(j)m^{q+1} + I_{ir}(j)m^q + \cdots + J_{ir}(j)m + H_{ir}(j), \tag{36}$$

while the matrix elements with $(i,r) = (1,3), (2,3), (3,1), (3,2), (3,3)$ are given by

$$K_{ir}^{(2q+1)}(m,j) = M_{ir}(j)m^q + N_{ir}(j)m^{q-1} + \cdots + P_{ir}(j)m + S_{ir}(j), \tag{37}$$

and the Frobenius number and genus have the asymptotic behavior: $F(n), G(n) = \mathcal{O}\left(n^{3q+2}\right)$.

Question 1 Keeping in mind

$$T_2 = 4, \quad T_3 = 18, \quad T_4 = 20 \ \text{if} \ n = 1 \ (\text{mod } 2) \ \text{and} \ T_4 = 40 \ \text{if} \ n = 0 \ (\text{mod } 2)$$

find T_k for the higher k.

Acknowledgements The research was supported by the Kamea Fellowship.

5 Appendix: Proof of Propositions 2

Proof Semigroup R_3^3 is symmetric due to requirement (4a). Find more n which satisfy (4a),

$$(n+1)^3 = e_1(n-1)^3 + e_2 n^3, \quad e_1, e_2 \in \mathbb{N}, \quad n > 3.$$

Simplifying the last equality we obtain the Diophantine equation

$$(e_1+e_2-1)t^3+3(2e_1+3e_2-4)t^2+3(4e_1+9e_2-16)t+8e_1+27e_2-64=0, \quad t=n-3. \tag{38}$$

Decompose the whole integer lattice $\mathbb{Z}_2^+ := \{e_1, e_2 \mid e_1, e_2 \geq 1\}$ in different sets,

$$\mathbb{Z}_2^+ = \bigcup_{j=1}^{5} \mathbb{E}_j, \quad \mathbb{E}_1 = \{e_1, e_2 \mid e_1 \geq 5;\ e_2 = 1\},$$
$$\mathbb{E}_2 = \{e_1, e_2 \mid e_1 \geq 2;\ e_2 = 2\}, \qquad \mathbb{E}_3 = \{e_1, e_2 \mid e_1 \geq 1;\ e_2 \geq 3\},$$
$$\mathbb{E}_4 = \{e_1, e_2 \mid 1 \leq e_1 \leq 4;\ e_2 = 1\}, \quad \mathbb{E}_5 = \{e_1 = e_2 = 1\}.$$

If $(e_1, e_2) \in \mathbb{E}_j$, $1 \leq j \leq 3$, then the sequence of coefficients in Eq. (38) has no changes of signs and therefore, by Descartes' rule of signs, Eq. (38) has no positive solutions in t. If $(e_1, e_2) \in \mathbb{E}_j$, $j = 4, 5$, a straightforward numerical verification shows that none of 5 cubic equations (38) has an integer positive solution in t.

Consider an alternative way to symmetrize R_n^3 by providing condition (4b), which may occur only when $n = 2q + 1$ and results in the Diophantine equation in $c_1, c_2 \in \mathbb{N}$, $q > 1$,

$$(2q+1)^3 = c_1 q^3 + c_2 (q+1)^3, \qquad \text{or}$$
$$(c_1+c_2-8)q^3 + 3(c_2-4)q^2 + 3(c_2-2)q + c_2 - 1 = 0. \tag{39}$$

Substituting $q = p + 1$, $p > 0$, into (39) we obtain the cubic Diophantine equation in p,

$$(c_1+c_2-8)p^3+3(c_1+2c_2-12)p^2+3(c_1+4c_2-18)p+c_1+8c_2-27=0, \tag{40}$$

which has no positive integer solutions p. Indeed, to prove this statement, we make use of Descartes' rule of signs for integer coefficients in Eq. (40). For this purpose decompose the whole integer lattice $\mathbb{Z}_2^+ := \{c_1, c_2 \mid c_1, c_2 \geq 1\}$ in different sets,

$$\mathbb{Z}_2^+ = \mathbb{C} \cup \overline{\mathbb{C}}, \qquad \mathbb{C} = \bigcup_{j=1}^{7} \mathbb{C}_j, \qquad \overline{\mathbb{C}} = \bigcup_{j=1}^{6} \overline{\mathbb{C}}_j,$$
$$\mathbb{C}_1 = \{c_1, c_2 \mid 1 \leq c_1 \leq 7,\ c_1 \geq 19;\ c_2 = 1\},$$
$$\mathbb{C}_2 = \{c_1, c_2 \mid 1 \leq c_1 \leq 6,\ c_1 \geq 11;\ c_2 = 2\},$$
$$\mathbb{C}_3 = \{c_1, c_2 \mid 1 \leq c_1 \leq 3,\ c_1 \geq 6;\ c_2 = 3\},$$
$$\mathbb{C}_4 = \{c_1, c_2 \mid c_1 \geq 4;\ c_2 = 4\}, \qquad \mathbb{C}_5 = \{c_1, c_2 \mid c_1 \geq 3;\ c_2 = 5\},$$
$$\mathbb{C}_6 = \{c_1, c_2 \mid c_1 \geq 2;\ c_2 = 6\}, \qquad \mathbb{C}_7 = \{c_1, c_2 \mid c_1 \geq 1;\ c_2 \geq 7\},$$
$$\overline{\mathbb{C}}_1 = \{c_1, c_2 \mid 8 \leq c_1 \leq 18;\ c_2 = 1\}, \quad \overline{\mathbb{C}}_2 = \{c_1, c_2 \mid 7 \leq c_1 \leq 10;\ c_2 = 2\},$$
$$\overline{\mathbb{C}}_3 = \{c_1, c_2 \mid 4 \leq c_1 \leq 5;\ c_2 = 3\}, \quad \overline{\mathbb{C}}_4 = \{c_1, c_2 \mid 1 \leq c_1 \leq 3;\ c_2 = 4\},$$
$$\overline{\mathbb{C}}_5 = \{c_1, c_2 \mid 1 \leq c_1 \leq 2;\ c_2 = 5\}, \quad \overline{\mathbb{C}}_6 = \{c_1 = 1;\ c_2 = 6\}.$$

If $(c_1, c_2) \in \mathbb{C}$ then the sequence of coefficients in Eq. (40) has no changes of signs and therefore, by Descartes' rule of signs, Eq. (40) has no positive solutions in p. Regarding the rest of the cases, when $(c_1, c_2) \in \overline{\mathbb{C}}$, a straightforward numerical verification shows that none of 23 cubic equations (40) has an integer positive solution in p. $\square$

References

1. A. Brauer, *On a problem of partitions*, Am. J. Math. **64**, 299–312 (1942)
2. L.G. Fel, *Frobenius problem for semigroups* $\mathsf{S}(d_1, d_2, d_3)$, Funct. Analysis and Other Math., **1**, 119–157 (2006)
3. L.G. Fel, *Symmetric Semigroups Generated by Fibonacci and Lucas Triples*, Integers, **9**, 107–116 (2009)
4. L.G. Fel, *Numerical semigroups generated by squares, cubes and quartics of three consecutive integers*, http://arxiv.org/pdf/1608.08693v1.pdf
5. J. Herzog, *Generators and Relations of Abelian Semigroups and Semigroup Rings*, Manuscripta Math., **3**, 175–193 (1970)
6. S.M. Johnson, *A linear Diophantine problem*, Canad. J. Math., **12**, 390–398 (1960)
7. M. Lepilov, J. O'Rourke, I. *Swanson, Frobenius numbers of numerical semigroups generated by three consecutive squares or cubes*, Semigroup Forum, **91**, 238–259 (2015)
8. J.M. Marin, J.L. Ramirez Alfonsin, M.P. Revuelta, *On the Frobenius number of Fibonacci numerical semigroups*, Integers, **7**, A14, 1–7 (2007)
9. D.C. Ong, V. Ponomarenko, *Frobenius number of geometric sequences*, Integers, **8**, A33 (2008)
10. J.L. Ramirez Alfonsin and Ö.J. Rödseth, *Numerical semigroups: Apéry sets and Hilbert series*, Semigroup Forum, **79**, 323–340 (2009)
11. J.B. Roberts, *Note on linear forms*, Proc. Am. Math. Soc. **7**, 465–469 (1956)
12. Ö.J. Rödseth, *A linear Diophantine problem of Frobenius*, J. Reine Angew. Math., **301**, 171–178 (1978)
13. K. Watanabe, *Examples of 1-dim Gorenstein Domains*, Nagoya Math. J., **49**, 101–109 (1973)

On Supra-SIM Sets of Natural Numbers

Isaac Goldbring and Steven Leth

Abstract We introduce the class of supra-SIM sets of natural numbers. We prove that this class is partition regular and closed under finite-embeddability. We also prove some results on sumsets and SIM sets motivated by their positive Banach density analogues.

1 Introduction

Ramsey theory on the integers can crudely be described as the study of *partition regular* properties of the integers, namely those properties $\mathscr{P}$ of integers such that, whenever $A \subseteq \mathbb{N}$ has $\mathscr{P}$ and $A = B \sqcup C$ (disjoint union), then at least one of B or C has property $\mathscr{P}$. Here are some of the more prominent examples of partition regular properties of the integers:

- having infinite cardinality (Pigeonhole principle);
- having arbitrary long arithmetic progressions (van der Waerden's theorem);
- containing a set of the form

$$\mathrm{FS}(X) := \{a_1 + \cdots + a_n \ : \ a_1, \ldots, a_n \in X \text{ distinct}, n \in \mathbb{N}\}$$

  for some infinite set X (Hindman's theorem);
- being piecewise syndetic;
- having positive Banach density.

I. Goldbring
Department of Mathematics, University of California, Irvine, 340 Rowland Hall (Bldg.# 400), Irvine, CA 92697-3875, USA
e-mail: isaac@math.uci.edu

S. Leth (✉)
School of Mathematical Sciences, University of Northern Colorado, Campus Box 122, 501 20th Street, Greeley, CO 80639, USA
e-mail: steven.leth@unco.edu

M. B. Nathanson (ed.), *Combinatorial and Additive Number Theory III*, Springer Proceedings in Mathematics & Statistics 297,
https://doi.org/10.1007/978-3-030-31106-3_9

In this paper, we introduce a new partition regular property of the natural numbers, namely that of being *supra-SIM*. SIM[1] sets were introduced by the second author in [6] in connection with Stewart and Tijdeman's result that intersections of difference sets of sets of positive density are syndetic. This property arises from an analogous natural property of *internal* subsets of the nonstandard natural numbers $^*\mathbb{N}$ in the sense of nonstandard analysis. While one can prove an analog of the aforementioned result of Stewart and Tijdeman by replacing the hypothesis of positive Banach density with the assumption of SIM, it was pointed out that the SIM property has some unusual features that should not lead one to view it simply as a notion of largeness. In particular, it was shown that a SIM set A has the property that all of its supersets are also SIM precisely when A is syndetic.

Thus, it is natural to consider the class of *supra-SIM* sets, which we define to be the class of sets which contain a SIM set. In this article, we show that the class of supra-SIM sets has better combinatorial features than the class of SIM sets itself. In particular, we show that this class is partition regular and is closed under *finite-embeddability*, neither of which are true for the class of SIM sets. We achieve these results by proving a simple nonstandard characterization of being supra-SIM.

In the final section, we continue the theme of proving analogues of results for positive Banach density with (supra-)SIM assumptions by considering results on sumsets. Indeed, we prove the SIM analogue of Jin's sumset theorem [5] as well as Nathanson's result from [8], which yielded partial progress on *Erdős' $B + C$ conjecture* (which was recently solved in [7]).

We assume that the reader is familiar with basic nonstandard analysis as it pertains to combinatorial number theory. Alternatively, one can consult the recent manuscript [2], which also contains a chapter on SIM sets. Nevertheless, we will recall the relevant definitions and facts about SIM sets in the next section.

We thank Mauro Di Nasso for useful conversations regarding this work.

2 Preliminaries

Let $I := [y, z]$ be an infinite, hyperfinite interval. Set $st_I := st_{[y,z]} : I \to [0, 1]$ to be the map $st_I(a) := st(\frac{a-y}{z-y})$. For $A \subseteq {}^*\mathbb{N}$ internal, we set $st_I(A) := st_I(A \cap I)$. We recall that $st_I(A)$ is a closed subset of $[0, 1]$ and we may thus consider $\lambda_I(A) := \lambda(st_I(A))$, where λ is Lebesgue measure on $[0, 1]$.

We also consider the quantity $g_A(I) := \frac{d-c}{|I|}$, where $[c, d] \subseteq I$ is maximal so that $[c, d] \cap A = \emptyset$.

The main idea in what is to follow is the desire to compare the notions of making $g_A(I)$ small (an internal notion) and making $\lambda_I(A)$ large (an external notion). There is always a connection in one direction, namely that if $\lambda_I(A) > 1 - \epsilon$, then $g_A(I) < \epsilon$. We now consider sets where there is also a relationship in the other direction.

[1] SIM stands for the *standard interval measure* property.

Definition 1 We say that A has the *interval-measure property* (or *IM property*) on I if for every $\epsilon > 0$, there is $\delta > 0$ such that, for all infinite $J \subseteq I$ with $g_A(J) \leq \delta$, we have $\lambda_J(A) \geq 1 - \epsilon$.

If A has the IM property on I, we let $\delta(A, I, \epsilon)$ denote the supremum of the δ's that witness the conclusion of the definition for the given ϵ.

It is clear from the definition that if A has the IM property on an interval, then it has the IM property on every infinite subinterval. Also note that it is possible that A has the IM property on I for a trivial reason, namely that there is $\delta > 0$ such that $g_A(J) > \delta$ for every infinite $J \subseteq I$. Let us temporarily say that A has the *nontrivial IM property* on I if this does *not* happen, that is, for every $\delta > 0$, there is an infinite interval $J \subseteq I$ such that $g_A(J) \leq \delta$. It will be useful to reformulate this in different terms. In order to do that, we recall an important standard tool that is often employed in the study of sets with the IM property, namely the *Lebesgue density theorem*. Recall that for a measurable set $E \subseteq [0, 1]$, a point $r \in E$ is a *(one-sided) point of density of* E if

$$\lim_{s \to r^+} \frac{\mu(E \cap [r, s])}{s - r} = 1.$$

The Lebesgue density theorem asserts that almost every point of E is a density point of E.

Fact 2.1 *Suppose that $A \subseteq {}^*\mathbb{N}$ is internal and I is an infinite, hyperfinite interval such that A has the IM property on I. Then the following are equivalent:*

1. *There is an infinite subinterval J of I such that A has the nontrivial IM property on J.*
2. *There is an infinite subinterval J of I such that $\lambda_J(A) > 0$.*

In practice, the latter property in the previous proposition is easier to work with. Consequently, let us say that A has the *enhanced IM property on I* if it has the IM property on I and $\lambda_I(A) > 0$.

In the proof of our main partition regularity result, the following *internal* partition regularity theorem will be essential.

Theorem 1 *Suppose that A has the enhanced IM property on I. Further suppose that $A \cap I = B_1 \cup \cdots \cup B_n$ with each B_i internal. Then there is i and infinite $J \subseteq I$ such that B_i has the enhanced IM property on J.*

Proof We prove the theorem by induction on n. The result is clear for $n = 1$. Now suppose that the result is true for $n - 1$ and suppose $A \cap I = B_1 \cup \cdots \cup B_n$ with each B_i internal. If there is an i and infinite $J \subseteq I$ such that $B_i \cap J = \emptyset$ and $\lambda_J(A) > 0$, then we are done by induction. We may thus assume that whenever $\lambda_J(A) > 0$, then each $B_i \cap J \neq \emptyset$. We claim that this implies that each of the B_i have the IM property on I. Since there must be an i such that $\lambda_I(B_i) > 0$, for such an i it follows that B_i has the enhanced IM property on I.

Fix i and set $B := B_i$. Suppose that $J \subseteq I$ is infinite, $\epsilon > 0$, and $g_B(J) \leq \delta(A, I, \epsilon)$; we show that $\lambda_J(B) \geq 1 - \epsilon$. Since $g_A(J) \leq g_B(J) \leq \delta(A, I, \epsilon)$, we have that $\lambda_J(A) \geq 1 - \epsilon$. Suppose that $[r, s] \subseteq [0, 1] \setminus st_J(B)$. Then $r = st_J(x)$ and $s = st_J(y)$ with $\frac{y-x}{|J|} \approx s - r$ and $B \cap [x, y] = \emptyset$. By our standing assumption, this implies that $\lambda_{[x,y]}(A) = 0$, whence it follows that $\lambda_J(A \cap [x, y]) = 0$. It follows that $\lambda_J(B) = \lambda_J(A) \geq 1 - \epsilon$, as desired. □

We will need two other facts about SIM sets, both of which are implicit in [6] but are spelled out in more detail in [2].

Fact 2.2 *If A is an internal set that has the IM property on I, then there is $w \in \mathbb{N}$ and a descending hyperfinite sequence $I = I_0, I_1, \ldots, I_K$ of hyperfinite subintervals of I such that:*

- $|I_K| \leq w$;
- $\frac{|I_{k+1}|}{|I_k|} \geq \frac{1}{w}$ *for all* $k < K$;
- *whenever I_k is infinite, we have* $\lambda_{I_k}(A) > 0$.

Fact 2.3 *Suppose that $A_1, \ldots, A_n$ are internal sets that satisfy the IM property on $I_1, \ldots, I_n$ respectively. Fix $\epsilon > 0$ such that $\epsilon < \frac{1}{n}$. Take $\delta > 0$ with $\delta < \min_{i=1,\ldots,n} \delta(A_i, I_i, \epsilon)$. Then there is $w \in \mathbb{N}$ such that, whenever $[a_i, a_i + b]$ satisfies*

$$[a_i, a_i + b] \subseteq I_i \text{ and } g_{A_i}([a_i, a_i + b]) \leq \delta \text{ for all } i = 1, \ldots, n,$$

then there is $c \in {}^*\mathbb{N}$ such that

$$A_i \cap [a_i + c, a_i + c + w] \neq \emptyset \text{ for all } i = 1, \ldots, n.$$

Finally, we recall the definition of SIM sets.

Definition 2 $A \subseteq \mathbb{N}$ has the *standard interval-measure property* (or *SIM property*) if:

- ${}^*A$ has the IM property on every infinite hyperfinite interval;
- ${}^*A$ has the enhanced IM property on some infinite hyperfinite interval.

It is possible to give a reformulation of SIM sets in completely standard terms; see [6] for the details.

3 Supra-SIM Sets and Their Properties

We begin by noting that the collection of SIM sets is not closed under the operation of taking supersets.

Example 3.1 Suppose that B has the SIM property but is not syndetic. Then as shown in [6], there is $A \supseteq B$ such that A is not SIM.

This implies that not all piecewise syndetic sets are SIM sets. As we will see below, the property of being a SIM set is also not partition regular. It is thus more interesting to consider the notion of a "supra-SIM" set, defined below.

Definition 3.2 $A \subseteq \mathbb{N}$ is *supra-SIM* if there is $B \subseteq A$ such that B has the SIM property.

Example 3.3 ([6]) Piecewise syndetic sets are supra-SIM.

In [6], SIM sets of Banach density 0 are constructed. This implies that there are supra-SIM sets that do not have positive Banach density, and thus also are not piecewise syndetic.

In order to prove our main results on supra-SIM sets, we use a convenient nonstandard reformulation. The next theorem is the core of the matter.

Theorem 3.4 *Suppose that $A \subseteq \mathbb{N}$ is such that $^*A$ has the enhanced IM property on some interval I. Then A is a supra-SIM set.*

Proof Without loss of generality, $I \subseteq {}^*\mathbb{N} \setminus \mathbb{N}$. For each $\epsilon > 0$, fix $\delta(\epsilon) < \min(\delta(A, I, \epsilon), \epsilon, \frac{1}{4})$.

For ease of notation, we set $\delta_k := \delta(\frac{1}{k})$. By underflow, for each $n, k \in \mathbb{N}$, there exists $M_{n,k} \in \mathbb{N}$ such that whenever a subinterval J of I satisfies $g_{^*A}(J) < \delta_k$ and $l(J) > M_{n,k}$, then it takes the sum of the lengths of at least n gaps of $^*A$ on J to add to $\frac{l(J)}{k}$. Since $\lambda_I(^*A) > 0$, for each n there exists an infinite subinterval J of I such that $g_{^*A}(J) < \frac{1}{n}$.

By transfer, we may inductively define a sequence of pairwise disjoint intervals (I_n) in $\mathbb{N}$ satisfying the following properties:

(i) Writing $I_n = [a_n, b_n]$, we have $a_n > nb_{n-1}$.
(ii) I_n has a subinterval of length at least n with $g_A(J) < \frac{1}{n}$.
(iii) For all $k \leq n$ and for all $J \subseteq I_n$, if $l(J) > M_{n,k}$ and $g_A(J) < \delta_k$, then at least n gaps of A on J are required to cover at least $\frac{l(J)}{k}$.

Set $B := \bigcup_n (A \cap I_n)$. We claim that B has the SIM property.

Let I' be an infinite hyperfinite interval. We show that $^*B$ has the IM property on I' as witnessed by the function $\delta'(\epsilon) := \frac{1}{2}\delta_k$, where $\frac{1}{k} < \epsilon$. Fix $\epsilon > 0$ and consider an infinite subinterval J of I' such that $g_{^*B}(J) \leq \frac{1}{2}\delta_k$.

By condition (i), If J intersects more than one of the I_K, with the largest such index being M, then every point in any $J \cap I_K$ with $K < M$ is less than $\frac{1}{M}a_M$, and so is infinitesimal compared to the length of J (which is at least $a_M - b_{M-1}$). Thus, all these points are mapped to 0 by the st_J mapping. Next note that a_M must be within the first δ_k portion of J, else $g_{^*B}(J) \geq \delta_k$. If the right endpoint of J is at most b_M, we then have that $l(J \cap I_M) \geq (1 - \delta_k)l(J)$. If J ends after I_M, then again we see that b_M must occur in the last δ_k portion of J, so $l(J \cap I_M) \geq (1 - 2\delta_k)l(J)$. In either case, we have $l(J \cap I_M) \geq (1 - 2\delta_k)l(J)$.

It follows that

$$g_{^*B}(J \cap I_M) \leq g_{^*B}(J) \cdot \frac{l(J)}{l(J \cap I_M)} \leq \frac{\delta_k}{2(1 - 2\delta_k)} \leq \delta_k.$$

Since $g_{^*B}(J \cap I_M) = g_{^*A}(J \cap I_M)$ and it requires M gaps of $^*A$ to add to $\frac{l(J)}{k}$, we see that $\lambda_J(^*B) \geq 1 - \frac{1}{k} > (1 - \epsilon)$, as desired.

It remains to show that ${}^*B$ has the enhanced IM property on some interval. To see that, observe that if N is of non-finite length, then I_N has a subinterval J of size at least N with $g_{{}^*A}(J) \leq \delta_N \approx 0$; since $g_{{}^*B}(J) = g_{{}^*A}(J)$, we see that ${}^*B$ has the enhanced IM property on J. □

Here is our promised nonstandard reformulation of supra-SIM sets.

Corollary 3.5 *A is supra-SIM if and only if there is $B \subseteq A$ and infinite hyperfinite I such that ${}^*B$ has the enhanced IM property on I.*

Proof If A is supra-SIM, then there is $B \subseteq A$ that is SIM. By definition of SIM, this B is as desired. Conversely, if B and I are as in the condition, then B is supra-SIM by the theorem, whence so is A. □

The partition regularity of supra-SIM now follows easily:

Corollary 3.6 *The notion of being a supra-SIM set is partition regular.*

Proof Suppose that A is supra-SIM and $A = C \sqcup D$. Take $B \subseteq A$ SIM. Take infinite I such that ${}^*B$ has the enhanced IM property on I. Then by Theorem 1, we have, without loss of generality, that ${}^*(B \cap C)$ has the enhanced IM property on some infinite subinterval of I. It follows from the previous corollary that C is supra-SIM. □

From this and Theorem 5.7 in [4] we obtain the following corollary.

Corollary 3.7 *Every supra-SIM set is contained in an ultrafilter consisting entirely of supra-SIM sets.*

Example 3.8 Being SIM is not partition regular. Indeed, consider

$$A := \{1, 3, 4, 7, 8, 9, 13, 14, 15, 16, \ldots\},$$

where A continues to consist of m elements in the set followed by m elements that are not in the set, with m increasing by 1 each time. Then, if k is large (but finite), on any infinite hyperfinite interval I that consists of k disjoint intervals that are in ${}^*A$ and k disjoint intervals that are not in ${}^*A$, we have that $g_{{}^*A}(I)$ and $g_{{}^*(\mathbb{N}\setminus A)}(I)$ are both roughly equal to $1/(2k)$, while $\lambda_I({}^*A)$ and $\lambda_I({}^*(\mathbb{N} \setminus A))$ are both $1/2$.

The argument in the proof of Theorem 3.4 is robust enough to allow us to adapt it to prove another desirable property of supra-SIM sets that is also possessed by sets of positive Banach density. Recall that A is said to be *finitely-embedded* in B if, given any finite $F \subseteq A$, there is $t \in \mathbb{N}$ such that $t + F \subseteq B$. (Equivalently, there is $t \in {}^*\mathbb{N}$ such that $t + A \subseteq {}^*B$.) Note that if A is finitely embedded in B and $(A) > 0$, then $(B) > 0$.

Theorem 3.9 *Suppose that A is finitely embedded in B and A is supra-SIM. Then B is supra-SIM.*

Proof Without loss of generality, we may assume that A is actually SIM. For $n \in \mathbb{N}$, let X_n be the set of intervals I in $^*\mathbb{N}$ of length at least n such that $g_{^*A}(I) \leq \frac{1}{n}$ and $t + (^*A \cap I) \subseteq {^*B}$ for some $t \in {^*\mathbb{N}}$. Since A is SIM and finitely-embeddable in B, each $X_n \neq \emptyset$. Thus, by overflow, there is $I \in \bigcap_n X_n$.

As in the proof of Theorem 3.4, we may use transfer to inductively define a sequence of pairwise disjoint intervals (I_n) in $\mathbb{N}$ and a sequence (t_n) from $\mathbb{N}$ satisfying the following properties:

(i) Writing $t_n + I_n = [a_n, b_n]$, we have $a_n > nb_{n-1}$.
(ii) I_n has a subinterval of length at least n with $g_A(J) < \frac{1}{n}$.
(iii) For all $k \leq n$ and for all $J \subseteq I_n$, if $|J| > M_{n,k}$ and $g_A(J) < \delta_k$, then at least n gaps of A on J are required to cover at least $\frac{l(J)}{k}$.
(iv) $t_n + (A \cap I_n) \subseteq B$.

Let $C := \bigcup_n (t_n + (A \cap I_n))$. As in the proof of Theorem 3.4, C has the SIM property. By (iv), $C \subseteq B$, so B is supra-SIM, as desired. □

Of course the previous proposition fails for SIM sets. As mentioned in the introduction, they are almost never even closed under taking supersets.

We end this section by mentioning arguably the most pressing open question concerning supra-SIM sets.

Question 1 Are sets of positive Banach density supra-SIM?

Our results from this section yield a *prima facie* simpler criterion for obtaining a positive solution to the previous question. First recall that, for $A \subseteq \mathbb{N}$, the *Shnirelmann density* of A is $\sigma(A) := \inf_{n \geq 1} \frac{|A \cap [1,n]|}{n}$.

Corollary 3.10 *Suppose there is $\epsilon > 0$ such that every set $A \subseteq \mathbb{N}$ with $\sigma(A) \geq 1 - \epsilon$ is supra-SIM. Then every set of positive Banach density is supra-SIM.*

Proof Suppose that ϵ is as in the hypothesis of the corollary and suppose that $(A) > 0$. Take a finite $F \subseteq \mathbb{N}$ such that $(A + F) \geq 1 - \epsilon$. Take $B \subseteq \mathbb{N}$ such that B is finitely embedded in A and $\sigma(B) \geq (A + F)$ (see, for example, [2, Corollary 12.12]). By assumption, B is supra-SIM. By Theorem 3.9, $A + F$ is supra-SIM. By Corollary 3.6, $A + i$ is supra-SIM for some $i \in F$. It remains to observe that being supra-SIM is translation invariant.

4 SIMsets and Sumsets

4.1 The Sumset Phenomenon

One of the first successes of nonstandard methods in combinatorial number theory was the following theorem of Renling Jin.

Fact 4.1 *Suppose that $A, B \subseteq \mathbb{N}$ are such that $(A), (B) > 0$. Then $A + B$ is piecewise syndetic.*

In this subsection, we prove the analogous result, replacing the positive Banach density assumption with a SIM assumption.

Proposition 4.2 *If A and B have the SIM property, then $A + B$ is piecewise syndetic.*

Proof By Fact 2.2 and the Lebesgue density theorem, we can obtain intervals I and J of the same infinite length such that $\lambda_I({}^*A) = \lambda_J({}^*B) = 1$. Let u be the left endpoint of I and v be the right endpoint of J. We apply Fact 2.3 with $A_1 := {}^*A - u$, $A_2 := v - {}^*B$, $I_1 := I - u$, and $I_2 := J - v$ to obtain a finite w as in the conclusion of that result. Now for any finite m, $g_{A_1}(I_1 + m) \approx 0$ and $g_{A_2}(I_2) \approx 0$. Thus, by the choice of w, there must exist $c \in {}^*\mathbb{N}$ such that

$$A_1 \cap [m + c, m + c + w] \neq \emptyset \text{ and } A_2 \cap [c, c + w] \neq \emptyset.$$

If we fix $x \in A_1 \cap [m + c, m + c + w]$ and $y \in A_2 \cap [c, c + w]$, then

$$x - y \in (A_1 - A_2) \cap [m - w, m + w] = \left(({}^*A - u) - (v - {}^*B)\right) \cap [m - w, m + w].$$

This shows that there is an element of ${}^*A + {}^*B$ in every interval of the form $[u + v + m - w, u + v + m + w]$. By overspill, there is an infinite interval starting at $u + v$ in which there is no gap of ${}^*A + {}^*B$ greater than $2w$, completing the proof. □

4.2 *Towards $B + C$ for SIMsets*

In [3], Erdős made the following conjecture.

Conjecture 4.3 Suppose that $A \subseteq \mathbb{N}$ is such that $\underline{d}(A) > 0$. Then there are infinite sets $B, C \subseteq \mathbb{N}$ such that $B + C \subseteq A$.

The first progress on this conjecture was due to Nathanson [8]:

Fact 4.4 *Suppose that $(A) > 0$. Then for any $n \in \mathbb{N}$, there are $B, C \subseteq \mathbb{N}$ such that B is infinite, $|C| = n$, and $B + C \subseteq A$.*

Nathanson's result follows immediately from repeated applications of the following fact, which he attributes to Kazhdan in [8].

Fact 4.5 *Suppose that $(A) > 0$. Then there are arbitrarily large $t \in \mathbb{N}$ such that $(A \cap (A - t)) > 0$.*

We remark in passing that the proof of Kazhdan's lemma appearing in [8] is quite complicated but that it is possible to give a very simple nonstandard proof as in [2]. In this subsection, we prove the supra-SIM version of Nathanson's result.

First, we should mention that, building somewhat upon ideas from [1], Moreira, Richter, and Robertson positively settle the Erdős conjecture in [7], even weakening the hypothesis to positive Banach density and also proving a version for countable amenable groups.

Here is the Kazhdan lemma for supra-SIM sets:

Proposition 4.6 (Kazhdan Lemma for supra-SIM sets) *Suppose that $A \subseteq \mathbb{N}$ is supra-SIM and set $\mathcal{T}_A := \{t \in \mathbb{N} \ : \ A \cap (A - t)$ is supra-SIM$\}$. Then $\mathcal{T}_A$ is syndetic.*

Proof Suppose that ${}^*A$ has the enhanced IM property on the interval I. Let $w \in \mathbb{N}$ be as in Fact 2.3 for $A_1 := A_2 := {}^*A$ and $I_1 := I_2 := I$, for some appropriately small ε and corresponding δ. We show that $\mathcal{T}_A$ has no gaps of length larger than w. Towards this end, fix $t \in \mathbb{N}$ and set

$$B_t := \bigcup_{k=0}^{w} \left({}^*A - (t+k)\right).$$

Claim: If J is any subinterval of I on which $\lambda_J({}^*A) > 0$,then

$${}^*A \cap B_t \cap J \neq \emptyset.$$

Proof of the Claim: By the Lebesgue density theorem, we may choose $[a_1, b] \subset J$ with sufficiently small gap that we may apply Fact 2.3 with $a_2 := a_1 + t$. This allows us to find a c with $c + w \leq b$ such that ${}^*A \cap [a_1 + c, a_1 + c + w] \neq \emptyset$
${}^*A \cap [a_1 + t + c, a_1 + t + c + w] \neq \emptyset$. This is equivalent to: ${}^*A \cap [a_1 + c, a_1 + c + w] \neq \emptyset$ $({}^*A - t) \cap [a_1 + c, a_1 + c + w] \neq \emptyset$.

Let d be an element in ${}^*A \cap [a_1 + c, a_1 + c + w]$. That same d must then be in B_t since it is within w of an element in $({}^*A - t)$, and this completes the proof of the claim. □

The claim implies that, for any infinite subinterval J of I, we have that

$$\lambda_J({}^*A \cap B_t) = \lambda_J({}^*A),$$

as J cannot contain any infinite intervals in the complement of ${}^*A \cap B_t$ that have positive ${}^*A$ measure. It follows immediately that ${}^*A \cap B_t$ has the enhanced IM property on I. By Theorem 1, it follows that for some $k = 0, \ldots, w$, we have that

$${}^*A \cap ({}^*A - (t+k))$$

has the enhanced IM property on some infinite subinterval of I. For this k, it follows that $A \cap (A - (t+k))$ is supra-SIM. □

As in the case of the original Nathanson result, repeated application of the previous proposition implies.

Corollary 4.7 (Nathanson's theorem for supra-SIM sets) *Suppose that A is supra-SIM. Then for any $n \in \mathbb{N}$, there is an infinite $B \subseteq A$ and $C \subseteq \mathbb{N}$ with $|C| = n$ such that $B + C \subseteq A$.*

Of course, we should ask.

Question 2 Suppose that A is supra-SIM. Do there exist infinite $B, C \subseteq \mathbb{N}$ such that $B + C \subseteq A$?

Acknowledgements Goldbring's work was partially supported by NSF CAREER grant DMS-1349399.

References

1. M. Di Nasso, I. Goldbring, R. Jin, S. Leth, M. Lupini, and K. Mahlburg, *On a sumset conjecture of Erdős*, Canadian Journal of Mathematics **67** (2015), 795–809.
2. M. Di Nasso, I. Goldbring, and M. Lupini, *Nonstandard methods in Ramsey theory and combinatorial number theory*, volume 2239 of Lecture Notes in Mathematics, Springer International (2019)
3. P. Erdős and R. L. Graham, *Old and new problems and results in combinatorial number theory*, volume 28 of Monographies de L'Enseignement Mathématique. Universit de Genève, L'Enseignement Mathématique, Geneva, 1980.
4. N. Hindman and D. Strauss, *Algebra in the Stone-Čech Compactification*, 2nd edn., De Gruyter, Berlin/Boston (2010)
5. R. Jin, *The sumset phenomenon*, Proceedings of the American Mathematical Society **130**(2002), 855–861.
6. S. Leth, *Some nonstandard methods in combinatorial number theory*, Polish Academy of Sciences. Institute of Philosophy and Sociology. Studia Logica. An International Journal for Symbolic Logic, **47**(1988), 265-278.
7. J. Moreira, F. Richter, and D. Robertson, *A proof of the Erdős sumset conjecture*, Ann. of Math. 189 no. 2 (2019), 605–652.
8. M. B. Nathanson, *Sumsets contained in infinite sets of integers*, Journal of Combinatorial Theory, Series A, **28** (1980), 150–155.

Mean Row Values in (u, v)-Calkin–Wilf Trees

Sandie Han, Ariane M. Masuda, Satyanand Singh and Johann Thiel

Abstract We fix integers $u, v \geq 1$, and consider an infinite binary tree $\mathscr{T}^{(u,v)}(z)$ with a root node whose value is a positive rational number z. For every vertex a/b, we label the left child as $a/(ua + b)$ and right child as $(a + vb)/b$. The resulting tree is known as the (u, v)-Calkin–Wilf tree. As z runs over $[1/u, v] \cap \mathbb{Q}$, the vertex sets of $\mathscr{T}^{(u,v)}(z)$ form a partition of $\mathbb{Q}^+$. When $u = v = 1$, the mean row value converges to $3/2$ as the row depth increases. Our goal is to extend this result for any $u, v \geq 1$. We show that, when $z \in [1/u, v] \cap \mathbb{Q}$, the mean row value in $\mathscr{T}^{(u,v)}(z)$ converges to a value close to $v + \log 2/u$ uniformly on z.

1 Introduction

In [8], Nathanson defines an infinite binary tree generated by the following rules:

1. fix two positive integers u and v,
2. label the root of the tree by a rational z, and
3. for any vertex labeled $\dfrac{a}{b}$, label its left and right children by $\dfrac{a}{ua+b}$ and $\dfrac{a+vb}{b}$, respectively.

In the case where u, v, and z are equal to 1, the tree generated is the well-known Calkin–Wilf tree [3] (see Fig. 1). Since Nathanson's definition represents a gener-

S. Han · A. M. Masuda · S. Singh (✉) · J. Thiel
Department of Mathematics, New York City college of Technology, CUNY, 300 Jay Street, Brooklyn 11201, USA
e-mail: ssingh@citytech.cuny.edu

S. Han
e-mail: shan@citytech.cuny.edu

A. M. Masuda
e-mail: amasuda@citytech.cuny.edu

J. Thiel
e-mail: jthiel@citytech.cuny.edu

M. B. Nathanson (ed.), *Combinatorial and Additive Number Theory III*, Springer Proceedings in Mathematics & Statistics 297,
https://doi.org/10.1007/978-3-030-31106-3_10

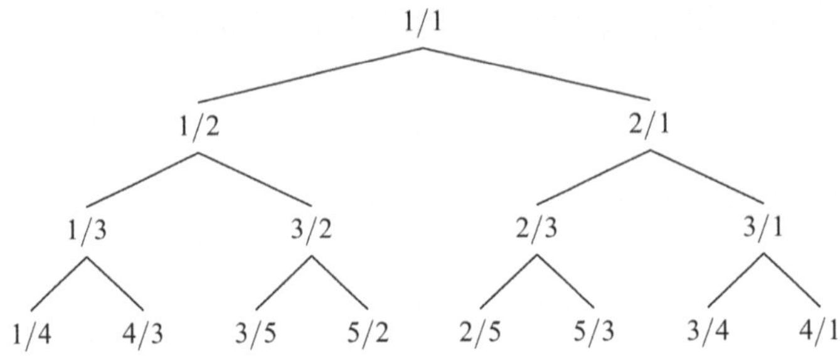

Fig. 1 The first four rows of the Calkin–Wilf tree

Fig. 2 The first three rows of $\mathscr{T}^{(u,v)}(z)$

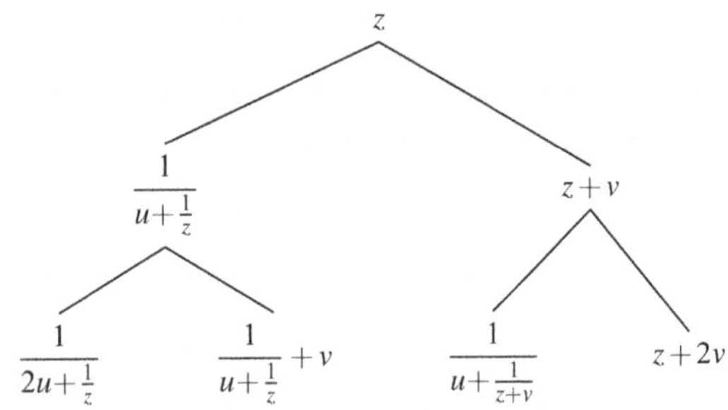

alization[1] of the Calkin–Wilf tree, we refer to trees defined in the above manner as (u, v)-Calkin–Wilf trees, and we denote them by $\mathscr{T}^{(u,v)}(z)$ (see Fig. 2). The set of depth n vertices of $\mathscr{T}^{(u,v)}(z)$ is denoted by $\mathscr{T}^{(u,v)}(z; n)$. For example, we see from Fig. 1 that $\mathscr{T}^{(1,1)}(1; 1) = \{1/2, 2\}$.

The vertices of $\mathscr{T}^{(1,1)}(1)$ are all positive rational numbers without any repetition [3]. More generally, the trees $\mathscr{T}^{(u,v)}(z)$ form a partition of $\mathbb{Q}^+$ as z runs over $[1/u, v] \cap \mathbb{Q}$; see [8]. The Calkin–Wilf tree has many other interesting properties [3, 5, 6, 8, 9], one of which is the fact that the mean value of vertices of depth n converges to $3/2$ as $n \to \infty$ [1, 10]. Our main result generalizes this property for all (u, v)-Calkin–Wilf trees.

The proof that the mean value of vertices of depth n converges to $3/2$ is not difficult and only makes use of one property of the Calkin–Wilf tree; namely, both a/b and b/a appear (in symmetric positions) on every row; see Fig. 1.

Proposition 1 *If* $\frac{a}{b} \in \mathscr{T}^{(1,1)}(1; n)$, *then* $\frac{b}{a} \in \mathscr{T}^{(1,1)}(1; n)$.

The proof of Proposition 1 follows quickly from induction on the depth n. We omit the details.

Theorem 1 *For* $n \geq 0$, *let* $A(n) = \frac{1}{2^n} \sum_{y \in \mathscr{T}^{(1,1)}(1;n)} y$. *Then* $\lim_{n\to\infty} A(n) = \frac{3}{2}$.

[1]For other generalizations, see [2, 7].

Proof Let $S(n) = \sum_{y \in \mathcal{T}^{(1,1)}(1;n)} y$. Rewriting y as a/b and using both the definition of the Calkin–Wilf tree and Proposition 1, we see that, for $n \geq 1$,

$$\begin{aligned} 2S(n) &= \sum_{\frac{a}{b} \in \mathcal{T}^{(1,1)}(1;n-1)} \left(\frac{a}{a+b} + \frac{a}{b} + 1 + \frac{b}{b+a} + \frac{b}{a} + 1 \right) \\ &= \sum_{\frac{a}{b} \in \mathcal{T}^{(1,1)}(1;n-1)} \left(\frac{a}{b} + \frac{b}{a} + 3 \right) \\ &= 2S(n-1) + 3 \cdot 2^{n-1}. \end{aligned}$$

This gives the recurrence relation $S(0) = 1$ and $S(n) = S(n-1) + 3 \cdot 2^{n-2}$ for $n \geq 1$. Solving the recurrence relation gives that $S(n) = \frac{3}{2} \cdot 2^n - \frac{1}{2}$ for $n \geq 0$. The desired result follows immediately since $A(n) = S(n)/2^n$. □

Let $S^{(u,v)}(z;n) = \sum_{y \in \mathcal{T}^{(u,v)}(z;n)} y$ and $A^{(u,v)}(z;n) = S^{(u,v)}(z;n)/2^n$. Suppose $uv > 1$. As a consequence of Lemma 5 and Theorem 2, we show that if $z \in [1/u, v] \cap \mathbb{Q}$, then $\lim_{n\to\infty} A^{(u,v)}(z;n)$ exists,[2] that the limit is independent of the value of z, and that the limit has a value close to $v + \log 2/u$. Unfortunately, Proposition 1 does not generalize to other (u, v)-Calkin–Wilf trees by Lemma 3, so a different approach is needed in this broader setting.

At first the value $v + \log 2/u$ may seem surprising, but a simple heuristic argument quickly leads to this quantity. Note that if a/b is a vertex in a (u, v)-Calkin–Wilf tree, then its children are given by

$$\frac{a}{ua+b} = \frac{1}{u + \frac{b}{a}} < \frac{1}{u} \quad \text{and} \quad \frac{a+vb}{b} = \frac{a}{b} + v > v.$$

Following this pattern from depth n to depth $n+1$ suggests that a quarter of all elements of a fixed (large) depth have integer part of roughly size v, an eighth have integer part of roughly size $2v$, etc. Similarly, half of all elements have a fractional part of roughly size $1/u$, a quarter have a fractional part of roughly size $1/(2u)$, etc. So we expect that

$$\begin{aligned} A^{(u,v)}(z;n) &\approx \frac{1}{2^n}\left(\frac{2^n}{4}\left(v + \frac{2}{u}\right) + \frac{2^n}{8}\left(2v + \frac{2}{2u}\right) + \frac{2^n}{16}\left(3v + \frac{2}{3u}\right) + \cdots \right) \\ &= \frac{v}{4}\sum_{k=0}^{\infty} \frac{k+1}{2^k} + \frac{1}{u}\sum_{k=1}^{\infty} \frac{1}{k2^k} \\ &= v + \frac{\log 2}{u}, \end{aligned}$$

[2]The reason for limiting our choice of roots to $[1/u, v] \cap \mathbb{Q}$ is that these rationals are the "orphan" roots in the sense that they are not the children of *any* rational in *any* (u, v)-Calkin–Wilf tree [8].

where the last equality follows from the Taylor series expansions for $1/(1-x)^2$ and $\log(1-x)$.

This heuristic throws away a lot of information from the denominator in the fractional part of each element. We would therefore expect the true value of $A^{(u,v)}(z;n)$ to be smaller than $v + \log 2/u$.

As for the independence of the limit of $A^{(u,v)}(z;n)$ from $z \in [1/u, v] \cap \mathbb{Q}$, we note that if a/b is a vertex in a (u, v)-Calkin–Wilf tree with continued fraction representation $a/b = [q_0, q_1, \ldots, q_r]$, then the children of a/b have easily computable continued fractions, as the next result shows.

Lemma 1 ([5, Lemma 5]) *Let a/b be a positive rational number with continued fraction representation $a/b = [q_0, q_1, \ldots, q_r]$. It follows that*

(a) if $q_0 = 0$, then $a/(ua+b) = [0, u+q_1, \ldots, q_r]$;
(b) if $q_0 \neq 0$, then $a/(ua+b) = [0, u, q_0, q_1, \ldots, q_r]$;
(c) and $(a+vb)/b = [v+q_0, q_1, \ldots, q_r]$.

It follows from the result above that, for large n, most vertices of depth n will have approximately $n/2$ coefficients in their continued fraction expansions. This lowers the influence of the root on the value of $A^{(u,v)}(z;n)$ as it is quickly buried by the above process. We will make this notion precise in Lemma 7.

2 Main Result

We show that for $z \in [1/u, v] \cap \mathbb{Q}$, the limit of $A^{(u,v)}(z;n)$ exists as $n \to \infty$ in two main steps:

(A) First we show that, for $z = 1/u$ or $z = v$, the mean $A^{(u,v)}(z;n)$ is monotonic increasing and bounded above as $n \to \infty$.
(B) Second we show that $A^{(u,v)}(z_1;n) - A^{(u,v)}(z_2;n) \to 0$ as $n \to \infty$ for any $z_1, z_2 \in [1/u, v] \cap \mathbb{Q}$.

We begin with a useful lemma for comparing rational numbers based on their continued fraction coefficients.

Lemma 2 ([11, p. 101]) *Suppose that $\alpha, \beta \in \mathbb{Q}$ are distinct with $\alpha = [p_0, p_1, \ldots, p_s]$ and $\beta = [q_0, q_1, \ldots, q_r]$. Let k be the smallest index such that $p_k \neq q_k$. Then $\alpha < \beta$ if and only if $p_k < q_k$ when k is even and $p_k > q_k$ when k is odd. If no such k exists and $n < m$, then $\alpha < \beta$ if and only if n is even.*

We note here two useful results from [5] that will be used to obtain our main result. Lemma 3 and Corollary 4 show two things: that there is a very close relationship between two vertices in the same (u, v)-Calkin–Wilf tree via their continued fraction representations if one is the descendant of the other, and that the continued fraction representation of a vertex in a (u, v)-Calkin–Wilf tree encodes its depth in the tree.

Lemma 3 ([5, Theorem 3]) *Suppose that z and z' are positive rational numbers with continued fraction representations $z = [q_0, q_1, \ldots, q_r]$ and $z' = [p_0, p_1, \ldots, p_s]$. Then z' is a descendant of z in the (u, v)-Calkin–Wilf tree with root z if and only if the following conditions all hold:*

(a) $s \geq r$ *and* $2 \mid (s - r)$*;*
(b) *for* $0 \leq j \leq s - r - 1$, $v \mid p_j$ *when* j *is even and* $u \mid p_j$ *when* j *is odd;*
(c) *for* $2 \leq i \leq r$, $p_{s-r+i} = q_i$*;*
(d) *and*

(i) *if* $q_0 \neq 0$, *then* $p_{s-r} \geq q_0$, $v \mid (p_{s-r} - q_0)$ *and* $p_{s-r+1} = q_1$*;*
(ii) *otherwise, if* $q_0 = 0$, *then* $v \mid p_{s-r}$, $p_{s-r+1} \geq q_1$, *and* $u \mid (p_{s-r+1} - q_1)$.

Lemma 4 ([5, Corollary 3]) *Using the same hypothesis as Lemma 3, if n is the depth of z', then*

$$n = \frac{1}{v}\Bigg(\sum_{\substack{0 \leq j \leq s-r-1 \\ j \text{ even}}} p_j + \sum_{\substack{0 \leq i \leq r \\ i \text{ even}}} (p_{s-r+i} - q_i)\Bigg) + \frac{1}{u}\Bigg(\sum_{\substack{0 \leq j \leq s-r-1 \\ j \text{ odd}}} p_j + \sum_{\substack{0 \leq i \leq r \\ i \text{ odd}}} (p_{s-r+i} - q_i)\Bigg).$$

The following lemma gives us the desired monotonicity for $A^{(u,v)}(z; n)$ when $z = 1/u$ or $z = v$.

Lemma 5 *For any $n \geq 0$, if $z = 1/u$ or $z = v$, then $S^{(u,v)}(z; n + 1) > 2S^{(u,v)}(z; n)$.*

Proof Let $n \geq 0$ be given. Enumerate the elements in $\mathscr{T}^{(u,v)}(z; n)$ and $\mathscr{T}^{(u,v)}(z; n + 1)$ as they appear from left to right in the (u, v)-Calkin–Wilf tree by $s_0, s_1, \ldots, s_{2^n-1}$ and $t_0, t_1, \ldots, t_{2^{n+1}-1}$, respectively. Clearly, for $0 \leq i \leq 2^n - 1$, t_{2i} and t_{2i+1} are the left and right children of s_i. Our goal is therefore to show that

$$2\sum_{i=0}^{2^n-1} s_i < \sum_{i=0}^{2^{n+1}-1} t_i.$$

This desired inequality can be reduced further by noting that $t_{2i+1} = s_i + v$. In other words, we obtain the desired result if we can show that

$$\sum_{i=0}^{2^n-1} s_i < 2^n v + \sum_{i=0}^{2^n-1} t_{2i}.$$

Let $\mathscr{I}_n = \sum_{i=0}^{2^n-1} \lfloor s_i \rfloor$. That is, $\mathscr{I}_n$ is the sum of the integer parts of all of the depth n elements of the (u, v)-Calkin–Wilf tree.

Claim: $\mathscr{I}_n = (2^n - 1)v + [w]$ for $n \geq 0$.

We prove the above claim by induction. Clearly $\mathscr{I}_0 = [w]$. Suppose that the claim holds for some $k \geq 1$. Since the left child of any number appearing in the (u, v)-Calkin–Wilf tree is smaller than $1/u$ and the right child of any element is always the original element plus v, it follows that $\mathscr{I}_{k+1} = \mathscr{I}_k + 2^k v$. By assumption, $\mathscr{I}_k = (2^k - 1)v + [w]$, from which the desired result immediately follows.

Our previous claim shows that we obtain the desired result if we can show that

$$[w] + \sum_{i=0}^{2^n-1} \{s_i\} < v + \sum_{i=0}^{2^n-1} t_{2i}. \tag{1}$$

If we take $w = 1/u$, then $[w] = 0$ and, by Lemma 4, the short continued fraction representation of $\{s_i\}$ must be of the form $[0, \alpha_1 u, \alpha_2 v, \ldots, \alpha_k u]$ with $m := m(s_i) = n + 2 - \sum_{i=1}^{k} \alpha_i > 0$. Since $\{s_{2^n-1}\} = [0, u]$ and $t_0 = [0, (n+2)u]$, we see that, in this case, (1) reduces further to the inequality

$$\sum_{i=0}^{2^n-2} \{s_i\} < \sum_{i=1}^{2^n-1} t_{2i}. \tag{2}$$

If $\alpha_k = 1$, then there is an $1 \leq i^* \leq 2^n - 1$ such that

$$t_{2i^*} = [0, \alpha_1 u, \alpha_2 v, \ldots, (\alpha_{k-1} + 1)v, mu].$$

If $\alpha_k > 1$, then there is an $1 \leq i^* \leq 2^n - 1$ such that

$$t_{2i^*} = [0, \alpha_1 u, \alpha_2 v, \ldots, (\alpha_k - 1)u, v, mu].$$

In either case, it follows that $\{s_i\} < t_{2i^*}$ by Lemma 2. Note that the above association between $\{\{s_i\}\}_{i=0}^{2^n-2}$ and $\{t_{2i}\}_{i=1}^{2^n-1}$ is bijective, from which (1) follows in this case.

If we take $w = v$, then $[w] = v$ and, by Lemma 4, the short continued fraction representation of $\{s_i\}$ must be of the form $[0, \alpha_1 u, \alpha_2 v, \ldots, \alpha_k v]$ with m defined as in the previous case. Since $\{s_{2^n-1}\} = 0$ and $t_0 = [0, (n+1)u, v]$, we see that, in this case, (1) also reduces to (2). If $m = 1$, then there is an $1 \leq i^* \leq 2^n - 1$ such that

$$t_{2i^*} = [0, \alpha_1 u, \alpha_2 v, \ldots, (\alpha_k + 1)v].$$

If $m > 1$, then there is an $1 \leq i^* \leq 2^n - 1$ such that

$$t_{2i^*} = [0, \alpha_1 u, \alpha_2 v, \ldots, \alpha_k v, (m-1)u, v].$$

As in the previous case, (1) follows, completing the proof of the lemma. □

The following theorem establishes $v + \log 2/u$ as an upper bound of $A^{(u,v)}(z; n)$. Note that by $f(x) = O(g(x))$ we mean that $|f(x)| \leq C|g(x)|$ for some constant C (which may differ depending on context) and all sufficiently large x.

Theorem 2 *If u and v are positive integers with $uv > 1$ and $z \in \mathbb{Q}$, then $A^{(u,v)}(z; n)$ is bounded above for all $n \geq 0$. In particular,*

$$v + \frac{\log 2}{u} - \lim_{n\to\infty} A^{(u,v)}(z; n) = O\left(\frac{1}{u^2 v}\right).$$

Proof For brevity, we let $S(n) := S^{(u,v)}(z; n)$, $A(n) := A^{(u,v)}(z; n)$, and $\mathcal{T}(n) := \mathcal{T}^{(u,v)}(z; n)$.

For $n \geq 1$, every rational number in the set $\mathcal{T}(n)$ is either the left-child or right-child of a rational number in the set $\mathcal{T}(n-1)$. In particular, for every $y \in \mathcal{T}(n-1)$, there is a unique $x \in \mathcal{T}(n)$ that is *the* right-child y. By definition, $x = y + v$. Likewise, there is a unique $z \in \mathcal{T}(n)$ that is *the* left-child y, making $z = \frac{1}{u+\frac{1}{y}}$. It follows that

$$S(n) = S(n-1) + 2^{n-1}v + \sum_{y\in\mathcal{T}(n-1)} \frac{1}{u + \frac{1}{y}}. \tag{3}$$

By dividing both sides of (3) by 2^n, we immediately obtain the equality

$$A(n) = \frac{1}{2}A(n-1) + \frac{v}{2} + \frac{1}{2^n} \sum_{y\in\mathcal{T}(n-1)} \frac{1}{u + \frac{1}{y}}. \tag{4}$$

By induction on (4), we can express $A(n)$ as

$$\begin{aligned} A(n) &= \frac{1}{2^n}A(0) + v\sum_{k=1}^{n} \frac{1}{2^k} + \frac{1}{2^n}\sum_{k=1}^{n} \sum_{y\in\mathcal{T}(n-k)} \frac{1}{u + \frac{1}{y}} \\ &= \frac{z}{2^n} + v\left(1 - \frac{1}{2^n}\right) + \frac{1}{2^n}\sum_{k=1}^{n} \sum_{y\in\mathcal{T}(n-k)} \frac{1}{u + \frac{1}{y}} \end{aligned} \tag{5}$$

Taking the limit as $n \to \infty$ of both sides of (5) shows that, to complete the proof, it is enough to prove that

$$\lim_{n\to\infty} \frac{1}{2^n}\sum_{k=1}^{n} \sum_{y\in\mathcal{T}(n-k)} \frac{1}{u + \frac{1}{y}} = \frac{\log 2}{u} + O\left(\frac{1}{u^2 v}\right). \tag{6}$$

Let $m = \lfloor n/2 \rfloor$. We split the double sum in (6) into two parts,

$$\sum_{k=1}^{n} \sum_{y\in\mathscr{T}(n-k)} \frac{1}{u+\frac{1}{y}} = \sum_{k=1}^{m} \sum_{y\in\mathscr{T}(n-k)} \frac{1}{u+\frac{1}{y}} + \sum_{k=m+1}^{n} \sum_{y\in\mathscr{T}(n-k)} \frac{1}{u+\frac{1}{y}}. \tag{7}$$

For $m < k \leq n$, we apply the following simple upper bound in (7),

$$\sum_{y\in\mathscr{T}(n-k)} \frac{1}{u+\frac{1}{y}} \leq \frac{2^{n-k}}{u}.$$

It follows that

$$\begin{aligned}\sum_{k=m+1}^{n} \sum_{y\in\mathscr{T}(n-k)} \frac{1}{u+\frac{1}{y}} &\leq \sum_{k=m+1}^{n} \frac{2^{n-k}}{u} \\ &= \frac{1}{u} \sum_{i=0}^{n-(m+1)} 2^i \\ &= \frac{2^{n-m}-1}{u}.\end{aligned} \tag{8}$$

Since $m \to \infty$ as $n \to \infty$, if we apply (8) to (7), then, by (6), we have reduced the problem to showing that

$$\lim_{n\to\infty} \frac{1}{2^n} \sum_{k=1}^{m} \sum_{y\in\mathscr{T}(n-k)} \frac{1}{u+\frac{1}{y}} = \frac{\log 2}{u} + O\left(\frac{1}{u^2 v}\right). \tag{9}$$

Using the same reasoning on the sum $\sum_{y\in\mathscr{T}(n-k)} \frac{1}{u+\frac{1}{y}}$ that led to (3), we see that, for $n-k > 2$,

$$\sum_{y\in\mathscr{T}(n-k)} \frac{1}{u+\frac{1}{y}} = \sum_{y\in\mathscr{T}(n-(k+1))} \frac{1}{2u+\frac{1}{y}} + \sum_{y\in\mathscr{T}(n-(k+1))} \frac{1}{u+\frac{1}{v+y}}. \tag{10}$$

We convert the rightmost sum on the right-hand side of (10) into a sum of geometric series,

$$\begin{aligned}\sum_{y\in\mathscr{T}(n-(k+1))} \frac{1}{u+\frac{1}{v+y}} &= \frac{1}{u} \sum_{y\in\mathscr{T}(n-(k+1))} \frac{1}{1+\frac{1}{u(v+y)}} \\ &= \frac{1}{u} \sum_{y\in\mathscr{T}(n-(k+1))} \sum_{j=0}^{\infty} \left(\frac{-1}{u(v+y)}\right)^j.\end{aligned} \tag{11}$$

The justification for (11) follows from the fact that $0 < \frac{1}{u(v+y)} \leq \frac{1}{uv} \leq \frac{1}{2}$ for *any* positive rational y. So

$$\sum_{y\in\mathcal{T}(n-(k+1))} \frac{1}{u+\frac{1}{v+y}} = \frac{1}{u} \sum_{y\in\mathcal{T}(n-(k+1))} \left(1+O\left(\frac{1}{uv}\right)\right)$$
$$= \frac{2^{n-(k+2)}}{u}\left(1+O\left(\frac{1}{uv}\right)\right) \tag{12}$$

Combining (12) with (10), we see that

$$\sum_{y\in\mathcal{T}(n-k)} \frac{1}{u+\frac{1}{y}} = \sum_{y\in\mathcal{T}(n-(k+1))} \frac{1}{2u+\frac{1}{y}} + \frac{2^{n-(k+2)}}{u}\left(1+O\left(\frac{1}{uv}\right)\right).$$

We can now repeat all of the above steps starting from (10) with the sum

$$\sum_{y\in\mathcal{T}(n-(k+1))} \frac{1}{2u+\frac{1}{y}}.$$

Inductively, for any positive integer $j < n-k$, it follows that

$$\sum_{y\in\mathcal{T}(n-k)} \frac{1}{u+\frac{1}{y}} = \sum_{y\in\mathcal{T}(n-(k+j))} \frac{1}{(j+1)u+\frac{1}{y}} + \sum_{i=1}^{j} \frac{2^{n-(k+i+1)}}{iu}\left(1+O\left(\frac{1}{uv}\right)\right) \tag{13}$$

where the constant associated with the big-oh term is uniform for all of the sums.

Let $m' = \lfloor n/4 \rfloor$. Then, from (13), for $1 \leq k \leq m$,

$$\sum_{y\in\mathcal{T}(n-k)} \frac{1}{u+\frac{1}{y}} = \sum_{y\in\mathcal{T}(n-(k+m'))} \frac{1}{(m'+1)u+\frac{1}{y}} + \sum_{i=1}^{m'} \frac{2^{n-(k+i+1)}}{iu}\left(1+O\left(\frac{1}{uv}\right)\right)$$
$$= O\left(\frac{2^{n-(k+m'+1)}}{(m'+1)u}\right) + \sum_{i=1}^{m'} \frac{2^{n-(k+i+1)}}{iu}\left(1+O\left(\frac{1}{uv}\right)\right). \tag{14}$$

(Note that for n sufficiently large, since $k \leq m$, then $k + m' \leq 3n/4$, so $n - (k + m') \geq 1$. In particular, we can apply (13) with $j = m'$.)

Using the Taylor series expansion of $\log(1-x)$ for $|x| < 1$, we see that

$$\sum_{i=1}^{m'} \frac{2^{n-(k+i+1)}}{iu} = \frac{2^{n-(k+1)}}{u} \sum_{i=1}^{m'} \frac{1}{i2^i}$$

$$= \frac{2^{n-(k+1)}}{u}\left(\log 2 - \sum_{i>m'} \frac{1}{i2^i}\right). \tag{15}$$

Combining (14) and (15) with the double sum from (9), it follows that

$$\frac{1}{2^{n-1}} \sum_{k=1}^{m} \sum_{y \in \mathcal{T}(n-k)} \frac{1}{u+\frac{1}{y}}$$
$$= \frac{1}{u} \sum_{k=1}^{m} \frac{1}{2^k}\left(\log 2 - \sum_{i>m'} \frac{1}{i2^i}\right)\left(1 + O\left(\frac{1}{uv}\right)\right) + O\left(\frac{1}{(m'+1)u}\right) \tag{16}$$

The result (9) now follows from taking the limit of (16) as $n \to \infty$. □

Lemma 5 and Theorem 2 immediately give (A). To show (B), we give a crude estimate of the difference between two rational numbers based on their short continued fraction representations.

Lemma 6 *Suppose that $\alpha, \beta \in \mathbb{Q}$ are distinct with $\alpha = [p_0, p_1, \ldots, p_s]$ and $\beta = [q_0, q_1, \ldots, q_r]$. Let k be the largest index such that $p_k = q_k$. Then*

$$|\alpha - \beta| \le \prod_{j=1}^{k} \frac{1}{p_j^2}.$$

Proof We rewrite the continued fraction representations of α and β as

$$\alpha = [p_0, p_1, \ldots, p_k, p_{k+1}, \ldots, p_s] \quad \text{and} \quad \beta = [p_0, p_1, \ldots, p_k, q_{k+1}, \ldots, q_r].$$

(Note that we cannot have $k = r = s$ and that if $k = r$ or $k = s$, the estimates below still apply.) Now, for $A_i = [p_i, \ldots, p_s]$ and $B_i = [q_i, \ldots, q_r]$ with $1 \le i \le k+1$,

$$\begin{aligned}
|\alpha - \beta| &= \left| p_0 + \frac{1}{p_1 + A_1} - p_0 - \frac{1}{p_1 + B_1} \right| \\
&= \left| \frac{1}{p_1 + A_1} - \frac{1}{p_1 + B_1} \right| \\
&\le \left| p_1 + \frac{1}{p_2 + A_2} - p_1 - \frac{1}{p_2 + B_2} \right| \cdot \frac{1}{p_1^2} \\
&\ \ \vdots \\
&\le \left| \frac{1}{p_{k+1} + A_{k+1}} - \frac{1}{q_{k+1} + B_{k+1}} \right| \cdot \prod_{j=1}^{k} \frac{1}{p_j^2} \\
&\le \prod_{j=1}^{k} \frac{1}{p_j^2}.
\end{aligned}$$

□

In the case where the rationals from Lemma 6 are vertices of possibly two different (u, v)-Calkin–Wilf trees, we get the following corollary.

Corollary 1 *With α and β as in Lemma 6 and, additionally, suppose that α and β are vertices of possibly two different (u, v)-Calkin–Wilf trees, then*

$$\alpha - \beta = O\left(\frac{\max\{u, v\}}{2^k}\right).$$

Proof The corollary follows from the fact that if the two rationals α and β are vertices on (u, v)-Calkin–Wilf trees, then p_i is divisible by v for even i and divisible by u for odd i by Lemma 3. □

Before we begin our proof of (B), we need one additional lemma.

Lemma 7 *Let $y = [q_0, q_1, \ldots, q_r]$ with $q_r \neq 1$ when $y \neq 1$ and $r = 0$ when $y = 1$ and define $\ell(y) = r$. Let $f_z(n, m) = \#\{y \in \mathcal{T}^{(u,v)}(z; n) : \ell(y) = m + \ell(z)\}$, then for $m \geq 0$,*

$$f_z(n, m) = \begin{cases} \binom{n+1}{m} & \text{if } 2 \nmid m \text{ and } z > 1 \\ \binom{n+1}{m+1} & \text{if } 2 \nmid m \text{ and } z < 1 \\ \binom{n}{m} & \text{if } z = 1 \\ 0 & \text{otherwise.} \end{cases}$$

Proof The desired result can be shown to be true for $n < 2$ by inspection.

Assume that the statement is true for all $0 \leq j \leq k$ for some $k \geq 2$ and let $y \in \mathcal{T}^{(u,v)}(z; k+1)$ be such that $\ell(y) = m + \ell(z)$. That is, we assume y is a rational number counted by $f_z(k+1, m)$. There is a sequence of rational numbers $z_0 = z, z_1, \ldots, z_{k+1} = y$ such that z_{i+1} is a descendant of z_i for $0 \leq i < k+1$. By Lemma 3, we see that $\ell(z_{i+1}) - \ell(z_i) \in \{0, 1, 2\}$. In fact, for $i \geq 1$, $\ell(z_{i+1}) - \ell(z_i) = 2$ if and only if z_{i+1} is a left child of z_i and z_i is a right child of z_{i-1}, $\ell(z_1) - \ell(z_0) = 2$ if and only if z_1 is a left child of z_0 with $z_0 > 1$, and $\ell(z_1) - \ell(z_0) = 1$ if and only if z_1 is a left child of z_0 with $z_0 = 1$.

We now consider the following three cases:

Case 1: z_2 is a right child of z_1 and z_1 is a right child of z_0.

In this case we have that $y \in \mathcal{T}^{(u,v)}(z_2; k-1)$ with $\ell(y) = m + \ell(z_2)$.

Case 2: z_2 is a left child of z_1 and z_1 is a right child of z_0.

In this case we have that $y \in \mathcal{T}^{(u,v)}(z_2; k-1)$ with $\ell(y) = m - 2 + \ell(z_2)$.

Case 3: z_1 is a left child of z_0.

In this case we have that $y \in \mathcal{T}^{(u,v)}(z_1; k)$ with

$$\ell(y) = \begin{cases} m - 2 + \ell(z_1) & \text{if } z_0 > 1 \\ m + \ell(z_1) & \text{if} z_0 < 1 \\ m - 1 + \ell(z_1) & \text{if} z_0 = 1. \end{cases}$$

It follows from the three cases above that,

$$f_z(k+1, m) = \begin{cases} f_{z'}(k-1, m) + f_{z''}(k-1, m-2) + f_{z'''}(k, m-2) & \text{if} z_0 > 1 \\ f_{z'}(k-1, m) + f_{z''}(k-1, m-2) + f_{z'''}(k, m) & \text{if} z_0 < 1 \\ f_{z'}(k-1, m) + f_{z''}(k-1, m-2) + f_{z'''}(k, m-1) & \text{if} z_0 = 1. \end{cases} \tag{17}$$

where $z' = z_0 + 2v > 1$, $z'' = \frac{1}{u+\frac{1}{v+z_0}} < 1$, and $z''' = \frac{1}{u+\frac{1}{z_0}} < 1$.

We will now make heavy use of the well-known binomial coefficient identity $\binom{n}{m} = \binom{n-1}{m} + \binom{n-1}{m-1}$ to complete the proof.

For $z_0 > 1$, the desired result is trivially true when $2 \mid m$, so we assume otherwise. Therefore, by assumption

$$\begin{aligned} f_z(k+1, m) &= \binom{k}{m} + \binom{k}{m-1} + \binom{k+1}{m-1} \\ &= \binom{k+1}{m} + \binom{k+1}{m-1} \\ &= \binom{k+2}{m}. \end{aligned}$$

Similarly, for $z_0 < 1$, the desired result is also trivially true when $2 \mid m$, so we assume otherwise. Therefore, by assumption

$$\begin{aligned} f_z(k+1, m) &= \binom{k}{m} + \binom{k}{m-1} + \binom{k+1}{m+1} \\ &= \binom{k+1}{m} + \binom{k+1}{m+1} \\ &= \binom{k+2}{m+1}. \end{aligned}$$

Finally, for $z_0 = 1$, by assumption, when m is odd,

$$\begin{aligned} f_z(k+1, m) &= \binom{k}{m} + \binom{k}{m-1} + 0 \\ &= \binom{k}{m} + \binom{k}{m-1} \\ &= \binom{k+1}{m} \end{aligned}$$

and when m is even,

$$f_z(k+1, m) = 0 + 0 + \binom{k+1}{m}$$
$$= \binom{k+1}{m}.$$

Having exhausted all possibilities, we complete the proof by induction. □

An application of the de Moivre-Laplace limit theorem [4, p. 186] shows that the number of continued fraction coefficients in depth n elements is normally distributed with mean approximately $n/2$.

Corollary 1 and Lemma 7 can now be used to compare the difference between rationals in different (u, v)-Calkin–Wilf trees that are in the same position relative to the root, showing that the mean values of the rows for different trees are asymptotically the same.

Proposition 2 *For any* $z_1, z_2 \in [1/u, v] \cap \mathbb{Q}$*, we have that*

$$A^{(u,v)}(z_1; n) - A^{(u,v)}(z_2; n) \to 0$$

as $n \to \infty$.

Proof We begin by considering the case where $z_1 = 1/u$ and $z_2 = v$. Let $y \in \mathcal{T}^{(u,v)}(v; n)$. Then by Lemmas 3 and 4, y has a continued fraction representation of the form $y = [\alpha_0 v, \alpha_1 u, \ldots, \alpha_k v]$ with $\sum_{i=0}^{k} \alpha_i = n + 1$. Consider the map $f : \mathcal{T}^{(u,v)}(v; n) \to \mathcal{T}^{(u,v)}(1/u; n)$ given by

$$f(y) = \begin{cases} [\alpha_0 v, \alpha_1 u, \ldots, (\alpha_{k-1} + 1)u] & \text{if } \alpha_k = 1 \\ [\alpha_0 v, \alpha_1 u, \ldots, (\alpha_k - 1)v, u] & \text{otherwise.} \end{cases}$$

It is clear that f represents a well-defined bijection. In particular, by Corollary 1 and Lemma 7,

$$A^{(u,v)}\left(\frac{1}{u}; n\right) - A^{(u,v)}(v; n)$$
$$= \frac{1}{2^n} \sum_{y \in \mathcal{T}^{(u,v)}(v;n)} f(y) - y$$
$$= O\left(\frac{\max\{u, v\}}{2^n}\left(\sum_{y \in \mathcal{T}^{(u,v)}(v;n), \alpha_k = 1} \frac{1}{2^{k-1}} + \sum_{y \in \mathcal{T}^{(u,v)}(v;n), \alpha_k > 1} \frac{1}{2^k}\right)\right)$$
$$= O\left(\frac{\max\{u, v\}}{2^n} \sum_{y \in \mathcal{T}^{(u,v)}(v;n)} \frac{1}{2^k}\right)$$

$$= O\left(\frac{\max\{u, v\}}{2^n}\sum_{k=0}^{n+1}\binom{n+1}{k}\frac{1}{2^k}\right)$$

$$= O\left(\max\{u, v\}\cdot\left(\frac{3}{4}\right)^n\right),$$

which goes to 0 as $n \to \infty$.

The cases $z_1 = 1/u$ and $z_2 \in (1/u, 1] \cap \mathbb{Q}$ and $z_1 = v$ and $z_2 \in [1, v) \cap \mathbb{Q}$ can be handled in a similar way. These three cases complete the proof of the proposition. □

Proposition 2 completes the proof of (B), giving the desired result.

Acknowledgements The second author received support for this project provided by a PSC-CUNY Award, #611157-00 49, jointly funded by The Professional Staff Congress and The City University of New York.

References

1. Alkauskas, G.: The moments of Minkowski question mark function: the dyadic period function. Glasg. Math. J. **52**(1), 41–64 (2010).
2. Bates, B., Mansour T.: The q-Calkin-Wilf tree. J. Combin. Theory Ser. A **118**, no. 3, 1143–1151 (2011).
3. Calkin, N., Wilf, H.S.: Recounting the rationals. Amer. Math. Monthly **107**, no. 4, 360–363 (2000).
4. Feller, W.: An Introduction to Probability Theory and its Applications, Vol. I, 3rd edition, John Wiley & Sons Inc., 1968.
5. Han, S., Masuda, A.M., Singh, S., Thiel, J.: The (u, v)-Calkin-Wilf forest. Int. J. Number Theory **12**, no. 5, 1311–1328 (2016).
6. Han, S., Masuda, A.M., Singh, S., Thiel, J.: Orphans in forests of linear fractional transformations. Electron. J. Combin. **23**, no. 3, Paper 3.6, 24pp (2016).
7. Mansour, T., Shattuck, M.: Two further generalizations of the Calkin-Wilf tree, J. Comb. **2**, no. 4, 507–524 (2011).
8. Nathanson, M.B.: A forest of linear fractional transformations, Int. J. Number Theory **11**, no. 4, 1275–1299 (2015).
9. Newman, M.: Recounting the rationals, continued, solution to problem 10906, Amer. Math. Monthly 110, 642–643 (2003).
10. Reznick, B.: Regularity properties of the Stern enumeration of the rationals, J. Integer Seq. **11**, Article 08.4.1, 17pp (2008).
11. J. Roberts, Elementary Number Theory: A Problem Oriented, MIT Press, 1977.

Dimensions of Monomial Varieties

Melvyn B. Nathanson

Abstract The dimensions of certain varieties defined by monomials are computed using only high school algebra.

2010 Mathematics Subject Classification 13C15 · 12D99 · 12-01 · 13-01.

1 Krull Dimension and Varieties

In this paper, a ring R is a commutative ring with a multiplicative identity, and a field $\mathbf{F}$ is an infinite field of any characteristic.

Let S be a nonempty set of polynomials in $\mathbf{F}[t_1, \ldots, t_n]$. The *variety* (also called the *algebraic set*) V determined by S is the set of points in $\mathbf{F}^n$ that are common zeros of the polynomials in S, that is,

$$V = V(S) = \left\{(x_1, \ldots, x_n) \in \mathbf{F}^n : f(x_1, \ldots, x_n) = 0 \text{ for all } f \in S\right\}.$$

The *vanishing ideal* $\mathfrak{I}(V)$ is the set of polynomials that vanish on the variety V, that is,

$$\mathfrak{I}(V) = \{f \in \mathbf{F}[t_1, \ldots, t_n] : f(x_1, \ldots, x_n) = 0 \text{ for all } (x_1, \ldots, x_n) \in V\}.$$

We have $S \subseteq \mathfrak{I}(V)$, and so $\mathfrak{I}(V)$ contains the ideal generated by S. The quotient ring

$$\mathbf{F}(V) = \mathbf{F}[t_1, \ldots, t_n]/\mathfrak{I}(V)$$

is called the *coordinate ring* of V.

A *prime ideal chain of length n* in the ring R is a strictly increasing sequence of $n + 1$ prime ideals of R. The *Krull dimension* of R is the supremum of the lengths

M. B. Nathanson (✉)
Lehman College (CUNY), Bronx, NY 10468, USA
e-mail: melvyn.nathanson@lehman.cuny.edu

M. B. Nathanson (ed.), *Combinatorial and Additive Number Theory III*,
Springer Proceedings in Mathematics & Statistics 297,
https://doi.org/10.1007/978-3-030-31106-3_11

of prime ideal chains in R. We define the dimension of the variety V as the Krull dimension of its coordinate ring $\mathbf{F}(V)$.

It is a basic theorem in commutative algebra that the polynomial ring $\mathbf{F}[t_1, \ldots, t_n]$ has Krull dimension n. (Nathanson [4] gives an elementary proof. Other references are Atiyah and Macdonald [1, Chap. 11], Cox et al. [2, Chap. 9], and Kunz [3, Chap. 2]). If $S = \{0\} \subseteq \mathbf{F}[t_1, \ldots, t_n]$ is the set whose only element is the zero polynomial, then $V = V(\{0\}) = \mathbf{F}^n$. By Lemma 2, the vanishing ideal of V is $\mathcal{I}(V) = \mathcal{I}(\mathbf{F}^n) = \{0\}$. We obtain the coordinate ring

$$\mathbf{F}(V) = \mathbf{F}[t_1, \ldots, t_n]/\mathfrak{I}(V) \cong \mathbf{F}[t_1, \ldots, t_n],$$

and so the variety $\mathbf{F}^n$ has dimension n.

We adopt standard polynomial notation. Let $\mathbf{N}_0$ denote the set of nonnegative integers. Associated to every n-tuple $I = (i_1, \ldots, i_n) \in \mathbf{N}_0^n$ is the monomial

$$t^I = t_1^{i_1} \cdots t_n^{i_n}.$$

Every polynomial $f \in R[t_1, \ldots, t_n]$ can be represented uniquely in the form

$$f = \sum_{I \in \mathbf{N}_0^n} c_I t^I$$

where $c_I \in R$ and $c_I \neq 0$ for only finitely many $I \in \mathbf{N}_0^n$.

In this paper, two results about polynomials from high school algebra will enable us to compute the dimensions of certain varieties defined by monomials. The first result is a factorization formula, and the second result follows from the fact that a polynomial of degree d has at most d roots in a field.

Lemma 1 *For every nonnegative integer i, there is the polynomial identity*

$$u^i - v^i = (u - v)\Delta_i(u, v). \tag{1}$$

where

$$\Delta_i(u, v) = \sum_{j=0}^{i-1} u^{i-1-j} v^j. \tag{2}$$

Proof We have

$$\begin{aligned}(u - v)\sum_{j=0}^{i-1} u^{i-1-j} v^j &= \sum_{j=0}^{i-1} u^{i-j} v^j - \sum_{j=0}^{i-1} u^{i-1-j} v^{j+1} \\ &= \sum_{j=0}^{i-1} u^{i-j} v^j - \sum_{j=1}^{i} u^{i-j} v^j \\ &= u^i - v^i.\end{aligned}$$

Lemma 2 *Let* $\mathbf{F}$ *be an infinite field. A polynomial* $f \in \mathbf{F}[t_1, \ldots, t_n]$ *satisfies* $f(x_1, \ldots, x_n) = 0$ *for all* $(x_1, \ldots, x_n) \in \mathbf{F}^n$ *if and only if* $f = 0$.

Proof The proof is by induction on n. Let $n = 1$. A nonzero polynomial $f \in \mathbf{F}[t_1]$ of degree d has at most d roots in $\mathbf{F}$, and so $f(x_1) \neq 0$ for some $x_1 \in \mathbf{F}$. Thus, if $f(x_1) = 0$ for all $x_1 \in \mathbf{F}$, then $f = 0$.

Let $n \geq 1$, and assume that the Lemma holds for polynomials in n variables. Let $f \in \mathbf{F}[t_1, \ldots, t_n, t_{n+1}]$ have degree d in the variable t_{n+1}. There exist polynomials $f_i \in \mathbf{F}[t_1, \ldots, t_n]$ such that

$$f = f(t_1, \ldots, t_n, t_{n+1}) = \sum_{i=0}^{d} f_i(t_1, \ldots, t_n) t_{n+1}^i$$

For all $(x_1, \ldots, x_n) \in \mathbf{F}^n$, the polynomial

$$g(t_{n+1}) = f(x_1, \ldots, x_n, t_{n+1}) = \sum_{i=0}^{d} f_i(x_1, \ldots, x_n) t_{n+1}^i \in \mathbf{F}[t_{n+1}]$$

satisfies $g(x_{n+1}) = 0$ for all $x_{n+1} \in \mathbf{F}$, and so $g = 0$. Therefore, $f_i(x_1, \ldots, x_n) = 0$ for all $(x_1, \ldots, x_n) \in \mathbf{F}^n$ and $i = 0, 1, \ldots, d$. By the induction hypothesis, $f_i = 0$ for all $i = 0, 1, \ldots, d$, and so $f = 0$.

2 An Example of a Plane Curve

A *hypersurface* is a variety that is the set of zeros of one nonzero polynomial. A *plane algebraic curve* is a hypersurface in $\mathbf{F}^2$. In this section we compute the dimension of a hypersurface V in $\mathbf{F}^{m+1}$ defined by a polynomial of the form

$$f^* = t_{m+1} - \lambda t_1^{a_1} t_2^{a_2} \cdots t_m^{a_m}$$

where $\lambda \in \mathbf{F}$ and $(a_1, \ldots, a_m) \in \mathbf{N}_0^m$. We shall prove that the *monomial hypersurface*

$$\begin{aligned} V &= \{(x_1, \ldots, x_m, x_{m+1}) \in \mathbf{F}^{m+1} : f^*(x_1, \ldots, x_m, x_{m+1}) = 0\} \\ &= \{(x_1, \ldots, x_m, x_{m+1}) \in \mathbf{F}^{m+1} : x_{m+1} = \lambda x_1^{a_1} x_2^{a_2} \cdots x_m^{a_m}\} \end{aligned}$$

has dimension m.

We begin with an example. Consider the monomial $4t_1^3$ and the curve in $\mathbf{F}^2$ defined by the polynomial

$$f^* = t_2 - 4t_1^3 \in \mathbf{F}[t_1, t_2].$$

Let

$$V = \left\{(x_1, x_2) \in \mathbf{F}^2 : f^*(x_1, x_2) = 0\right\} = \left\{(x_1, x_2) \in \mathbf{F}^2 : x_2 = 4x_1^3\right\}.$$

We shall prove that the vanishing ideal $\mathfrak{I}(V)$ is the principal ideal generated by f^*. Because $f^* \in \mathfrak{I}(V)$, it suffices to show that every polynomial in $\mathfrak{I}(V)$ is divisible by f^*.

For $I = (i_1, i_2) \in \mathbf{N}_0^2$ and $t^I = t_1^{i_1} t_2^{i_2}$, let

$$b_1 = i_1 + 3i_2$$

and let Δ_{i_2} be the polynomial defined by (2) in Lemma 1. We have

$$\begin{aligned} t^I - 4^{i_2} t_1^{b_1} &= t_1^{i_1} t_2^{i_2} - 4^{i_2} t_1^{i_1+3i_2} = t_1^{i_1}\left(t_2^{i_2} - 4^{i_2} t_1^{3i_2}\right) \\ &= t_1^{i_1}\left(t_2^{i_2} - \left(4t_1^3\right)^{i_2}\right) = t_1^{i_1} \Delta_{i_2}(t_2, 4t_1^3)\left(t_2 - 4t_1^3\right) \\ &= g_I f^* \end{aligned}$$

where $g_I = t_1^{i_1} \Delta_{i_2}(t_2, 4t_1^3) \in \mathbf{F}[t_1, t_2]$.

Every polynomial $f \in \mathbf{F}[t_1, t_2]$ can be represented uniquely in the form

$$f = \sum_{I=(i_1,t_2)\in\mathbf{N}_0^2} c_I t_1^{i_1} t_2^{i_2} = \sum_{b_1\in\mathbf{N}_0} \sum_{\substack{I=(i_1,i_2)\in\mathbf{N}_0^2 \\ i_1+3i_2=b_1}} c_I t_1^{i_1} t^{i_2}.$$

A polynomial $f \in \mathbf{F}[t_1,\ t_2]$ is in the vanishing ideal $\mathfrak{I}(V)$ if and only if, for all $x_1 \in \mathbf{F}$,

$$\begin{aligned} 0 = f(x_1, 4x_1^3) &= \sum_{b_1\in\mathbf{N}_0} \sum_{\substack{I=(i_1,i_2)\in\mathbf{N}_0^2 \\ i_1+3i_2=b_1}} c_I x_1^{i_1}\left(4x_1^3\right)^{i_2} \\ &= \sum_{b_1\in\mathbf{N}_0} \sum_{\substack{I=(i_1,i_2)\in\mathbf{N}_0^2 \\ i_1+3i_2=b_1}} c_I 4^{i_2} x_1^{i_1+3i_2} \\ &= \sum_{b_1\in\mathbf{N}_0} \left(\sum_{\substack{I=(i_1,i_2)\in\mathbf{N}_0^2 \\ i_1+3i_2=b_1}} c_I 4^{i_2} \right) x_1^{b_1}. \end{aligned}$$

By Lemma 2, because $\mathbf{F}$ is an infinite field, the coefficients of this polynomial are zero, and so

$$\sum_{\substack{I=(i_1,i_2)\in\mathbf{N}_0^2 \\ i_1+3i_2=b_1}} c_I 4^{i_2} = 0$$

for all $b_1 \in \mathbf{N}_0$. The ordered pair $I = (b_1, 0)$ is one of the terms in this sum, and so

$$-c_{(b_1,0)} = \sum_{\substack{I=(i_1,i_)\in\mathbf{N}_0^2,\\ i_\ell+3i_{m+1}=b_1\\ I\neq(b_1,0)}} c_I 4^{i_2}.$$

Therefore, $f \in \mathfrak{I}(V)$ implies

$$\begin{aligned}
f &= \sum_{b_1\in\mathbf{N}_0} \left(c_{(b_1,0)} t_1^{b_1} + \sum_{\substack{I=(i_1,i_2)\in\mathbf{N}_0^2\\ i_1+3i_2=b_1\\ I\neq(b_1,0)}} c_I t_1^{i_1} t_2^{i_2} \right)\\
&= \sum_{b_1\in\mathbf{N}_0} \left(\sum_{\substack{I=(i_1,i_2)\in\mathbf{N}_0^2\\ i_1+3i_2=b_1\\ I\neq(b_1,0)}} c_I t_1^{i_1} t_2^{i_2} - \sum_{\substack{I=(i_1,i_)\in\mathbf{N}_0^2,\\ i_\ell+3i_{m+1}=b_1\\ I\neq(b_1,0)}} c_I 4^{i_2} t_1^{b_1} \right)\\
&= \sum_{b_1\in\mathbf{N}_0} \sum_{\substack{I=(i_1,i_2)\in\mathbf{N}_0^2\\ i_1+3i_2-b_1\\ I\neq(b_1,0)}} c_I \left(t_1^{i_1} t_2^{i_2} - 4^{i_2} t_1^{b_1} \right)\\
&= \sum_{b_1\in\mathbf{N}_0} \sum_{\substack{I=(i_1,i_2)\in\mathbf{N}_0^2\\ i_1+3i_2=b_1\\ I\neq(b_1,0)}} c_I g^I f^*
\end{aligned}$$

and so f^* divides f. Thus, every polynomial $f \in \mathfrak{I}(V)$ is contained in the principal ideal generated by f^*.

The function

$$\varphi : \mathbf{F}[t_1, t_2] \to \mathbf{F}[t_1]$$

defined by

$$\varphi(t_1) = t_1$$

and

$$\varphi(t_2) = 4t_1^3$$

is a surjective ring homomorphism with

$$\text{kernel}(\varphi) = \left\{ f \in \mathbf{F}[t_1, t_2] : f(t_1, 4^{i_2} t_1^3) = 0 \right\} = \mathfrak{I}(V).$$

Therefore,

$$\mathbf{F}[V] = \mathbf{F}[t_1, t_2]/\mathcal{I}(V) \cong \mathbf{F}[t_1].$$

The polynomial ring $\mathbf{F}[t_1]$ has Krull dimension 1, and so the coordinate ring $\mathbf{F}(V)$ of the curve has Krull dimension 1 and the curve has dimension 1.

3 Dimension of a Monomial Hypersurface

We shall prove that every monomial hypersurface in $\mathbf{F}^{m+1}$ has dimension m. The proof is elementary, like the proof in Sect. 2, but a bit more technical.

Lemma 3 *For $\lambda \in \mathbf{F}$ and $(a_1, \ldots, a_m) \in \mathbf{N}_0^m$, consider the polynomial*

$$f^* = t_{m+1} - \lambda t_1^{a_1} \cdots t_m^{a_m} \in \mathbf{F}[t_1, \ldots, t_{m+1}].$$

For $I = (i_1, \ldots, i_m, i_{m+1}) \in \mathbf{N}_0^{m+1}$ and $\ell \in \{1, \ldots, m\}$, let

$$b_\ell = i_\ell + a_\ell i_{m+1}.$$

There exists a polynomial $g_I \in \mathbf{F}[t_1, \ldots, t_{m+1}]$ such that

$$t^I - \lambda^{i_{m+1}} t_1^{b_1} \cdots t_m^{b_m} = g_I f^*.$$

Proof Let $\Delta_i(u, v)$ be the polynomial defined by (2). We have

$$\begin{aligned} t^I - &\lambda^{i_{m+1}} t_1^{b_1} \cdots t_m^{b_m} \\ &= t_1^{i_1} \cdots t_m^{i_m} t_{m+1}^{i_{m+1}} - \lambda^{i_{m+1}} t_1^{i_1 + a_1 i_{m+1}} \cdots t_m^{i_m + a_m i_{m+1}} \\ &= t_1^{i_1} \cdots t_m^{i_m} \left(t_{m+1}^{i_{m+1}} - \lambda^{i_{m+1}} t_1^{a_1 i_{m+1}} \cdots t_m^{a_m i_{m+1}} \right) \\ &= t_1^{i_1} \cdots t_m^{i_m} \left(t_{m+1}^{i_{m+1}} - \left(\lambda t_1^{a_1} \cdots t_m^{a_m} \right)^{i_{m+1}} \right) \\ &= t_1^{i_1} \cdots t_m^{i_m} \Delta_{i_{m+1}}(t_{m+1}, \lambda t_1^{a_1} \cdots t_m^{a_m}) \left(t_{m+1} - \lambda t_1^{a_1} \cdots t_m^{a_m} \right) \\ &= g_I f^* \end{aligned}$$

where

$$g_I = t_1^{i_1} \cdots t_m^{i_m} \Delta_{i_{m+1}}(t_{m+1}, \lambda t_1^{a_1} \cdots t_m^{a_m}) \in \mathbf{F}[t_1, \ldots, t_{m+1}].$$

This completes the proof. □

Theorem 1 *Let $\mathbf{F}$ be an infinite field. For $\lambda \in \mathbf{F}$ and $(a_1, \ldots, a_m) \in \mathbf{N}_0^m$, consider the polynomial*

$$f^* = t_{m+1} - \lambda t_1^{a_1} t_2^{a_2} \cdots t_m^{a_m} \in \mathbf{F}[t_1, \ldots, t_m, t_{m+1}]$$

and the associated hypersurface

$$
\begin{aligned}
V &= \left\{(x_1, \ldots, x_m, x_{m+1}) \in \mathbf{F}^{m+1} : f^*(x_1, \ldots, x_m, x_{m+1}) = 0\right\} \\
&= \left\{\left(x_1, \ldots, x_m, \lambda x_1^{a_1} x_2^{a_2} \cdots x_m^{a_m}\right) \in \mathbf{F}^{m+1} : (x_1, \ldots, x_m) \in \mathbf{F}^m\right\}.
\end{aligned}
$$

The vanishing ideal $\mathfrak{I}(V)$ is the principal ideal generated by f^*.

Proof The vanishing ideal $\mathfrak{I}(V)$ contains f^*, and so $\mathfrak{I}(V)$ contains the principal ideal generated by f^*. Therefore, it suffices to prove that $\mathfrak{I}(V)$ is contained in the principal ideal generated by f^*.

For every $(m+1)$-tuple $I = (i_1, \ldots, i_m, i_{m+1}) \in \mathbf{N}_0^{m+1}$, there is a unique m-tuple $(b_1, \ldots, b_m) \in \mathbf{N}_0^m$ defined by

$$b_\ell = i_\ell + a_\ell i_{m+1}$$

for $\ell = 1, \ldots, m$. Thus, every polynomial $f \in \mathbf{F}[t_1, \ldots, t_m, t_{m+1}]$ can be represented uniquely in the form

$$
\begin{aligned}
f &= \sum_{I \in \mathbf{N}_0^{m+1}} c_I t^I \\
&= \sum_{(b_1,\ldots,b_m) \in \mathbf{N}_0^m} \sum_{\substack{I=(i_1,\ldots,i_m,i_{m+1}) \in \mathbf{N}_0^{m+1} \\ i_\ell + a_\ell i_{m+1} = b_\ell \\ \text{for } \ell=1,\ldots,m}} c_I t_1^{i_1} \cdots t_m^{i_m} t_{m+1}^{i_{m+1}}.
\end{aligned}
$$

A polynomial $f \in \mathbf{F}[t_1, \ldots, t_m, t_{m+1}]$ is in the vanishing ideal $\mathfrak{I}(V)$ if and only if, for all $(x_1, \ldots, x_m) \in \mathbf{F}^m$,

$$
\begin{aligned}
0 &= f(x_1, \ldots, x_m, \lambda x_1^{a_1} x_2^{a_2} \cdots x_m^{a_m}) \\
&= \sum_{(b_1,\ldots,b_m) \in \mathbf{N}_0^m} \sum_{\substack{I=(i_1,\ldots,i_m,i_{m+1}) \in \mathbf{N}_0^{m+1} \\ i_\ell + a_\ell i_{m+1} = b_\ell \\ \text{for } \ell=1,\ldots,m}} c_I x_1^{i_1} \cdots x_m^{i_m} \left(\lambda x_1^{a_1} x_2^{a_2} \cdots x_m^{a_m}\right)^{i_{m+1}} \\
&= \sum_{(b_1,\ldots,b_m) \in \mathbf{N}_0^m} \sum_{\substack{I=(i_1,\ldots,i_m,i_{m+1}) \in \mathbf{N}_0^{m+1} \\ i_\ell + a_\ell i_{m+1} = b_\ell \\ \text{for } \ell=1,\ldots,m}} c_I \lambda^{i_{m+1}} x_1^{i_1 + a_1 i_{m+1}} \cdots x_m^{i_m + a_m i_{m+1}} \\
&= \sum_{(b_1,\ldots,b_m) \in \mathbf{N}_0^m} \left(\sum_{\substack{I=(i_1,\ldots,i_m,i_{m+1}) \in \mathbf{N}_0^{m+1} \\ i_\ell + a_\ell i_{m+1} = b_\ell \\ \text{for } \ell=1,\ldots,m}} c_I \lambda^{i_{m+1}} \right) x_1^{b_1} \cdots x_m^{b_m}.
\end{aligned}
$$

By Lemma 2, the coefficients of this polynomial are zero, and so

$$\sum_{\substack{I=(i_1,\ldots,i_m,i_{m+1})\in \mathbf{N}_0^{m+1} \\ i_\ell + a_\ell i_{m+1} = b_\ell \\ \text{for } \ell=1,\ldots,m}} c_I \lambda^{i_{m+1}} = 0 \tag{3}$$

for all $(b_1, \ldots, b_m) \in \mathbf{N}_0^m$. The $(m+1)$-tuple $I = (b_1, \ldots, b_m, 0)$ is one of the terms in the sum (3), and so

$$-c_{(b_1,\ldots,b_m,0)} = \sum_{\substack{I=(i_1,\ldots,i_m,i_{m+1})\in \mathbf{N}_0^{m+1}, \\ i_\ell + a_\ell i_{m+1} = b_\ell \\ \text{for } \ell=1,\ldots,m, \\ I \neq (b_1,\ldots,b_m,0)}} c_I \lambda^{i_{m+1}}.$$

Therefore, $f \in \mathfrak{I}(V)$ implies

$$\begin{aligned}
f &= \sum_{(b_1,\ldots,b_m)\in \mathbf{N}_0^m} \sum_{\substack{I=(i_1,\ldots,i_m,i_{m+1})\in \mathbf{N}_0^{m+1} \\ i_\ell + a_\ell i_{m+1} = b_\ell \\ \text{for } \ell=1,\ldots,m}} c_I t^I \\
&= \sum_{(b_1,\ldots,b_m)\in \mathbf{N}_0^m} \left(\sum_{\substack{I=(i_1,\ldots,i_m,i_{m+1})\in \mathbf{N}_0^{m+1} \\ i_\ell + a_\ell i_{m+1} = b_\ell \\ \text{for } \ell=1,\ldots,m \\ I \neq (b_1,\ldots,b_m,0)}} c_I t^I + c_{(b_1,\ldots,b_m,0)} t_1^{b_1} \cdots t_m^{b_m} \right) \\
&= \sum_{(b_1,\ldots,b_m)\in \mathbf{N}_0^m} \left(\sum_{\substack{I=(i_1,\ldots,i_m,i_{m+1})\in \mathbf{N}_0^{m+1} \\ i_\ell + a_\ell i_{m+1} = b_\ell \\ \text{for } \ell=1,\ldots,m \\ I \neq (b_1,\ldots,b_m,0)}} c_I t^I - \sum_{\substack{I=(i_1,\ldots,i_m,i_{m+1})\in \mathbf{N}_0^{m+1} \\ i_\ell + a_\ell i_{m+1} = b_\ell \\ \text{for } \ell=1,\ldots,m \\ I \neq (b_1,\ldots,b_m,0)}} c_I \lambda^{i_{m+1}} t_1^{b_1} \cdots t_m^{b_m} \right) \\
&= \sum_{(b_1,\ldots,b_m)\in \mathbf{N}_0^m} \sum_{\substack{I=(i_1,\ldots,i_m,i_{m+1})\in \mathbf{N}_0^{m+1} \\ i_\ell + a_\ell i_{m+1} = b_\ell \\ \text{for } \ell=1,\ldots,m \\ I \neq (b_1,\ldots,b_m,0)}} c_I \left(t^I - \lambda^{i_{m+1}} t_1^{b_1} \cdots t_m^{b_m} \right) \\
&= \sum_{(b_1,\ldots,b_m)\in \mathbf{N}_0^m} \sum_{\substack{I=(i_1,\ldots,i_m,i_{m+1})\in \mathbf{N}_0^{m+1} \\ i_\ell + a_\ell i_{m+1} = b_\ell \\ \text{for } \ell=1,\ldots,m \\ I \neq (b_1,\ldots,b_m,0)}} c_I g_I f^*
\end{aligned}$$

by Lemma 3, and so f is in the principal ideal generated by f^*. This completes the proof. □

Theorem 2 *Let* $\mathbf{F}$ *be an infinite field. For* $\lambda \in \mathbf{F}$ *and* $(a_1, \ldots, a_m) \in \mathbf{N}_0^m$*, the hypersurface*

$$V = \left\{ \left(x_1, \ldots, x_m, \lambda x_1^{a_1} x_2^{a_2} \cdots x_m^{a_m}\right) \in \mathbf{F}^{m+1} : (x_1, \ldots, x_m) \in \mathbf{F}^m \right\}$$

has dimension m.

Proof The function

$$\varphi : \mathbf{F}[t_1, \ldots, t_{m+1}] \to \mathbf{F}[t_1, \ldots, t_m]$$

defined by

$$\varphi(t_\ell) = t_\ell \qquad \text{for } \ell = 1, \ldots, m$$

and

$$\varphi(t_{m+1}) = \lambda t_1^{a_1} \cdots t_m^{a_m}$$

is a surjective ring homomorphism with

$$\text{kernel}(\varphi) = \left\{ f \in \mathbf{F}[t_1, \ldots, t_m] : f(t_1, \ldots, t_m, \lambda^{i_{m+1}} t_1^{a_1} \cdots t_m^{a_m} = 0 \right\} = \mathfrak{I}(V).$$

Therefore,

$$\mathbf{F}[V] = \mathbf{F}[t_1, \ldots, t_m, t_{m+1}]/\mathcal{I}(V) \cong \mathbf{F}[t_1, \ldots, t_m].$$

The polynomial ring $\mathbf{F}[t_1, \ldots, t_m]$ has Krull dimension m, and so the coordinate ring $\mathbf{F}[V]$ has Krull dimension m and the hypersurface V has dimension m. This completes the proof. □

4 Varieties Defined by Several Monomials

Let m and k be positive integers, and let $n = m + k$. For $j = 1, 2, \ldots, k$, let $\lambda_j \in \mathbf{F}$ and $(a_{1,j}, a_{2,j}, \ldots, a_{m,j}) \in \mathbf{N}_0^m$. Consider the polynomials

$$f_j^* = t_{m+j} - \lambda_j t_1^{a_{1,j}} t_2^{a_{2,j}} \cdots t_m^{a_{m,j}} \in \mathbf{F}[t_1, \ldots, t_n]. \tag{4}$$

Let V be the variety in $\mathbf{F}^n$ determined by the set of polynomials

$$S = \{f_j^* : j = 1, \ldots, k\}$$

and let $\mathfrak{I}(V)$ be the vanishing ideal of V. We shall prove that the coordinate ring $\mathbf{F}[V] = \mathbf{F}[t_1, \ldots, t_n]/\mathfrak{I}(V)$ is isomorphic to the polynomial ring $\mathbf{F}[t_1, \ldots, t_m]$, and so V has dimension m.

Lemma 4 *Let* $\mathbf{F}$ *be an infinite field. For* $j = 1, 2, \ldots, k$*, let* $\lambda_j \in \mathbf{F}$ *and* $(a_{1,j}, a_{2,j}, \ldots, a_{m,j}) \in \mathbf{N}_0^m$*. Define the polynomial* f_j^* *by* (4)*. For* $I = (i_1, \ldots, i_m, i_{m+1}, \ldots,$

$i_{m+k}) \in \mathbf{N}_0^{m+k}$ and $\ell \in \{1, \ldots, m\}$, let

$$b_\ell = i_\ell + \sum_{j=1}^{k} a_{\ell,j} i_{m+j}.$$

There exist polynomials $g_{I,1}, \ldots, g_{I,k} \in \mathbf{F}[t_1, \ldots, t_{m+k}]$ such that

$$t^I - \prod_{j=1}^{k} \lambda_j^{i_{m+j}} t_1^{b_1} \cdots t_m^{b_m} = \sum_{j=1}^{k} g_{I,j} f_j^*. \tag{5}$$

Proof The proof is by induction on k. The case $k = 1$ is Lemma 3. Assume that Lemma 4 is true for the positive integer k. We shall prove the Lemma for $k + 1$.

For

$$I = (i_1, \ldots, i_{m+k}) \in \mathbf{N}_0^{m+k}$$

and

$$I' = (i_1, \ldots, i_{m+k+1}) \in \mathbf{N}_0^{m+k+1}$$

we have

$$t^I = \prod_{\ell=1}^{m} t_\ell^{i_\ell} \prod_{j=1}^{k} t_{m+j}^{i_{m+j}}$$

and

$$t^{I'} = \prod_{\ell=1}^{m} t_\ell^{i_\ell} \prod_{j=1}^{k+1} t_{m+j}^{i_{m+j}} = t^I \, t_{m+k+1}^{i_{m+k+1}}.$$

For $\ell = 1, \ldots, m$, define

$$b_\ell = i_\ell + \sum_{j=1}^{k} a_{\ell,j} i_{m+j}$$

and

$$b'_\ell = i_\ell + \sum_{j=1}^{k+1} a_{\ell,j} i_{m+j} = b_\ell + a_{\ell,k+1} i_{m+k+1}.$$

We have

$$t^{I'} - \prod_{j=1}^{k+1} \lambda_j^{i_{m+j}} t_1^{b'_1} \cdots t_m^{b'_m}$$

$$= t_{m+k+1}^{i_{m+k+1}} \left(t^I - \prod_{j=1}^{k} \lambda_j^{i_{m+j}} \, t_1^{b_1} \cdots t_m^{b_m} \right)$$

$$+ t_{m+k+1}^{i_{m+k+1}} \left(\prod_{j=1}^{k} \lambda_j^{i_{m+j}} \, t_1^{b_1} \cdots t_m^{b_m} \right) - \prod_{j=1}^{k+1} \lambda_j^{i_{m+j}} \, t_1^{b'_1} \cdots t_m^{b'_m}$$

By the induction hypothesis, there exist polynomials $g_{I,1}, \ldots, g_{I,k} \in \mathbf{F}[t_1, \ldots, t_{m+k}]$ that satisfy (5), and so

$$t_{m+k+1}^{i_{m+k+1}} \left(t^I - \prod_{j=1}^{k} \lambda_j^{i_{m+j}} \, t_1^{b_1} \cdots t_m^{b_m} \right) = t_{m+k+1}^{i_{m+k+1}} \sum_{j=1}^{k} g_{I,j} \, f_j^* .$$

Applying the factorization formula (1), we obtain

$$t_{m+k+1}^{i_{m+k+1}} \left(\prod_{j=1}^{k} \lambda_j^{i_{m+j}} \, t_1^{b_1} \cdots t_m^{b_m} \right) - \prod_{j=1}^{k+1} \lambda_j^{i_{m+j}} \, t_1^{b'_1} \cdots t_m^{b'_m}$$

$$= \left(\prod_{j=1}^{k} \lambda_j^{i_{m+j}} \, t_1^{b_1} \cdots t_m^{b_m} \right) \left(t_{m+k+1}^{i_{m+k+1}} - \lambda_{k+1}^{i_{m+k+1}} \prod_{\ell=1}^{m} t_\ell^{a_{\ell,k+1} i_{m+k+1}} \right)$$

$$= \left(\prod_{j=1}^{k} \lambda_j^{i_{m+j}} \, t_1^{b_1} \cdots t_m^{b_m} \right) \left(t_{m+k+1}^{i_{m+k+1}} - \left(\lambda_{k+1} \prod_{\ell=1}^{m} t_\ell^{a_{\ell,k+1}} \right)^{i_{m+k+1}} \right)$$

$$= \left(\prod_{j=1}^{k} \lambda_j^{i_{m+j}} \, t_1^{b_1} \cdots t_m^{b_m} \right) \Delta_{i_{m+k+1}} \left(t_{m+k+1}, \lambda_{k+1} \prod_{\ell=1}^{m} t_\ell^{a_{\ell,k+1}} \right) \left(t_{m+k+1} - \lambda_{k+1} \prod_{\ell=1}^{m} t_\ell^{a_{\ell,k+1}} \right)$$

$$= g_{I,k+1} \, f_{k+1}^*$$

where

$$g_{I,k+1} = \left(\prod_{j=1}^{k} \lambda_j^{i_{m+j}} \, t_1^{b_1} \cdots t_m^{b_m} \right) \Delta_{i_{m+k+1}} \left(t_{m+k+1}, \ \lambda_{k+1} \prod_{\ell=1}^{m} t_\ell^{a_{\ell,k+1}} \right).$$

This completes the proof. □

Theorem 3 *Let* $\mathbf{F}$ *be an infinite field. Let* $n = m + k$. *For* $j = 1, 2, \ldots, k$, *let* $\lambda_j \in \mathbf{F}$ *and* $(a_{1,j}, a_{2,j}, \ldots, a_{m,j}) \in \mathbf{N}_0^m$, *and let* $f_j^* \in \mathbf{F}[t_1, \ldots, t_n]$ *be the polynomial*

$$f_j^*(t_1, \ldots, t_m, t_{m+1}, \ldots, t_{m+k}) = t_{m+j} - \lambda_j t_1^{a_{1,j}} t_2^{a_{2,j}} \cdots t_m^{a_{m,j}} .$$

Let $V \subseteq \mathbf{F}^n$ *be the variety determined by the set* $S = \{f_1^*, \ldots, f_k^*\} \subseteq \mathbf{F}[t_1, \ldots, t_n]$. *The vanishing ideal* $\mathfrak{I}(V)$ *is the ideal generated by* S.

Proof The ideal $\mathfrak{I}(V)$ contains S, and so $\mathfrak{I}(V)$ contains the ideal generated by S. Thus, it suffices to prove that ideal generated by S contains every polynomial in $\mathfrak{I}(V)$.

The variety determined by S is

$$V = \left\{ \left(x_1, \ldots, x_m, \lambda_1 x_1^{a_{1,1}} \cdots x_m^{a_{m,1}}, \ldots, \lambda_k x_1^{a_{1,k}} \cdots x_m^{a_{m,k}} \right) : (x_1, \ldots, x_m) \in \mathbf{F}^m \right\}.$$

For every $(m+k)$-tuple $I = (i_1, \ldots, i_m, i_{m+1}, \ldots, i_{m+k}) \in \mathbf{N}_0^{m+k}$, there is a unique m-tuple $(b_1, \ldots, b_m) \in \mathbf{N}_0^m$ such that

$$b_\ell = i_\ell + \sum_{j=1}^{k} a_{\ell,j} i_{m+j}$$

for $\ell = 1, \ldots, m$. Let

$$\sum_{I(b_1,\ldots,b_m)} = \sum_{\substack{I=(i_1,\ldots,i_m,i_{m+1},\ldots,i_{m+k})\in\mathbf{N}_0^n \\ i_\ell+\sum_{j=1}^k a_{\ell,j} i_{m+j} = b_\ell \\ \text{for } \ell=1,\ldots,m}}.$$

Every polynomial $f \in \mathbf{F}[t_1, \ldots, t_n]$ has a unique representation in the form

$$f = \sum_{I \in \mathbf{N}_0^n} c_I t^I = \sum_{(b_1,\ldots,b_m)\in\mathbf{N}_0^m} \sum_{I(b_1,\ldots,b_m)} c_I t^I$$

where $c_I \in \mathbf{F}$ and $c_I \neq 0$ for only finitely many n-tuples I. If $f \in \mathfrak{I}(V)$, then

$$\begin{aligned}
0 &= f\left(x_1, \ldots, x_m, \lambda_1 x_1^{a_{1,1}} \cdots x_m^{a_{m,1}}, \ldots, \lambda_k x_1^{a_{1,k}} \cdots x_m^{a_{m,k}}\right) \\
&= \sum_{(b_1,\ldots,b_m)\in\mathbf{N}_0^m} \sum_{I(b_1,\ldots,b_m)} c_I x_1^{i_1} \cdots x_m^{i_m} \left(\lambda_1 x_1^{a_{1,1}} \cdots x_m^{a_{m,1}}\right)^{i_{m+1}} \cdots \left(\lambda_k x_1^{a_{1,k}} \cdots x_m^{a_{m,k}}\right)^{i_{m+k}} \\
&= \sum_{(b_1,\ldots,b_m)\in\mathbf{N}_0^m} \sum_{I(b_1,\ldots,b_m)} c_I \prod_{j=1}^{k} \lambda_j^{i_{m+j}} \prod_{\ell=1}^{m} x_\ell^{i_\ell + \sum_{j=1}^k a_{\ell,j} i_{m+j}} \\
&= \sum_{(b_1,\ldots,b_m)\in\mathbf{N}_0^m} \left(\sum_{I(b_1,\ldots,b_m)} c_I \prod_{j=1}^{k} \lambda_j^{i_{m+j}} \right) x_1^{b_1} \cdots x_m^{b_m}
\end{aligned}$$

for all $(x_1, \ldots, x_m) \in \mathbf{F}^m$. Note that $I = (b_1, \ldots, b_m, 0, \ldots, 0) \in I_{(b_1,\ldots,b_m)}$. It follows from Lemma 2 that

$$0 = \sum_{I_{(b_1,\ldots,b_m)}} c_I \prod_{j=1}^{k} \lambda_j^{i_{m+j}} = c_{(b_1,\ldots,b_m,0,\ldots,0)} + \sum_{\substack{I_{(b_1,\ldots,b_m)} \\ I \neq (b_1,\ldots,b_m,0,\ldots,0)}} c_I \prod_{j=1}^{k} \lambda_j^{i_{m+j}}$$

for all $(b_1, \ldots, b_m) \in \mathbf{N}_0^m$, and so

$$
\begin{aligned}
f &= \sum_{(b_1,\ldots,b_m)\in\mathbf{N}_0^m} \sum_{I(b_1,\ldots,b_m)} c_I t^I \\
&= \sum_{(b_1,\ldots,b_m)\in\mathbf{N}_0^m} \left(\sum_{\substack{I_{(b_1,\ldots,b_m)} \\ I\neq(b_1,\ldots,b_m,0,\ldots,0)}} c_I t^I + c_{(b_1,\ldots,b_m,0,\ldots,0)} t_1^{b_1}\cdots t_m^{b_m} \right) \\
&= \sum_{(b_1,\ldots,b_m)\in\mathbf{N}_0^m} \left(\sum_{\substack{I_{(b_1,\ldots,b_m)} \\ I\neq(b_1,\ldots,b_m,0,\ldots,0)}} c_I t^I - \sum_{\substack{I_{(b_1,\ldots,b_m)} \\ I\neq(b_1,\ldots,b_m,0,\ldots,0)}} c_I \prod_{j=1}^{k} \lambda_j^{i_{m+j}} t_1^{b_1}\cdots t_m^{b_m} \right) \\
&= \sum_{(b_1,\ldots,b_m)\in\mathbf{N}_0^m} \sum_{\substack{I_{(b_1,\ldots,b_m)} \\ I\neq(b_1,\ldots,b_m,0,\ldots,0)}} c_I \left(t^I - \prod_{j=1}^{k} \lambda_j^{i_{m+j}} t_1^{b_1}\cdots t_m^{b_m} \right).
\end{aligned}
$$

Lemma 4 immediately implies that f is in the ideal generated by S. This completes the proof. □

Theorem 4 *The variety V has dimension m.*

Proof The function

$$\varphi : \mathbf{F}[t_1, \ldots, t_{m+1}, \ldots, t_{m+k}] \to \mathbf{F}[t_1, \ldots, t_m]$$

defined by

$$\varphi(t_\ell) = t_\ell \qquad \text{for } \ell = 1, \ldots, m$$

and

$$\varphi(t_{m+j}) = \lambda_j t_1^{a_{1,j}} \cdots t_m^{a_{m,j}} \qquad \text{for } j = 1, \ldots, k$$

is a surjective ring homomorphism with

$$
\begin{aligned}
&\operatorname{kernel}(\varphi) \\
&= \left\{ f \in \mathbf{F}[t_1, \ldots, t_n] : f\left(t_1, \ldots, t_m, \lambda_1 t_1^{a_{1,1}} \cdots t_m^{a_{m,1}}, \ldots, \lambda_k t_1^{a_{1,k}} \cdots t_m^{a_{m,k}}\right) = 0 \right\} \\
&= \mathfrak{I}(V).
\end{aligned}
$$

Therefore,

$$\mathbf{F}[V] = \mathbf{F}[t_1, \ldots, t_n]/\mathcal{I}(V) \cong \mathbf{F}[t_1, \ldots, t_m]$$

and the coordinate ring of $\mathfrak{I}(V)$ has Krull dimension m. This completes the proof. □

Acknowledgements Supported in part by a grant from the PSC-CUNY Research Award Program.

References

1. M. F. Atiyah and I. G. Macdonald, *Introduction to Commutative Algebra*, Addison-Wesley, Reading, MA, 1969.
2. D. Cox, J. Little, and D. O'Shea, *Ideals, Varieties, and Algorithms*, 3rd edn. Springer, New York, 2007.
3. E. Kunz, *Introduction to Commutative Algebra and Algebraic Geometry*, Modern Birkhäuser Classics, Birkhäuser/Springer, New York, 2013.
4. M. B. Nathanson, *An elementary proof for the Krull dimension of a polynomial ring*, The American Mathematical Monthly **125** (2018), 623–637.

Matrix Scaling Limits in Finitely Many Iterations

Melvyn B. Nathanson

Abstract The alternate row and column scaling algorithm applied to a positive $n \times n$ matrix A converges to a doubly stochastic matrix $S(A)$, sometimes called the *Sinkhorn limit* of A. For every positive integer n, a two parameter family of row but not column stochastic $n \times n$ positive matrices is constructed that become doubly stochastic after exactly one column scaling.

2010 Mathematics Subject Classification 11C20 · 11B75 · 11J68 · 11J70

1 The Alternate Scaling Algorithm

A *positive matrix* is a matrix with positive coordinates. A *nonnegative matrix* is a matrix with nonnegative coordinates. Let $D = \text{diag}(x_1, \ldots, x_n)$ denote the $n \times n$ diagonal matrix with coordinates $x_1, \ldots, x_n$ on the main diagonal. The diagonal matrix D is *positive* if its coordinates $x_1, \ldots, x_n$ are positive. If $A = (a_{i,j})$ is an $m \times n$ positive matrix, if $X = \text{diag}(x_1, \ldots, x_m)$ is an $m \times m$ positive diagonal matrix, and if $Y = \text{diag}(y_1, \ldots, y_n)$ is an $n \times n$ positive diagonal matrix, then $XA = (x_i a_{i,j})$, $AY = (a_{i,j} y_j)$, $XAY = (x_i a_{i,j} y_j)$ are $m \times n$ positive matrices.

Let $A = (a_{i,j})$ be an $n \times n$ matrix. The ith *row sum* of A is

$$\text{rowsum}_i(A) = \sum_{j=1}^{n} a_{i,j}.$$

The jth *column sum* of A is

$$\text{colsum}_j(A) = \sum_{i=1}^{n} a_{i,j}.$$

M. B. Nathanson (✉)
Lehman College (CUNY), Bronx, NY 10468, USA
e-mail: melvyn.nathanson@lehman.cuny.edu

M. B. Nathanson (ed.), *Combinatorial and Additive Number Theory III*, Springer Proceedings in Mathematics & Statistics 297,
https://doi.org/10.1007/978-3-030-31106-3_12

The matrix A is *row stochastic* if it is nonnegative and $\text{rowsum}_i(A) = 1$ for all $i \in \{1, \ldots, n\}$. The matrix A is *column stochastic* if it is nonnegative and $\text{colsum}_j(A) = 1$ for all $j \in \{1, \ldots, n\}$. The matrix A is *doubly stochastic* if it is both row stochastic and column stochastic.

Let $A = (a_{i,j})$ be a nonnegative $n \times n$ matrix such that $\text{rowsum}_i(A) > 0$ and $\text{colsum}_j(A) > 0$ for all $i, j \in \{1, \ldots, n\}$. Define the $n \times n$ positive diagonal matrix

$$X(A) = \text{diag}\left(\frac{1}{\text{rowsum}_1(A)}, \frac{1}{\text{rowsum}_2(A)}, \ldots, \frac{1}{\text{rowsum}_n(A)}\right).$$

Multiplying A on the left by $X(A)$ multiplies each coordinate in the ith row of A by $1/\text{rowsum}_i(A)$, and so

$$(X(A)A)_{i,j} = \frac{a_{i,j}}{\text{rowsum}_i(A)}$$

and

$$\begin{aligned}\text{rowsum}_i\,(X(A)A) &= \sum_{j=1}^{n} (X(A)A)_{i,j} = \sum_{j=1}^{n} \frac{a_{i,j}}{\text{rowsum}_i(A)} \\ &= \frac{\text{rowsum}_i(A)}{\text{rowsum}_i(A)} = 1\end{aligned}$$

for all $i \in \{1, 2, \ldots, n\}$. The process of multiplying A on the left by $X(A)$ to obtain the row stochastic matrix $X(A)A$ is called *row scaling*. We have $X(A)A = A$ if and only if A is row stochastic if and only if $X(A) = I$. Note that the row stochastic matrix $X(A)A$ is not necessarily column stochastic.

Similarly, we define the $n \times n$ positive diagonal matrix

$$Y(A) = \text{diag}\left(\frac{1}{\text{colsum}_1(A)}, \frac{1}{\text{colsum}_2(A)}, \ldots, \frac{1}{\text{colsum}_n(A)}\right).$$

Multiplying A on the right by $Y(A)$ multiplies each coordinate in the jth column of A by $1/\text{colsum}_j(A)$, and so

$$(AY(A))_{i,j} = \frac{a_{i,j}}{\text{colsum}_j(A)}$$

and

$$\begin{aligned}\text{colsum}_j(AY(A)) &= \sum_{i=1}^{n} (AY(A))_{i,j} = \sum_{i=1}^{n} \frac{a_{i,j}}{\text{colsum}_j(A)} \\ &= \frac{\text{colsum}_j(A)}{\text{colsum}_j(A)} = 1\end{aligned}$$

for all $j \in \{1, 2, \ldots, n\}$. The process of multiplying A on the right by $Y(A)$ to obtain a column stochastic matrix $AY(A)$ is called *column scaling*. We have $AY(A) = A$ if and only if $Y(A) = I$ if and only if A is column stochastic. The column stochastic matrix $AY(A)$ is not necessarily row stochastic.

Let A be a positive $n \times n$ matrix. Alternately row scaling and column scaling the matrix A produces an infinite sequence of matrices that converges to a doubly stochastic matrix This result (due to Brualdi, Parter, and Schnieder [1], Letac [3], Menon [4], Sinkhorn [7], Sinkhorn–Knopp [8], Tverberg [9], and others) is classical.

Nathanson [5, 6] proved that if A is a 2×2 positive matrix that is not doubly stochastic but becomes doubly stochastic after a finite number L of scalings, then L is at most 2, and the 2×2 row stochastic matrices that become doubly stochastic after exactly one column scaling were computed explicitly. An open question was to describe $n \times n$ matrices with $n \geq 3$ that are not doubly stochastic but become doubly stochastic after finitely many scalings. Ekhad and Zeilberger [2] discovered the following row-stochastic but not column stochastic 3×3 matrix, which requires exactly one column scaling to become doubly stochastic:

$$A = \begin{pmatrix} 1/5 & 1/5 & 3/5 \\ 2/5 & 1/5 & 2/5 \\ 3/5 & 1/5 & 1/5 \end{pmatrix}. \tag{1}$$

Column scaling A produces the doubly stochastic matrix

$$AY(A) = \begin{pmatrix} 1/6 & 1/3 & 3/6 \\ 2/6 & 1/3 & 2/6 \\ 3/6 & 1/3 & 1/6 \end{pmatrix}.$$

The following construction generalizes this example. For every $n \geq 3$, there is a two parameter family of row-stochastic $n \times n$ matrices that require exactly one column scaling to become doubly stochastic

Let $A = \left(a_{i,j}\right)$ be an $m \times n$ matrix. For $i = 1, \ldots, m$, we denote the ith row of A by

$$\text{row}_i(A) = \left(a_{i,1}, a_{i,2}, \ldots, a_{i,n}\right).$$

Theorem 1 *Let k and ℓ be positive integers, and let $n > \max(2k, 2\ell)$. Let x and z be positive real numbers such that*

$$0 < x + z < \frac{1}{k} \quad \textit{and} \quad x + z \neq \frac{2}{n} \tag{2}$$

and let

$$y = \frac{x+z}{2} \quad \textit{and} \quad w = \frac{1 - k(x+z)}{n - 2k}. \tag{3}$$

The $n \times n$ matrix A such that

$$row_i(A) = \begin{cases} (\underbrace{x, x, \ldots, x}_{k} \underbrace{w, w, \ldots, w}_{n-2k} \underbrace{z, z, \ldots, z}_{k} & if\, i \in \{1, 2, \ldots, \ell\} \\ (\underbrace{y, y, \ldots, y}_{k} \underbrace{w, w, \ldots, w}_{n-2k} \underbrace{y, y, \ldots, y}_{k} & if\, i \in \{\ell+1, \ell+2, \ldots, n-\ell\} \\ (\underbrace{z, z, \ldots, z}_{k} \underbrace{w, w, \ldots, w}_{n-2k} \underbrace{x, x, \ldots, x}_{k} & if\, i \in \{n-\ell+1, n-\ell+2, \ldots, n\} \end{cases}$$

is row stochastic but not column stochastic. The matrix obtained from A after one column scaling is doubly stochastic.

Proof If

$$i \in \{1, 2, \ldots, \ell\} \cup \{n - \ell + 1, n - \ell + 2, \ldots, n\}$$

then

$$\text{rowsum}_i(A) = k(x + z) + (n - 2k)w = 1.$$

If

$$i \in \{\ell + 1, \ell + 2, \ldots, n - \ell\}$$

then

$$\text{rowsum}_i(A) = 2ky + (n - 2k)w = 1.$$

Thus, the matrix A is row stochastic.

If

$$j \in \{1, 2, \ldots, k\} \cup \{n - k + 1, n - k + 2, \ldots, n\}$$

then

$$\text{colsum}_j(A) = \ell x + (n - 2\ell)y + \ell z = ny = \frac{n}{2}(x + z) \neq 1.$$

If

$$j \in \{k + 1, k + 2, \ldots, n - k\}$$

then

$$\text{colsum}_j(A) = nw \neq 1.$$

Thus, matrix A is not column stochastic.

The column scaling matrix for A is the positive diagonal matrix

$$Y(A) = \text{diag}\left(\underbrace{\frac{1}{ny}, \ldots, \frac{1}{ny}}_{k}, \underbrace{\frac{1}{nw}, \ldots, \frac{1}{nw}}_{n-2k}, \underbrace{\frac{1}{ny}, \ldots, \frac{1}{ny}}_{k} \right).$$

For the column scaled matrix $AY(A)$, we have the following row sums. If

$$i \in \{1, 2, \ldots, \ell\} \cup \{n - \ell + 1, n - \ell + 2, \ldots, n\}$$

then

$$\text{rowsum}_i(AY(A)) = \frac{kx}{ny} + \frac{(n-2k)w}{nw} + \frac{kz}{ny} = \frac{k(x+z)}{ny} + 1 - \frac{2k}{n} = 1.$$

If

$$i \in \{\ell + 1, \ell + 2, \ldots, n - \ell\}$$

then

$$\text{rowsum}_i(A) = \frac{2ky}{ny} + \frac{(n-2k)w}{nw} = \frac{2k}{n} + 1 - \frac{2k}{n} = 1.$$

Thus, the matrix $AY(A)$ is row stochastic. This completes the proof. □

For example, let $k = \ell = 1$ and $n = 3$, and let w, x, y, z be positive real numbers such that

$$0 < x + z < 1, \qquad x + z \neq \frac{2}{3}$$

$$y = \frac{x+z}{2} \quad \text{and} \quad w = 1 - x - z.$$

The matrix

$$A = \begin{pmatrix} x & w & z \\ y & w & y \\ z & w & x \end{pmatrix}, \tag{4}$$

is row stochastic but not column stochastic. By Theorem 1, column scaling A produces a doubly stochastic matrix. Choosing $x = 1/5$ and $z = 3/5$, we obtain the matrix (1).

Here is another example. Let $k = 2$, $\ell = 3$, and $n = 7$. Choosing

$$x = \frac{1}{4}, \quad y = \frac{3}{16}, \quad z = \frac{1}{8}, \quad w = \frac{1}{12}$$

we obtain the row but not column stochastic matrix

$$A = \begin{pmatrix} 1/4 & 1/4 & 1/12 & 1/12 & 1/12 & 1/8 & 1/8 \\ 1/4 & 1/4 & 1/12 & 1/12 & 1/12 & 1/8 & 1/8 \\ 1/4 & 1/4 & 1/12 & 1/12 & 1/12 & 1/8 & 1/8 \\ 3/16 & 3/16 & 1/12 & 1/12 & 1/12 & 3/16 & 3/16 \\ 1/8 & 1/8 & 1/12 & 1/12 & 1/12 & 1/4 & 1/4 \\ 1/8 & 1/8 & 1/12 & 1/12 & 1/12 & 1/4 & 1/4 \\ 1/8 & 1/8 & 1/12 & 1/12 & 1/12 & 1/4 & 1/4 \end{pmatrix}.$$

Column scaling produces the doubly stochastic matrix

$$AY(A) = \begin{pmatrix} 4/21 & 4/21 & 1/7 & 1/7 & 1/7 & 2/21 & 2/21 \\ 4/21 & 4/21 & 1/7 & 1/7 & 1/7 & 2/21 & 2/21 \\ 4/21 & 4/21 & 1/7 & 1/7 & 1/7 & 2/21 & 2/21 \\ 1/7 & 1/7 & 1/7 & 1/7 & 1/7 & 1/7 & 1/7 \\ 2/21 & 2/21 & 1/7 & 1/7 & 1/7 & 4/21 & 4/21 \\ 2/21 & 2/21 & 1/7 & 1/7 & 1/7 & 4/21 & 4/21 \\ 2/21 & 2/21 & 1/7 & 1/7 & 1/7 & 4/21 & 4/21 \end{pmatrix}.$$

Theorem 2 *Every $n \times n$ matrix A constructed in Theorem 1 satisfies* $\det(A) = 0$.

Proof There are three cases.

If $k > 1$ or $n - 2k > 1$, then A has two equal columns and $\det(A) = 0$.

If $\ell > 1$ or $n - 2\ell > 1$, then A has two equal rows and $\det(A) = 0$.

If $k = \ell = 1$ and $n = 3$, then

$$A = \begin{pmatrix} x & w & z \\ y & w & y \\ z & w & x \end{pmatrix}$$

and

$$\det(A) = w(x - z)(x + z - 2y) = 0.$$

This completes the proof. □

Theorem 2 is of interest for the following reason. Let $A = \left(a_{i,j}\right)$ be an $n \times n$ matrix. If $\det(A) \neq 0$, then the system of linear equations

$$\begin{aligned} a_{1,1}t_1 + a_{2,1}t_2 + \cdots + a_{n,1}t_n &= 1 \\ a_{1,2}t_1 + a_{2,2}t_2 + \cdots + a_{n,2}t_n &= 1 \\ &\vdots \\ a_{1,n}t_1 + a_{2,n}t_2 + \cdots + a_{n,n}t_n &= 1 \end{aligned}$$

has a unique solution. Equivalently, if $\det(A) \neq 0$, then there exists a unique $n \times n$ diagonal matrix $T = \operatorname{diag}(t_1, \ldots, t_n)$ such that the matrix $B = TA$ is column stochastic.

Suppose that the matrix A is positive and row stochastic. If $t_i > 0$ for all $i \in \{1, \ldots, n\}$, then T is invertible and $B = TA$ is a positive column stochastic matrix. Setting $X = T^{-1}$, we have $XB = A$. Moreover, X is the row scaling matrix associated to B. Thus, if A is a row stochastic matrix such that column scaling A produces a doubly stochastic matrix, then we have pulled A back to a column stochastic matrix B, and we have increased by 1 the number of scalings needed to get a doubly stochastic matrix.

Unfortunately, the matrices constructed in Theorem 1 have determinant 0.

2 Open Problems

1. Does there exist a positive 3×3 row stochastic but not column stochastic matrix A with nonzero determinant such that A becomes doubly stochastic after one column scaling?
2. Let A be a positive 3×3 row stochastic but not column stochastic matrix that becomes doubly stochastic after one column scaling. Does $\det(A) = 0$ imply that A has the shape of matrix (4)?
3. Here is the inverse problem: Let A be an $n \times n$ row-stochastic matrix. Does there exist a column stochastic matrix B such that row scaling B produces A (equivalently, such that $X(B)B = A$)? Compute B.
4. Modify the above problems so that the matrices are required to have rational coordinates.
5. Determine if, for positive integers $L \geq 3$ and $n \geq 3$, there exists a positive $n \times n$ matrix that requires exactly L scalings to reach a doubly stochastic matrix.
6. Classify all matrices for which the alternate scaling algorithm terminates in finitely many steps.

Acknowledgements Supported in part by a grant from the PSC-CUNY Research Award Program.

References

1. R. A. Brualdi, S. V. Parter, and H. Schneider, *The diagonal equivalence of a nonnegative matrix to a stochastic matrix*, J. Math. Anal. Appl. 16 (1966), 31–50.
2. S. B. Ekhad and D. Zeilberger, *Answers to some questions about explicit Sinkhorn limits posed by Mel Nathanson*, arXiv:1902.10783, 2019.
3. G. Letac, *A unified treatment of some theorems on positive matrices*, Proc. Amer. Math. Soc. **43** (1974), 11–17.
4. M. V. Menon, *Reduction of a matrix with positive elements to a doubly stochastic matrix*, Proc. Amer. Math. Soc. 18 (1967), 244–247.
5. M. B. Nathanson, *Alternate minimization and doubly stochastic matrices*, arXiv:1812.11935, 2018.
6. M. B. Nathanson, *Matrix scaling and explicit doubly stochastic limits*, Linear Algebra and its Applications 578 (2019), 111–132; https://doi.org/10.1016/j.laa.2019.05.004
7. R. Sinkhorn, *A relationship between arbitrary positive matrices and doubly stochastic matrices*, Ann. Math. Statist. **35** (1964), 876–879.
8. R. Sinkhorn and P. Knopp, *Concerning nonnegative matrices and doubly stochastic matrices*, Pacific J. Math. **21** (1967), 343–348.
9. H. Tverberg, *On Sinkhorn's representation of nonnegative matrices*, J. Math. Anal. Appl. 54 (1976), no. 3, 674–677.

Not All Groups Are LEF Groups, or Can You Know If a Group Is Infinite?

Melvyn B. Nathanson

Abstract This is an introduction to the class of groups that are locally embeddable into finite groups.

2010 Mathematics Subject Classification: 20E25 · 20F05 · 20-02

1 Finite or Infinite?

A simple question: Do the finite subsets of a group tell us if the group is infinite? Assume that we can only see the finite subsets of a group, and, also, that we can determine if a finite subset is a subset of some finite group. This means that we can answer the following question. Let A be a finite subset of a group G. Does there exist a finite group H and a *partial homomorphism* $f : A \to H$ that is one-to-one. A partial homomorphism from a subset A of a group to a group H is a function $f : A \to H$ such that, if $a, b \in A$ and $ab \in A$, then $f(ab) = f(a)f(b)$. A one-to-one partial homomorphism is also called a *local embedding*. Of course, if the group G is finite, then, for every subset A of G, the restriction of the identity homomorphism on G to the subset A is a local embedding into a finite group.

Does there exist an infinite group G such that every finite subset of G looks like (equivalently, can be partially embedded into) a subset of a finite group? Does there exist an infinite group G in which some finite subset of G is not also a subset of a finite group?

Theorem 3 answers the second question. The following example answers the first question. Let A be a nonempty finite subset of the infinite abelian group $\mathbf{Z}$. Choose an integer

$$m > \max\{|a - b| : a, b \in A\} = \max(A) - \min(A).$$

M. B. Nathanson (✉)
Lehman College (CUNY), Bronx, NY 10468, USA
e-mail: melvyn.nathanson@lehman.cuny.edu

M. B. Nathanson (ed.), *Combinatorial and Additive Number Theory III*,
Springer Proceedings in Mathematics & Statistics 297,
https://doi.org/10.1007/978-3-030-31106-3_13

Consider the function $f : A \to \mathbf{Z}/m\mathbf{Z}$ defined by $f(a) = a + m\mathbf{Z}$ for all $a \in A$. This is a partial homomorphism because it is the restriction of the canonical homomorphism $a \mapsto a + m\mathbf{Z}$ from $\mathbf{Z}$ to $\mathbf{Z}/m\mathbf{Z}$. For $a, b \in A$, we have $f(a) = f(b)$ if and only if $a \equiv b \pmod{m}$ if and only if m divides $|a - b|$. The inequality $|a - b| < m$ implies that $f(a) = f(b)$ if and only if $a = b$, and so f is a local embedding. Thus, every finite subset of the infinite group $\mathbf{Z}$ can be embedded into a finite cyclic group. By looking only at finite subsets, we cannot decide if $\mathbf{Z}$ is infinite.

Let us call a group G *locally embeddable into finite groups*, or an *LEF group*, if every finite subset of G can be embedded into a finite group. Mal'cev [5] introduced this concept in general algebraic structures. Vershik and Gordon [8] extended it to groups, and obtained many fundamental results.

Here are two classes of LEF groups.

Theorem 1 *Every locally finite group is an LEF group. Every abelian group is an LEF group.*

Proof A group is *locally finite* if every finite subset generates a finite group. For such groups, the proof is immediate from the definition.

For abelian groups, the proof follows easily from the structure theorem for finitely generated abelian groups, and an easy modification of the preceding argument that $\mathbf{Z}$ is an LEF group. □

It is natural to ask: Is every infinite group an LEF group, or does there exist an infinite group that is not an LEF group?

2 Finitely Presented Groups

Let W be a group with identity e, and let X be a subset of W that generates W. We assume that $e \notin X$. The *length* of an element $w \in W$ with $w \neq e$ is the smallest positive integer $k = \ell(w)$ such that there is a representation of w in the form

$$w = x_1^{\varepsilon_1} x_2^{\varepsilon_2} \cdots x_k^{\varepsilon_k} \tag{1}$$

where

$$x_i \in X \text{ and } \varepsilon_i \in \{1, -1\} \text{ for } i = 1, \ldots, k. \tag{2}$$

We define $\ell(e) = 0$. Note that $\ell(w) = 1$ if and only if $w = x$ or $w = x^{-1}$ for some $x \in X$.

For every nonnegative integer L, we define the "closed ball"

$$B_L = \{w \in W : \ell(w) \leq L\}.$$

We have

$$B_0 = \{e\} \qquad \text{and} \qquad B_1 = \{e\} \cup \{x^{\varepsilon} : x \in X \text{ and } \varepsilon \in \{1, -1\}\}.$$

If the generating set X is finite, then, for every L, the group W contains only finitely many elements w of length $\ell(w) \leq L$, and so the B_L is a finite subset of W.

If $w \in B_L$, then w satisfies (1) and (2) for some $k \leq L$. For all $j = 1, \ldots, k$, the partial product

$$w_j = x_1^{\varepsilon_1} x_2^{\varepsilon_2} \cdots x_j^{\varepsilon_j}$$

has length $\ell(w_j) \leq j \leq L$, and so $w_j \in B_L$. (We observe that if $\ell(w_j) < j$, then $\ell(w) < k$, which is absurd. Therefore, $\ell(w_j) = j$ for all $j \in \{1, \ldots, k\}$.) Let $w_0 = e$. Note that $w = w_k$ and that

$$w_j = w_{j-1} x_j^{\varepsilon_j}$$

for all $j \in \{1, \ldots, k\}$. If $f : B_L \to H$ is a partial homomorphism, then

$$\begin{aligned} f(w) &= f(w_{k-1} x_k^{\varepsilon_k}) = f(w_{k-1}) f(x_k^{\varepsilon_k}) \\ &= f(w_{k-2} x_{k-1}^{\varepsilon_{k-1}}) f(x_k^{\varepsilon_k}) = f(w_{k-2}) f\left(x_{k-1}^{\varepsilon_{k-1}}\right) f(x_k^{\varepsilon_k}) \\ &= \cdots \\ &= f\left(x_1^{\varepsilon_1}\right) f(x_2^{\varepsilon_2}) \cdots f\left(x_{k-1}^{\varepsilon_{k-1}}\right) f(x_k^{\varepsilon_k}) \end{aligned}$$

For partial products in finite groups, see Nathanson [6].

Let X be a nonempty set, and let $F(X)$ be the free group generated by X. Let R be a nonempty subset of $F(X)$. The *normal closure* of R in $F(X)$, denoted $N(R)$, is the smallest normal subgroup of $F(X)$ that contains R. The subgroup $N(R)$ is generated by the set

$$\left\{ w r^{\varepsilon} w^{-1} : w \in F(X),\ r \in R,\ \varepsilon \in \{1, -1\} \right\}.$$

A group G is *finitely presented* if

$$G = \langle X; R \rangle = F(X)/N(R)$$

where $F(X)$ is the free group generated by a finite set X and the subgroup $N(R)$ is the normal closure of a finite subset R of $F(X)$. If $\pi : F(X) \to G$ is the canonical homomorphism, then the set

$$X^* = \pi(X) = \{x N(R) : x \in X\}$$

generates G.

The following result is Proposition 1.10 in Pestov and Kwiatkowska [7].

Theorem 2 *Let G be a finitely presented infinite group. If G is an LEF group, then G contains a nontrivial proper normal subgroup. Equivalently, a finitely presented infinite simple group is not an LEF group.*

Proof Let $G = \langle X; R \rangle = F(X)/N$ be a finitely presented infinite group, where $F(X)$ is the free group generated by a finite set X, and $N = N(R)$ is the normal closure of a finite subset R of $F(X)$. Let e_F be the identity in $F(X)$. The identity in G is $e_G = e_F N = N$. The canonical homomorphism $\pi : F(X) \to G$ is defined by $\pi(w) = wN$ for all $w \in F(X)$.

Choose an integer L such that

$$L \geq \max\{\ell(w) : w \in X \cup R\}.$$

The closed ball

$$B_L = \{w \in F(X) : \ell(w) \leq L\}$$

is a finite subset of $F(X)$. We have

$$\{e_F\} \cup X \cup X^{-1} \cup R \subseteq B_L.$$

The set

$$A = \pi(B_L) \subseteq G$$

is a finite subset of G that contains $X^* = \pi(X)$. Also, $e_G = \pi(e_F) = N \in A$.

If G is an LEF group, then there exist a finite group H and a local embedding f of A into H. Let e_H be the identity in H. For all $x \in X$, we have $\pi(x) \in A$ and so

$$f\pi(x) \in H.$$

By the universal property of a free group, there exists a unique homomorphism

$$f^* : F(X) \to H$$

such that

$$f^*(x) = f\pi(x)$$

for all $x \in X$. The subgroup

$$N^* = \text{kernel}(f^*)$$

is a normal subgroup of $F(X)$. We shall prove that

$$N \subsetneq N^* \subsetneq F(X). \tag{3}$$

The diagram is

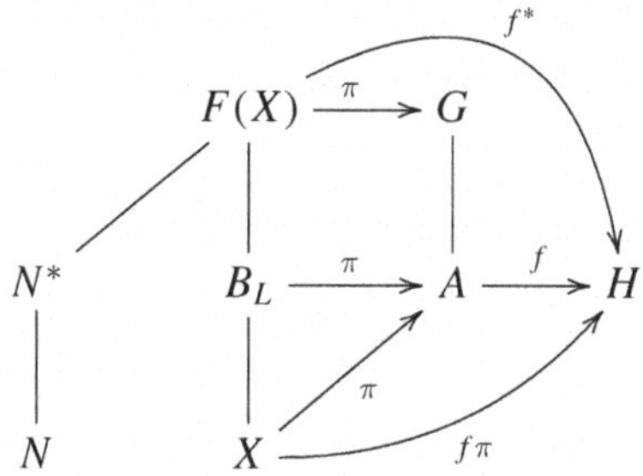

If $N^* = F(X)$, then $x \in N^*$ for all $x \in X$. Because $X \subseteq B_L$ and $\pi(x) = xN \in A$, we have

$$f(xN) = f\pi(x) = f^*(x) = e_H = f(N).$$

Because f is one-to-one and $f(xN) = f(N)$, it follows that $\pi(x) = xN = N$ for all $x \in X$. The set $\pi(X)$ generates G, and so $G = \{N\}$ is the trivial group, which is absurd. Therefore, N^* is a proper normal subgroup of $F(X)$.

Next we prove that N^* contains N. Let $r \in R$. There is a nonnegative integer $k = \ell(r) \le L$ such that

$$r = \prod_{i=1}^{k} x_i^{\varepsilon_i}$$

where $(x_i)_{i=1}^k$ is a sequence of elements of X and $(\varepsilon_i)_{i=1}^k$ is a sequence of elements of $\{1, -1\}$.

Because $r \in R \subseteq N$, we have $rN = N$ and

$$\begin{aligned} f^*(r) &= f^*\left(\prod_{i=1}^{k} x_i^{\varepsilon_i}\right) = \prod_{i=1}^{k} f^*(x_i)^{\varepsilon_i} = \prod_{i=1}^{k} f\pi(x_i)^{\varepsilon_i} \\ &= \prod_{i=1}^{k} f(x_i N)^{\varepsilon_i} = f\left(\prod_{i=1}^{k} x_i^{\varepsilon_i} N\right) = f(rN) \\ &= f(N) = e_H. \end{aligned}$$

Therefore, $r \in N^*$. Because $R \subseteq N^*$ and N^* is a normal subgroup of $F(X)$, it follows that N^* contains N, which is the normal closure of R, and so $N \subseteq N^*$.

Finally, if $N = N^* = \text{kernel}(f^*)$, then $G = F(X)/N = F(X)/N^*$ is isomorphic to a subgroup of the finite group H, and so G is finite, which is absurd. Therefore, N is a proper subgroup of N^*.

This proves relation (3). The correspondence theorem in group theory implies that N^*/N is a nontrivial proper normal subgroup of G, and so G is not a simple group. It follows that no finitely presented infinite simple group is an LEF group. This completes the proof. □

Theorem 3 *There exist infinite groups that are not LEF groups. In particular, the Thompson groups T and V are not LEF groups.*

Proof The Thompson groups T and V are finitely presented infinite simple groups (Cannon, Floyd, and Parry [2],Cannon and Floyd [1]). □

For recent work, including other examples of groups that are not LEF groups, see [3, 4].

Acknowledgements Supported in part by a grant from the PSC-CUNY Research Award Program.

References

1. J. W. Cannon and W. J. Floyd, What is ... Thompson's group?, Notices Amer. Math. Soc. **58** (2011), no. 8, 1112–1113.
2. J. W. Cannon, W. J. Floyd, and W. R. Parry, *Introductory notes on Richard Thompson's groups*, Enseign. Math. (2) **42** (1996), no. 3-4, 215–256.
3. Y. Cornulier, Sofic profile and computability of Cremona groups, Michigan Math. J. **62** (2013), no. 4, 823–841.
4. Y. de Cornulier, L. Guyot, and W. Pitsch, On the isolated points in the space of groups, J. Algebra **307** (2007), no. 1, 254–277.
5. A. Malcev, *On isomorphic matrix representations of infinite groups*, Rec. Math. [Mat. Sbornik] N.S. **8 (50)** (1940), 405–422.
6. M. B. Nathanson, Partial products in finite groups, Discrete Math. **15** (1976), no. 2, 201–203.
7. V. G. Pestov and A. Kwiatkowska, *An introduction to hyperlinear and sofic groups*, Appalachian Set Theory 2006–2012, London Math. Soc. Lecture Note Ser., vol. 406, Cambridge Univ. Press, Cambridge, 2013, pp. 145–185.
8. A. M. Vershik and E. I. Gordon, Groups that are locally embeddable in the class of finite groups, Algebra i Analiz **9** (1997), no. 1, 71–97.

Binary Quadratic Forms in Difference Sets

Alex Rice

Abstract We show that if $h(x, y) = ax^2 + bxy + cy^2 \in \mathbb{Z}[x, y]$ satisfies $\Delta(h) = b^2 - 4ac \neq 0$, then any subset of $\{1, 2, \ldots, N\}$ lacking nonzero differences in the image of h has size at most a constant depending on h times $N \exp(-c\sqrt{\log N})$, where $c = c(h) > 0$. We achieve this goal by adapting an L^2 density increment strategy previously used to establish analogous results for sums of one or more single-variable polynomials. Our exposition is thorough and self-contained, in order to serve as an accessible gateway for readers who are unfamiliar with previous implementations of these techniques.

MSC 2010 11B30

1 Introduction

Established independently by Sárközy and Furstenberg during the 1970s, settling a question of Lovász, it is a well-studied fact that any set of integers of positive upper density necessarily contains two distinct elements that differ by a perfect square. Equivalently, if $A \subseteq \mathbb{N}$ contains no such pair, then

$$\lim_{N\to\infty} \frac{|A \cap [1, N]|}{N} = 0.$$

Here we use $[1, N]$ to denote $\{1, 2, \ldots, N\}$ and $|X|$ to denote the size of a finite set X. Furstenberg [2] achieved this result qualitatively via ergodic theory, specifically his correspondence principle, but obtained no information on the rate at which the density must decay, while Sárközy [20] employed a Fourier analytic density increment strategy to show that if $A \subseteq [1, N]$ has no square differences, then

A. Rice (✉)
Department of Mathematics, Millsaps College, Jackson, MS 39210, USA
e-mail: riceaj@millsaps.edu

M. B. Nathanson (ed.), *Combinatorial and Additive Number Theory III*,
Springer Proceedings in Mathematics & Statistics 297,
https://doi.org/10.1007/978-3-030-31106-3_14

$$\frac{|A|}{N} \ll \left(\frac{(\log\log N)^2}{\log N}\right)^{1/3}. \tag{1}$$

Throughout the paper we use log to denote the natural logarithm, and we use "$\ll$" to denote "less than a constant times", with subscripts indicating on what parameters, if any, the implied constant depends. Sárközy's argument was driven by the Hardy–Littlewood circle method, and was inspired by Roth's [14] proof that sets of positive upper density contain three-term arithmetic progressions.

Using a more intricate Fourier analytic argument, Pintz, Steiger, and Szemerédi [13] improved (1) to

$$|A| \ll N(\log N)^{-c\log\log\log\log N}, \tag{2}$$

with $c = 1/12$. While more elementary Fourier analytic proofs [3, 10] and a Fourier-free density increment proof [4] have also been discovered, it is versions of these two Fourier analytic attacks that have yielded the best quantitative information. In the ensuing decades, these two methods have been refined and applied to other sets of prohibited differences, such as more general polynomial images [1, 5, 9, 22], shifted primes [8, 19, 21], polynomial curves in higher-dimensional integer lattices [11], and images of the primes under polynomials [7, 17].

With regard to sums of one or more single-variable polynomials, the author [15] pushed these two methods to their breaking points. In the case of one single-variable polynomial, if $h \in \mathbb{Z}[x]$ has degree $k \geq 2$ and $h(\mathbb{N})$ contains a multiple of q for every $q \in \mathbb{N}$, known as an *intersective polynomial*, then any set $A \subseteq [1, N]$ with no nonzero differences in the image of h satisfies (2) for any $c < (\log((k^2+k)/2))^{-1}$, with the implied constant depending on h and c. The intersective condition is necessary to force any density decay, as otherwise one can take $A = q\mathbb{N}$ if $h(\mathbb{N})$ misses $q\mathbb{Z}$, and in that sense this is a maximal extension of the elaborate techniques developed in [1, 13].

Further, if we allow the additional degree of freedom of a second polynomial, then the more straightforward density increment approach yields density bounds that are even better than (2), as described below.

Theorem 1 ([15]) *Suppose $g, h \in \mathbb{Z}[x]$ are nonzero intersective polynomials and $A \subseteq [1, N]$. If*

$$a - a' \neq g(m) + h(n)$$

for all distinct pairs $a, a' \in A$ and all $m, n \in \mathbb{N}$, then

$$|A| \ll_{g,h} Ne^{-c(\log N)^{\mu}},$$

where $c = c(g, h) > 0$, $\mu = \mu(\deg(g), \deg(h)) > 0$, and $\mu(2, 2) = 1/2$.

As a notable example, Theorem 1 gives an upper bound of $\exp(-c\sqrt{\log N})$ for the density of subsets of $[1, N]$ lacking differences that are the sum of two squares. There is also a brief discussion of sums of three or more single-variable polynomials at the end of [15], but the improvements in density bounds are modest as $\exp(-c\sqrt{\log N})$ arises as a natural limit of the method, a fact that we discuss in Sect. 2.3.

While the generality of Theorem 1 is pleasing, prohibited differences of the form $g(m) + h(n)$ can be thought of as the diagonal special case of differences of the form $h(m,n)$ where $h \in \mathbb{Z}[x,y]$. Of course, if $h(x,y) = \tilde{h}(g(x,y))$ for some $g \in \mathbb{Z}[x,y]$ and $\tilde{h} \in \mathbb{Z}[x]$ with $\deg(\tilde{h}) \geq 2$, then the image of h is contained in the image of $\tilde{h}$, in which case we could not hope to improve on the original setting of one single-variable polynomial. However, in other cases, we expect that the freedom of two variables should allow for improved density bounds. It is with this expectation in mind that we gently wade into the arena of potentially non-diagonal two-variable polynomials by exploring the following natural generalization of the aforementioned special case $m^2 + n^2$.

Definition 1 $h \in \mathbb{Z}[x,y]$ is called a *binary quadratic form* if

$$h(x,y) = ax^2 + bxy + cy^2$$

for some $a, b, c \in \mathbb{Z}$. Further, we define the *discriminant* of h by

$$\Delta(h) = b^2 - 4ac,$$

noting that $h(x,y) = d(rx+sy)^2$ for some $d, r, s \in \mathbb{Z}$ if and only if $\Delta(h) = 0$.

Our main result is the following, which says that under the necessary restriction that a binary quadratic form does not collapse into a dilated perfect square, we achieve the same density bounds previously established in the diagonal case, which are likely the best possible for our chosen method.

Theorem 2 *Suppose $h \in \mathbb{Z}[x,y]$ is a binary quadratic form with $\Delta(h) \neq 0$. If $A \subseteq [1,N]$ with*

$$a - a' \neq h(m,n)$$

for all distinct pairs $a, a' \in A$ and all $m, n \in \mathbb{N}$, then

$$|A| \ll_h Ne^{-c\sqrt{\log N}},$$

where $c = c(h) > 0$.

We note that the image of every nonzero binary quadratic form contains a dilation of the squares, and hence our result is only material because the established density bound is better than (2). Our goal for the remainder of the paper is twofold: to establish Theorem 2, which we hope will serve as a starting point for the application of these methods to more general polynomials in several variables, and to provide thorough and self-contained exposition of all of the components of this iteration scheme for those unfamiliar with its previous applications, such as the original case of the squares.

2 Main Iteration Lemma: Deducing Theorem 2

The principle behind a density increment strategy is that a set which lacks the desired arithmetic structure should spawn a new, significantly denser subset of a slightly smaller interval with an inherited lack of arithmetic structure. Iterating this procedure enough times for the density to reach 1 provides an upper bound on the density of the original set.

For this section, we fix a binary quadratic form $h \in \mathbb{Z}[x, y]$ with $\Delta(h) \neq 0$, and we let

$$I(h) = \{h(m, n) : m, n \in \mathbb{N}\} \setminus \{0\}.$$

Our iteration scheme is encapsulated by the following lemma, from which we quickly deduce Theorem 2.

Lemma 1 *Suppose* $A \subseteq [1, N]$ *with* $|A| = \delta N$ *and* $\delta \geq N^{-1/20}$. *If* $(A - A) \cap I(h) = \emptyset$, *then there exists* $A' \subseteq [1, N']$ *with* $|A'| = \delta' N'$ *and a constant* $c = c(h) > 0$ *with*

$$N' \gg_h \delta^4 N, \quad \delta' \geq (1 + c)\delta, \quad \text{and} \quad (A' - A') \cap I(h) = \emptyset.$$

2.1 *Proof of Theorem 2*

Suppose $A \subseteq [1, N]$ with $|A| = \delta N$ and $(A - A) \cap I(h) = \emptyset$. Setting $A_0 = A$, $N_0 = N$, and $\delta_0 = \delta$, Lemma 1 yields, for each m, a set $A_m \subseteq [1, N_m]$ with $|A_m| = \delta_m N_m$ and $(A_m - A_m) \cap I(h) = \emptyset$ satisfying

$$N_m \geq c\delta^4 N_{m-1} \geq (c\delta^4)^m N \tag{3}$$

and

$$\delta_m \geq (1 + c)\delta_{m-1} \geq (1 + c)^m \delta \tag{4}$$

as long as

$$\delta_m \geq N_m^{-1/20}. \tag{5}$$

By (4), we see that the density δ_m will surpass 1, and hence (5) must fail, for $m = C \log(\delta^{-1})$. In particular, by (3) we have

$$\delta \leq (c\delta^4)^{-C \log(\delta^{-1})} N^{-1/20},$$

which can be rearranged to

$$N \leq e^{C \log^2(\delta^{-1})}$$

and hence implies

$$\delta \ll_h e^{-c\sqrt{\log N}},$$

as required. □

2.2 Roadmap for the Remainder of the Paper

Our task is is now completely reduced to a proof of Lemma 1, the two major components of which are described below.

i. The condition $(A - A) \cap I(h) = \emptyset$ represents unexpected, nonuniform behavior, which we expect to be detectable in the Fourier analytic behavior of A. More specifically, we use orthogonality of characters and adaptations of standard exponential sum estimates to locate a single small denominator q such that the Fourier transform of the characteristic function of A, translated to have mean value zero, has substantial L^2 concentration near rationals with denominator q. The Fourier analytic infrastructure is introduced in Sect. 3.1, the proof of this component is carried out in Sect. 4.2, and the required exponential sum estimates are exposed in great detail in Sect. 5.
ii. The substantial L^2 concentration of the transform of the translated characteristic function of A near rationals with a particular denominator q indicates a correlation of A with a linear phase function. In particular, we show that this implies that A has significantly increased relative density on a long arithmetic progression P of step size q. Since this implication has nothing to do with h, or any other assumptions about A, we prove a general version preemptively in Sect. 3.2. Finally, by shifting and rescaling the intersection of A with a subprogression of P of step size q^2, we obtain our new, denser set A' with $(A' - A') \cap I(h) = \emptyset$. The complete deduction of Lemma 1 from these two components is carried out in Sect. 4.1.

2.3 A Discussion of Novelty and Bounds

As indicated in the introduction, the procedure outlined in Sect. 2.2, though refined over the years, goes back to Sárközy in the 1970s. The improvement in bounds in Theorems 1 and 2, as compared to the case of one single-variable polynomial, comes from the details of the numerology in Lemma 1, most notably the size of the density increment $\delta' \geq (1 + c)\delta$. This effectively optimal increase in density is facilitated by the quality of the exponential sum estimates mentioned in item (i) above.

More specifically, the size of the density increment can be traced to the level of decay achieved in normalized complete local exponential sums. In the original setting of square differences, for example, the relevant decay comes from the standard estimate

$$\left| \frac{1}{q} \sum_{r=0}^{q-1} e^{2\pi i r^2 a/q} \right| \ll q^{-1/2} \tag{6}$$

for $(a, q) = 1$, which ultimately leads to a density increment $\delta' \geq \delta + c\delta^2$. Substituting this increment size, and other minor necessary modifications, into the proof in Sect. 2.1 leads to the upper bound

$$\delta \ll \frac{\log\log N}{\log N},$$

which is better than Sárközy's original result (1). The reader may refer to [12] or [16] for full expositions of this refinement in the cases of squares, shifted primes, and, in the latter case, intersective polynomials.

In the case of sums of two squares, covered in Theorem 1, the relevant decay comes from the analogous two-variable sum that then splits, allowing one to use the same estimate (6) to conclude

$$\left|\frac{1}{q^2}\sum_{r,s=0}^{q-1} e^{2\pi i(r^2+s^2)a/q}\right| = \left|\frac{1}{q}\sum_{r=0}^{q-1} e^{2\pi i r^2 a/q}\right|^2 \ll q^{-1}$$

for $(a,q)=1$, which is good enough to get the optimal density increment. The novelty of Theorem 2 is rooted in the fact that when $\Delta(h)\neq 0$, we get the same level of decay, namely

$$\left|\frac{1}{q^2}\sum_{r,s=0}^{q-1} e^{2\pi i h(r,s)a/q}\right| \ll_h q^{-1}$$

for $(a,q)=1$, even though the sum no longer necessarily splits.

In order to improve on the bound $\exp(-c\sqrt{\log N})$ using this approach, for any fixed set of prohibited differences, one of two components of the numerology of Lemma 1 must be improved: either the ratio N'/N must decay more slowly than any power of δ, or the ratio δ'/δ must tend to infinity, as $\delta\to 0$, neither of which appear feasible in any nontrivial context. However, the question of whether the known upper bounds are even remotely sharp remains completely open in all of the aforementioned cases. For a more detailed discussion of lower bounds, constructions, and conjectures, the reader may refer to Sect. 1.4 of [15].

3 Preliminaries

3.1 Fourier Analysis and the Circle Method on $\mathbb{Z}$

We embed our finite sets in $\mathbb{Z}$, on which we utilize the discrete Fourier transform. Specifically, for a function $F:\mathbb{Z}\to\mathbb{C}$ with finite support, we define $\widehat{F}:\mathbb{T}\to\mathbb{C}$, where $\mathbb{T}$ denotes the circle parametrized by the interval [0, 1] with 0 and 1 identified, by

$$\widehat{F}(\alpha)=\sum_{n\in\mathbb{Z}} F(n)e^{-2\pi i n\alpha}.$$

In this finite support context, Plancherel's Identity

$$\sum_{n\in\mathbb{Z}} |F(n)|^2 = \int_0^1 |\widehat{F}(\alpha)|^2 d\alpha \tag{7}$$

is a direct consequence of the orthogonality relation

$$\int_0^1 e^{2\pi i n\alpha} d\alpha = \begin{cases} 1 & \text{if } n = 0 \\ 0 & \text{if } n \in \mathbb{Z}\setminus\{0\}. \end{cases} \tag{8}$$

Given $N \in \mathbb{N}$ and a set $A \subseteq [1, N]$ with $|A| = \delta N$, we examine the Fourier analytic behavior of A by considering the *balanced function*, f_A, defined by

$$f_A = 1_A - \delta 1_{[1,N]}.$$

We analyze $\widehat{f_A}$, and other exponential sums, using the Hardy–Littlewood circle method, decomposing the frequency space into two components: the set of points on the circle that are close to rationals with small denominator, and the complement.

Definition 2 Given $N \in \mathbb{N}$ and $\eta > 0$, we define, for each $q \in \mathbb{N}$ and $a \in [1, q]$,

$$\mathbf{M}_{a/q} = \mathbf{M}_{a/q}(N, \eta) = \left\{\alpha \in \mathbb{T} : |\alpha - \frac{a}{q}| < \frac{1}{\eta^2 N}\right\}, \quad \mathbf{M}_q = \bigcup_{(a,q)=1} \mathbf{M}_{a/q},$$

and

$$\mathbf{M}'_q = \bigcup_{r|q} \mathbf{M}_q = \bigcup_{a=1}^{q} \mathbf{M}_{a/q}.$$

We then define $\mathfrak{M}$, the *major arcs* and $\mathfrak{m}$, the *minor arcs*, by

$$\mathfrak{M} = \bigcup_{q=1}^{\eta^{-1}} \mathbf{M}_q, \quad \mathfrak{m} = \mathbb{T}\setminus\mathfrak{M}.$$

We note that if $\eta^2 N > 2Q^2$, then

$$\mathbf{M}_{a/q} \cap \mathbf{M}_{b/r} = \emptyset \tag{9}$$

whenever $a/q \neq b/r$ and $q, r \leq Q$.

3.2 Density Increment Lemma

The following standard result shows that for $A \subseteq [1, N]$, L^2 concentration of $\widehat{f_A}$ near rationals with a particular denominator q implies increased relative density on a long arithmetic progression of step size q, as described in item (ii) in Sect. 2.2.

Lemma 2 *Suppose $A \subseteq [1, N]$ with $|A| = \delta N$. If $q \in \mathbb{N}$, $\sigma, \eta > 0$, and*

$$\int_{\mathbf{M}'_q} |\widehat{f_A}(\alpha)|^2 d\alpha \geq \sigma\delta^2 N,$$

then there exists an arithmetic progression

$$P = \{x + \ell q : 1 \leq \ell \leq L\}$$

with $qL \gg \min\{\sigma, \eta^2\}N$ and $|A \cap P| \geq (1 + \sigma/32)\delta L$.

Proof Suppose $A \subseteq [1, N]$ with $|A| = \delta N$, $\sigma, \eta > 0$. Suppose further that

$$\int_{\mathbf{M}'_q} |\widehat{f_A}(\alpha)|^2 d\alpha \geq \sigma\delta^2 N, \tag{10}$$

and let $P = \{q, 2q, \ldots, Lq\}$ with $L = \lfloor \min\{\sigma, \eta^2\}N/128q \rfloor$. We will show that some translate of P satisfies the conclusion of Lemma 2. We note that for $\alpha \in [0, 1]$,

$$|\widehat{1_P}(\alpha)| = \left|\sum_{\ell=1}^{L} e^{-2\pi i \ell q \alpha}\right| \geq L - \sum_{\ell=1}^{L} |1 - e^{-2\pi i \ell q \alpha}| \geq L - 2\pi L^2 \|q\alpha\|, \tag{11}$$

where $\|\cdot\|$ denotes the distance to the nearest integer. Further, if $\alpha \in \mathbf{M}'_q$, then

$$\|q\alpha\| \leq \frac{q}{\eta^2 N} \leq \frac{1}{4\pi L}. \tag{12}$$

Therefore, by (11) and (12) we have

$$|\widehat{1_P}(\alpha)| \geq L/2 \quad \text{for all} \quad \alpha \in \mathbf{M}'_q. \tag{13}$$

By (10), (13), and Plancherel's Identity (7) we see

$$\sigma\delta^2 N \leq \int_{\mathbf{M}'_q} |\widehat{f_A}(\alpha)|^2 d\alpha \leq \frac{4}{L^2} \int_0^1 |\widehat{f_A}(\alpha)|^2 |\widehat{1_P}(\alpha)|^2 d\alpha = \frac{4}{L^2} \sum_{n \in \mathbb{Z}} |f_A * \widetilde{1_P}(n)|^2, \tag{14}$$

where $\widetilde{1_P}(n) = 1_P(-n)$ and

$$f_A * \widetilde{1_P}(n) = \sum_{m\in\mathbb{Z}} f_A(m)1_P(m-n) = |A \cap (P+n)| - \delta|(P+n)\cap[1,N]|. \tag{15}$$

We now take advantage of the fact that f_A, and consequently $f_A * \widetilde{1_P}$, has mean value zero. In other words,

$$\sum_{n\in\mathbb{Z}} f_A * \widetilde{1_P}(n) = 0. \tag{16}$$

As with any real valued function, we can write

$$|f_A * \widetilde{1_P}| = 2(f_A * \widetilde{1_P})_+ - f_A * \widetilde{1_P}, \tag{17}$$

where $(f_A * \widetilde{1_P})_+ = \max\{f_A * \widetilde{1_P}, 0\}$.

For the purposes of proving Lemma 2, we can assume that $f_A * \widetilde{1_P}(n) \leq 2\delta L$ for all $n \in \mathbb{Z}$, as otherwise the progression $P+n$ would more than satisfy the conclusion. Combined with the trivial upper bound $f_A * \widetilde{1_P}(n) \geq -\delta L$, we can assume

$$|f_A * \widetilde{1_P}(n)| \leq 2\delta L \quad \text{for all} \quad n \in \mathbb{Z}. \tag{18}$$

By (14), (16)–(18), we have

$$\sum_{n\in\mathbb{Z}} (f_A * \widetilde{1_P})_+(n) = \frac{1}{2}\sum_{n\in\mathbb{Z}} |f_A * \widetilde{1_P}| \geq \frac{1}{4\delta L}\sum_{n\in\mathbb{Z}} |f_A * \widetilde{1_P}|^2 \geq \frac{\sigma\delta NL}{16}. \tag{19}$$

By (15), we see that $f_A * \widetilde{1_P}(n) = 0$ if $n \notin [-qL, N]$. Letting $E = \{n \in \mathbb{Z} : P + n \subseteq [1, N]\}$ and $F = [-qL, N] \setminus E$, we see that $|F| \leq 2qL$. Therefore, by (18), (19), and the bound $128qL \leq \sigma N$, we have

$$\sum_{n\in E} (f_A * \widetilde{1_P})_+(n) \geq \frac{\sigma\delta NL}{16} - 2\delta L|F| \geq \frac{\sigma\delta NL}{16} - 4q\delta L^2 > \frac{\sigma\delta NL}{32}. \tag{20}$$

Recalling that $|E| \leq N$ and $f_A * \widetilde{1_P}(n) = |A \cap (P+n)| - \delta L$ for all $n \in E$, we have that there exists $n \in \mathbb{Z}$ with

$$|A \cap (P+n)| \geq (1 + \sigma/32)\delta L,$$

as required.

4 L^2 Concentration

For this section, we once again fix a binary a quadratic form $h \in \mathbb{Z}[x, y]$ with $\Delta(h) \neq 0$, and let

$$I(h) = \{h(m, n) : m, n \in \mathbb{N}\} \setminus \{0\}.$$

The following result makes precise the implication outlined in item (i) in Sect. 2.2, in which the condition $(A - A) \cap I(h) = \emptyset$ forces substantial L^2 concentration of $\widehat{f_A}$ near rationals with a single small denominator. Combining this with Lemma 2, we then quickly deduce Lemma 1.

Lemma 3 *Suppose $A \subseteq [1, N]$ with $|A| = \delta N$, and let $\eta = c_0\delta$ for a sufficiently small constant $c_0 = c_0(h) > 0$. If $(A - A) \cap I(h) = \emptyset$, $\delta \geq N^{-1/20}$, and $|A \cap (N/9, 8N/9)| \geq 3\delta N/4$, then there exists $q \leq \eta^{-1}$ such that*

$$\int_{\mathbf{M}'_q} |\widehat{f_A}(\alpha)|^2 d\alpha \gg_h \delta^2 N.$$

4.1 *Proof of Lemma 1*

Suppose $A \subseteq [1, N]$, $|A| = \delta N$, $\delta \geq N^{-1/20}$, and $(A - A) \cap I(h) = \emptyset$.

If $|A \cap (N/9, 8N/9)| < 3\delta N/4$, then

$$\max\{|A \cap [1, N/9]|, |A \cap [8N/9, N]|\} > \delta N/8.$$

In other words, A has density at least $9\delta/8$ on one of these intervals.

Otherwise, Lemmas 3 and 2 apply, so in either case, letting $\eta = c_0\delta$, there exists $q \leq \eta^{-1}$ and an arithmetic progression

$$P = \{x + \ell q : 1 \leq \ell \leq L\}$$

with $qL \gg_h \delta^2 N$ and $|A \cap P| \geq (1 + c)\delta L$. Partitioning P into subprogressions of step size q^2, the pigeonhole principle yields a progression

$$P' = \{y + \ell q^2 : 1 \leq \ell \leq N'\} \subseteq P$$

with $N' \geq L/2q$ and $|A \cap P'| \geq (1 + c)\delta N'$. This allows us to define a set $A' \subseteq [1, N']$ by

$$A' = \{\ell \in [1, N'] : y + \ell q^2 \in A\},$$

which clearly satisfies $|A'| \geq (1 + c)\delta N'$ and $N' \gg_h \delta^2 N/q^2 \gg_h \delta^4 N$. Moreover, since $q^2 h(m, n) = h(qm, qn)$, A' inherits the lack of $h(m, n)$ differences from A. □

Our task is now completely reduced to a proof of Lemma 3.

4.2 Proof of Lemma 3

Suppose $A \subseteq [1, N]$ with $|A| = \delta N$, and let $\eta = c_0\delta$. We let $J = |b_1| + |b_2| + |b_3|$, $M = \sqrt{N/9J}$, $Z = \{(m, n) \in [1, M]^2 : h(m, n) = 0\}$, and $\Lambda = [1, M]^2 \setminus Z$.

We note that

$$|Z| \ll_h M. \tag{21}$$

If $(A - A) \cap I(h) = \emptyset$, then since $h(\Lambda) \subseteq [-N/9, N/9]$, we see that

$$\begin{aligned}\sum_{\substack{x\in\mathbb{Z}\\(m,n)\in\Lambda}} f_A(x)f_A(x+h(m,n)) = &\sum_{\substack{x\in\mathbb{Z}\\(m,n)\in\Lambda}} 1_A(x)1_A(x+h(m,n))\\ &-\delta\sum_{\substack{x\in\mathbb{Z}\\(m,n)\in\Lambda}} 1_A(x)1_{[1,N]}(x+h(m,n))\\ &-\delta\sum_{\substack{x\in\mathbb{Z}\\(m,n)\in\Lambda}} 1_{[1,N]}(x-h(m,n))1_A(x)\\ &+\delta^2\sum_{\substack{x\in\mathbb{Z}\\(m,n)\subset\Lambda}} 1_{[1,N]}(x)1_{[1,N]}(x+h(m,n))\\ &\le \Big(\delta^2 N - 2\delta|A\cap(N/9, 8N/9)|\Big)|\Lambda|.\end{aligned}$$

Therefore, if $|A \cap (N/9, 8N/9)| \ge 3\delta N/4$, we have

$$\sum_{\substack{n\in\mathbb{Z}\\1\le m\le M}} f_A(n)f_A(x+h(m,n)) \le -\delta^2 N|\Lambda|/2. \tag{22}$$

One can check using orthogonality (8) and Plancherel's Identity (7) that

$$\begin{aligned}&\sum_{\substack{x\in\mathbb{Z}\\(m,n)\in\Lambda}} f_A(x)f_A(x+h(m,n))\\ &=\sum_{\substack{x,y\in\mathbb{Z}\\(m,n)\in\Lambda}} f_A(x)f_A(y)\int_0^1 e^{2\pi i(x-y+h(m,n))\alpha}d\alpha\\ &=\int_0^1\left(\sum_{x\in\mathbb{Z}} f_A(x)e^{2\pi i x\alpha}\right)\left(\sum_{y\in\mathbb{Z}} f_A(y)e^{-2\pi i y\alpha}\right)\left(\sum_{(m,n)\in\Lambda} e^{2\pi i h(m,n)\alpha}\right)d\alpha\\ &=\int_0^1 |\widehat{f_A}(\alpha)|^2 S_M(\alpha)d\alpha + O(\delta N|Z|),\end{aligned}$$

where

$$S_x(\alpha) = \sum_{1\le m,n\le x} e^{2\pi i h(m,n)\alpha}.$$

Combined with (21), (22), and the triangle inequality, this yields

$$\int_0^1 |\widehat{f_A}(\alpha)|^2 |S_M(\alpha)| d\alpha \ge \delta^2 N M^2/4. \tag{23}$$

By adapting traditional exponential sum estimates to this two-variable setting, and at one point carefully exploiting that $\Delta(h) \neq 0$, we have that if $\delta \ge N^{-1/20}$, then

$$|S_M(\alpha)| \ll_h M^2/q \quad \text{for } \alpha \in \mathbf{M}_q,\ q \le \eta^{-1}, \tag{24}$$

and

$$|S_M(\alpha)| \le C\eta M^2 \le \delta M^2/8 \quad \text{for } \alpha \in \mathfrak{m}, \tag{25}$$

provided we choose $c_0 \le 1/8C$. We prove and discuss these estimates in detail in Sect. 5.

By (25) and Plancherel's Identity (7), we have

$$\int_{\mathfrak{m}} |\widehat{f_A}(\alpha)|^2 |S_M(\alpha)| d\alpha \le \delta^2 N M^2/8,$$

which by (23) yields

$$\int_{\mathfrak{M}} |\widehat{f_A}(\alpha)|^2 |S_M(\alpha)| d\alpha \ge \delta^2 N M^2/8. \tag{26}$$

By (24) and (26) we have

$$\delta^2 N M^2 \ll_h \sum_{q=1}^{\eta^{-1}} \frac{M^2}{q} \int_{\mathbf{M}_q} |\widehat{f_A}(\alpha)|^2 d\alpha. \tag{27}$$

We then make use of the following proposition, a more general version of which can be found in Proposition 5.6 of [15], which exploits the more inclusive definition of $\mathbf{M}'_q$ as compared with $\mathbf{M}_q$.

Proposition 1 *If $\eta^2 N > 2Q^2$, then*

$$\max_{q\le Q} \int_{\mathbf{M}'_q} |\widehat{f_A}(\alpha)|^2 d\alpha \ge \frac{1}{2} \sum_{q=1}^{Q} q^{-1} \int_{\mathbf{M}_q} |\widehat{f_A}(\alpha)|^2 d\alpha.$$

Proof By (9) we have

$$\begin{aligned} Q \max_{q\leq Q} \int_{\mathbf{M}'_q} |\widehat{f_A}(\alpha)|^2 d\alpha &\geq \sum_{q=1}^{Q} \int_{\mathbf{M}'_q} |\widehat{f_A}(\alpha)|^2 d\alpha \\ &= \sum_{q=1}^{Q} \sum_{r|q} \int_{\mathbf{M}_r} |\widehat{f_A}(\alpha)|^2 d\alpha \\ &= \sum_{r=1}^{Q} \lfloor Q/r \rfloor \int_{\mathbf{M}_r} |\widehat{f_A}(\alpha)|^2 d\alpha \\ &\geq \frac{Q}{2} \sum_{r=1}^{Q} r^{-1} \int_{\mathbf{M}_r} |\widehat{f_A}(\alpha)|^2 d\alpha, \end{aligned}$$

and the proposition follows.

Lemma 3 then follows immediately from (27) and Proposition 1. □

5 Exponential Sum Estimates

In this section, we carefully adapt traditional exponential sum estimates in order to establish the crucial upper bounds (24) and (25). For the entirety of the section, we fix a nonzero binary quadratic form

$$h(x, y) = b_1x^2 + b_2xy + b_3y^2 \in \mathbb{Z}[x, y].$$

Unlike in previous sections, we do not make the perpetual assumption that $\Delta(h) = b_2^2 - 4b_1b_3 \neq 0$, but rather enforce this condition only when necessary.

5.1 Major Arc Estimates

We begin our pursuit of (24) by establishing an asymptotic formula for the relevant exponential sum near rationals with small denominator. To achieve this goal, we make multiple appeals to the following standard formula, which is simply integration by parts applied to an appropriate Riemann–Stieltjes integral.

Lemma 4 (Abel's Partial Summation Formula) *If $\phi : \mathbb{R} \to \mathbb{C}$ is continuously differentiable, $f : \mathbb{N} \to \mathbb{C}$, $F(x) = \sum_{1\leq n\leq x} f(n)$, and $M > 0$, then*

$$\sum_{1\leq n\leq M} f(n)\phi(n) = F(M)\phi(M) - \int_0^M F(x)\phi'(x)dx.$$

We now proceed with the asymptotic formula, obtained by applying Lemma 4 one variable at a time.

Lemma 5 *If* $a, q \in \mathbb{N}$, $\alpha = a/q + \beta$, *and* $M > 0$, *then*

$$\begin{aligned} S_M(\alpha) &= \sum_{1 \leq m,n \leq M} e^{2\pi i h(m,n)\alpha} \\ &= q^{-2} G(a,q) \int_0^M \int_0^M e^{2\pi i h(x,y)\beta} dx dy + O(qM(1 + JM^2\beta)), \end{aligned}$$

where $J = |b_1| + |b_2| + |b_3|$ *and*

$$G(a,q) = \sum_{r,s=0}^{q-1} e^{2\pi i h(r,s)a/q}.$$

Proof For each fixed $1 \leq m \leq M$ and $y > 0$, we see that

$$\begin{aligned} S_y^m(a/q) &= \sum_{1 \leq n \leq y} e^{2\pi i h(m,n)a/q} \\ &= \sum_{s=0}^{q-1} e^{2\pi i h(m,s)a/q} \, |\{1 \leq n \leq y \; : \; n \equiv s \bmod q\}| \\ &= \frac{y}{q} G_m(a,q) + O(q), \end{aligned}$$

where

$$G_m(a,q) = \sum_{s=0}^{q-1} e^{2\pi i h(m,s)a/q}.$$

Then, letting $h_y = \frac{\partial h}{\partial y}$ and combining the above with Lemma 4 and integration by parts, we have

$$\begin{aligned} S_M^m(\alpha) &= \sum_{1 \leq n \leq M} e^{2\pi i h(m,n)a/q} e^{2\pi i h(m,n)\beta} \\ &= S_M^m(a/q) e^{2\pi i h(m,M)\beta} - \int_0^M S_y^m(a/q)(2\pi i h_y(m,y)\beta) e^{2\pi i h(m,y)\beta} dy \\ &= q^{-1} G_m(a,q) \left(M e^{2\pi i h(m,M)\beta} - \int_0^M y 2\pi i h_y(m,y)\beta e^{2\pi i h(m,y)\beta} dy \right) \\ &\quad + O(q(1 + JM^2\beta)) \\ &= q^{-1} G_m(a,q) \int_0^M e^{2\pi i h(m,y)\beta} dy + O(q(1 + JM^2\beta)). \end{aligned}$$

Similarly, summing in m we have

$$\begin{aligned}\tilde{S}_x(a/q) &= \sum_{1\leq m\leq x} G_m(a,q)\\ &= \sum_{r=0}^{q-1} G_r(a,q)\,|\{1\leq m\leq x \,:\, m\equiv r \bmod q\}|\\ &= \frac{x}{q}G(a,q)+O(q),\end{aligned}$$

and, letting $h_x=\frac{\partial h}{\partial x}$, we apply the same sequence of steps to see that $S_M(\alpha)$ equals

$$\begin{aligned}&q^{-1}\sum_{1\leq m\leq M} G_m(a,q)\int_0^M e^{2\pi i h(m,y)\beta}dy + O(qM(1+JM^2\beta))\\ =&q^{-1}\Big(\tilde{S}_M(a/q)\int_0^M e^{2\pi i h(M,y)\beta}dy\\ &-\int_0^M\int_0^M \tilde{S}_x(a/q)(2\pi i h_x(x,y)\beta)e^{2\pi i h(x,y)\beta}dxdy\Big)+O(qM(1+JM^2\beta))\\ =&q^{-2}G(a,q)\Big(M\int_0^M e^{2\pi i h(M,y)\beta}dy\\ &-\int_0^M\int_0^M x(2\pi i h_x(x,y)\beta)e^{2\pi i h(x,y)\beta}dxdy\Big)+O(qM(1+JM^2\beta))\\ =&q^{-2}G(a,q)\int_0^M\int_0^M e^{2\pi i h(x,y)\beta}dxdy+O(qM(1+JM^2\beta)),\end{aligned}$$

and the formula is established.

The crucial denominator q in (24) comes from the following result, previously discussed in Sect. 2.3, which is the one and only juncture at which we require $\Delta(h)\neq 0$. This key ingredient, as well as the standard proof we recreate for Lemma 8, rely on a technique known as *Weyl differencing*, in which we take the modulus squared of the exponential sum in order to reduce the quadratic dependence in each variable to a linear dependence.

Lemma 6 *If $\Delta(h)\neq 0$ and $a,q\in\mathbb{N}$ with $(a,q)=1$, then*

$$\left|\sum_{r,s=0}^{q-1} e^{2\pi i h(r,s)a/q}\right| \ll_h q.$$

Proof Fixing $a,q\in\mathbb{N}$ with $(a,q)=1$, exploiting that $|z|^2=z\bar{z}$ for any $z\in\mathbb{C}$, and changing variables $r'=r+t$, $s'=s+u$, we see that

$$\begin{aligned}
&\left|\sum_{r,s=0}^{q-1} e^{2\pi i h(r,s)a/q}\right|^2 \\
&= \sum_{r,r',s,s'=0}^{q-1} e^{2\pi i (h(r',s')-h(r,s))a/q} \\
&= \sum_{r,s,t,u=0}^{q-1} e^{2\pi i (h(r+t,s+u)-h(r,s))a/q} \\
&= \sum_{r,s,t,u=0}^{q-1} e^{2\pi i (2b_1rt+b_1t^2+b_2ru+b_2st+b_2tu+2b_3su+b_3u^2)a/q} \\
&= \sum_{t,u=0}^{q-1} e^{2\pi i h(t,u)a/q} \left(\sum_{r=0}^{q-1} e^{2\pi i (2b_1t+b_2u)ra/q}\right)\left(\sum_{s=0}^{q-1} e^{2\pi i (b_2t+2b_3u)sa/q}\right) \\
&= \sum_{t,u=0}^{q-1} e^{2\pi i h(t,u)a/q} \begin{cases} q^2 & \text{if } 2b_1t+b_2u \equiv b_2t+2b_3u \equiv 0 \bmod q \\ 0 & \text{else} \end{cases},
\end{aligned}$$

where the last equality follows from the orthogonality relation

$$\sum_{r=0}^{q-1} e^{2\pi i rb/q} = \begin{cases} q & \text{if } q \mid b \\ 0 & \text{else} \end{cases}.$$

Looking at the two congruence conditions above, multiplying the first expression by b_2, and multiplying the second expression by $2b_1$, we get the system

$$2b_1b_2t + b_2^2u \equiv 2b_1b_2t + 4b_1b_3u \equiv 0 \bmod q.$$

By subtracting the two resulting expressions we see that q must divide $\Delta(h)u$. Letting $d = \gcd(q, \Delta(h))$, we have that u must be one of the d multiples of q/d, which each yield at most $\gcd(q, 2b_1b_2)$ choices for t. In particular, if $\Delta(h) \neq 0$, then the number of simultaneous solutions is $O_h(1)$, and the lemma follows.

5.2 *Proof of (24)*

Returning to the setting of the proof of Lemma 3, if $\alpha \in \mathbf{M}_q$ with

$$q \le \eta^{-1} \ll_h \delta^{-1} \le N^{1/20} \ll M^{1/10},$$

then $\alpha = a/q + \beta$ with

$$|\beta| < \frac{1}{\eta^2 N} \ll_h N^{-9/10} \ll M^{-9/5}$$

for some a with $(a, q) = 1$. In this case, Lemma 5 tells us that

$$S_M(\alpha) = q^{-2}G(a,q)\int_0^M\int_0^M e^{2\pi i h(x,y)\beta}dxdy + O_h(M^{1.3}).$$

Applying Lemma 6 and trivially bounding the double integral by M^2, we have

$$|S_M(\alpha)| \ll_h M^2/q,$$

as claimed in (24). □

5.3 Minor Arc Estimates

We begin our pursuit of (25) with the following standard oscillatory integral estimate, which will allow us to exhibit (25) in the case that α is fairly close to a rational with small denominator, but not so close as to lie in the major arcs.

Lemma 7 (Van der Corput's Lemma for Quadratic Polynomials) *If $g(x) = x^2 + bx + c \in \mathbb{R}[x]$ and $I \subseteq \mathbb{R}$ is an interval, then*

$$\left|\int_I e^{2\pi i g(x)\beta}dx\right| \ll |\beta|^{-1/2}.$$

Proof Fix $g(x) = x^2 + bx + c \in \mathbb{R}[x]$ and an interval $I \subseteq \mathbb{R}$, and let $E = (I + b/2) \cap \{x : |x| \geq |\beta|^{-1/2}\}$, where $I + b/2$ denotes the translation of the interval I by $b/2$. We know that the measure of $(I + b/2) \setminus E$ is at most $2|\beta|^{-1/2}$, so we complete the square and change variables to see that

$$\begin{aligned}\left|\int_I e^{2\pi i g(x)\beta}dx\right| &= \left|\int_I e^{2\pi i((x+b/2)^2-b^2/4+c)\beta}dx\right| \\ &= \left|\int_I e^{2\pi i(x+b/2)^2\beta}dx\right| \\ &= \left|\int_{I+b/2} e^{2\pi i y^2\beta}dy\right| \\ &\ll |\beta|^{-1/2} + \left|\int_E e^{2\pi i y^2\beta}dy\right|.\end{aligned}$$

Writing

$$e^{2\pi i y^2\beta} = \frac{1}{4\pi i y\beta}\frac{d}{dx}(e^{2\pi i y^2\beta}),$$

we have by integration by parts that

$$\int_E e^{2\pi i y^2\beta}dy = \left[\frac{e^{2\pi i y^2\beta}}{4\pi i y\beta}\right] + \int_E \frac{e^{2\pi i y^2\beta}}{4\pi i y^2\beta}dy,$$

where the expression in brackets is appropriately evaluated at endpoints of E. By construction, $|y| \geq |\beta|^{-1/2}$ at each endpoint of E, and hence

$$\left|\int_E e^{2\pi i y^2\beta}dy\right| \ll |\beta|^{-1/2} + |\beta|^{-1}\int_{|y|\geq|\beta|^{-1/2}} \frac{1}{y^2}dy \ll |\beta|^{-1/2},$$

which establishes the desired estimate.

With regard to estimating the double integral in the conclusion of Lemma 5, since we assumed h was not identically zero, we can relabel or make a linear change of variables to reduce to the case where $b_1 \neq 0$. Then, by applying Lemma 7 to the integral in x for every fixed y, we immediately get the following estimate.

Corollary 1 *If $M > 0$, then*

$$\left|\int_0^M \int_0^M e^{2\pi i h(x,y)\beta}dxdy\right| \ll_h M|\beta|^{-1/2}. \tag{28}$$

For our final ingredient, we turn to the following traditional estimate, which we utilize to establish (25) when α is close to a denominator that is neither too small nor too large.

Lemma 8 (Weyl's Inequality for Quadratic Polynomials) *Suppose $g(x) = bx^2 + cx + d \in \mathbb{R}[x]$, $b \in \mathbb{N}$, $a, q \in \mathbb{N}$, $t \geq 1$, and $x > 0$. If $(a, q) = 1$ and $|\alpha - a/q| \leq tq^{-2}$, then*

$$\left|\sum_{1\leq n\leq x} e^{2\pi i g(n)\alpha}\right| \ll \left(bx\log q + tx + btx^2/q + q\log q\right)^{1/2}.$$

Proof Letting S denote the exponential sum we wish to estimate, we see that

$$|S|^2 = \sum_{1\leq n,n'\leq x} e^{2\pi i(h(n')-h(n))\alpha} = x + 2\Re\left(\sum_{1\leq n<n'\leq x} e^{2\pi i(h(n')-h(n))\alpha}\right), \tag{29}$$

where the x accounts for terms where $n = n'$, and $\Re$ denotes the real part. With a change of variables $n' = n + h$, we have

$$\begin{aligned}\sum_{1\le n<n'\le x} e^{2\pi i(h(n')-h(n))\alpha} &= \sum_{1\le n\le x-1}\ \sum_{1\le h\le x-n} e^{2\pi i(h(n+h)-h(n))\alpha}\\ &= \sum_{1\le n\le x-1}\ \sum_{1\le h\le x-n} e^{2\pi(2bnh+h^2+ch)\alpha}\\ &= \sum_{1\le h\le x-1} e^{2\pi i(h^2+ch)\alpha} \sum_{1\le n\le x-h} e^{2\pi i(2bhn)\alpha}.\end{aligned}$$

Applying the geometric series formula to the inner sum, and the triangle inequality, gives us

$$\left|\sum_{1\le n<n'\le x} e^{2\pi i(h(n')-h(n))\alpha}\right| \ll \sum_{1\le h\le 2bx} \min\left\{x, \|h\alpha\|^{-1}\right\}, \tag{30}$$

where $\|\cdot\|$ denotes the distance to the nearest integer.

Fixing $q\in\mathbb{N}$ and breaking the sum in h into intervals of length q, we have

$$\sum_{1\le h\le 2bx} \min\left\{x, \|h\alpha\|^{-1}\right\} \le \sum_{1\le j\le 2bx/q}\ \sum_{s=0}^{q-1} \min\left\{x, \|(qj+s)\alpha\|^{-1}\right\}. \tag{31}$$

If $a\in\mathbb{N}$ with $|\alpha-a/q|\le tq^{-2}$, we can write $\alpha=a/q+O(t/q^2)$, and hence

$$(qj+s)\alpha = qj\alpha + \frac{sa}{q} + O(t/q).$$

Further, if we let k be the nearest integer to $q^2j\alpha$, then $qj\alpha=k/q+O(t/q)$ and hence

$$(qj+s)\alpha = \frac{sa+k}{q} + O(t/q).$$

Combined with (31), this yields

$$\sum_{1\le h\le 2bx} \min\left\{x, \|h\alpha\|^{-1}\right\} \le \sum_{1\le j\le 2bx/q}\ \sum_{s=0}^{q-1} \min\left\{x, \left\|\frac{sa+k}{q}+O(t/q)\right\|^{-1}\right\}. \tag{32}$$

If $(a,q)=1$, then as s runs over all congruence classes modulo q, so does sa. In particular, the $O(t/q)$ error term dominates for at most $O(t)$ terms, and we have

$$\begin{aligned}\sum_{1\le j\le 2bx/q}\ \sum_{s=0}^{q-1} \min\left\{x, \left\|\frac{sa+k}{q}+O(t/q)\right\|^{-1}\right\} &\ll \sum_{1\le j\le 2bx/q}\left(tx+\sum_{s=1}^{q/2}\frac{q}{s}\right)\\ &\ll (2bx/q+1)(tx+q\log q),\end{aligned}$$

which combines with (29), (30), and (32) to yield the desired estimate.

In the same way we deduce Corollary 1 from Lemma 7, we reduce to the case of $b_1 \neq 0$ and apply Lemma 8 to the sum in m for every fixed n to immediately get the following estimate.

Corollary 2 *Suppose $a, q \in \mathbb{N}$, $\alpha \in [0, 1]$, and $x > 0$. If $(a, q) = 1$ and $|\alpha - a/q| \leq q^{-2}$, then*

$$\left| \sum_{1 \leq m,n \leq x} e^{2\pi i h(m,n)\alpha} \right| \ll_h x \left(x \log q + x^2/q + q \log q\right)^{1/2}. \tag{33}$$

Remark. We note that under the assumption $\Delta(h) \neq 0$, the estimates (28) and (33) can be improved to $|\beta|^{-1}$ and

$$\left(x^4/q^2 + (x^3/q + x^2 + qx) \log q\right)^{1/2},$$

respectively. For the former, since it is in a continuous setting, one can simply use that if $b^2 - 4ac \neq 0$, then

$$ax^2 + bxy + cy^2 = u^2 + v^2$$

after an invertible linear change of variables, and then apply Lemma 7 separately in u and v. The latter estimate can be established by mimicking the two-variable Weyl differencing process, and exploitation of nonzero discriminant, exhibited in the proof of Lemma 6. However, for the purposes of proving Theorem 2, we only require this sort of "optimal cancellation" on the major arcs, so for the sake of brevity, and for the sake of exposing the components used in previous applications of this method, we leave the details of these improvements as exercises for the reader.

5.4 *Proof of (25)*

Returning to the setting of the proof of Lemma 3, we consider $\alpha \in \mathfrak{m}$. By the pigeonhole principle, there exists $1 \leq q \leq M^{7/4}$ and $(a, q) = 1$ such that

$$|\alpha - a/q| \leq \frac{1}{qM^{7/4}} \leq \frac{1}{q^2}.$$

Writing $\alpha = a/q + \beta$, if $q \leq M^{1/4}$, then we have from Lemma 5 that

$$S_M(\alpha) = q^{-2}G(a, q) \int_0^M \int_0^M e^{2\pi i h(x,y)\beta} dx dy + O_h(M^{3/2}). \tag{34}$$

If $q \leq \eta^{-1}$, then it must be the case that $|\beta| > (\eta^2 N)^{-1}$, since otherwise we would have $\alpha \in \mathfrak{M}$. In this case, recalling that $N \ll_h M^2$ and $\eta \gg_h \delta \geq N^{-1/20} \gg_h M^{-1/10}$, it follows from (34), Corollary 1, and trivially bounding $G(a, q)$ by q^2 that

$$|S_M(\alpha)| \ll M|\beta|^{-1/2} + O_h(M^{3/2}) \ll_h \eta M^2.$$

If $\eta^{-1} \leq q \leq M^{1/4}$, then by (34), Lemma 6, and trivially bounding the double integral by M^2, we have

$$S_M(\alpha) \ll_h M^2/q + O_h(M^{3/2}) \ll_h \eta M^2.$$

Finally, if $M^{1/4} \leq q \leq M^{7/4}$, then by Corollary 2 we have

$$|S_M(\alpha)| \ll_h M(M \log q + M^2/q + q \log q)^{1/2} \ll M^{15/8} \ll_h \eta M^2,$$

and (25) is established in all cases. □

Acknowledgements The author would like to thank Neil Lyall who co-authored the expository note [12], in the context of squares and shifted primes, that served as a template for this paper.

References

1. A. Balog, J. Pelikán, J. Pintz, E. Szemerédi, *Difference sets without κ-th powers*, Acta. Math. Hungar. 65 (2) (1994), 165–187
2. H. Furstenberg, *Ergodic behavior of diagonal measures and a theorem of Szemerédi on arithmetic progressions*, J. d'Analyse Math, 71 (1977), 204–256
3. B. Green, *On arithmetic structures in dense sets of integers*, Duke Math. Jour. 114 (2002) no. 2, 215–238
4. B. Green, T. Tao, T. Ziegler, *A Fourier-free proof of the Furstenberg-Sárközy theorem*, https://terrytao.wordpress.com/2013/02/28/a-fourier-free-proof-of-the-furstenberg-sarkozy-theorem/
5. M. Hamel, N. Lyall, A. Rice, *Improved bounds on Sárközy's theorem for quadratic polynomials*, Int. Math. Res. Not. no. 8 (2013), 1761–1782
6. T. Kamae, M. Mendès France, *van der Corput's difference theorem*, Israel J. Math. 31, no. 3–4, (1978), pp. 335–342
7. H.- Z. Li, H. Pan, *Difference sets and polynomials of prime variables*, Acta. Arith. 138, no. 1 (2009), 25–52
8. J. Lucier, *Difference sets and shifted primes*, Acta. Math. Hungar. 120 (2008), 79–102
9. J. Lucier, *Intersective sets given by a polynomial*, Acta Arith. 123 (2006), 57–95
10. N. Lyall, *A new proof of Sárközy's theorem*, Proc. Amer. Math. Soc. 141 (2013), 2253–2264
11. N. Lyall, À. Magyar, *Polynomial configurations in difference sets*, J. Number Theory 129 (2009), 439–450
12. N. Lyall, A. Rice, *Two theorems of Sárközy*, http://alexricemath.com/wp-content/uploads/2013/06/DoubleSarkozy.pdf
13. J. Pintz, W. L. Steiger, E. Szemerédi, *On sets of natural numbers whose difference set contains no squares*, J. London Math. Soc. 37 (1988), 219–231
14. K. F. Roth, *On certain sets of integers*, J. London Math. Soc. 28 (1953), pp. 104–109
15. A. Rice, *A maximal extension of the best-known bounds for the Sárközy-Furstenberg Theorem*, Acta Arith. 187 (2019), 1–41

16. A. RICE, *Improvements and extensions of two theorems of Sárközy*, Ph.D. thesis, University of Georgia, 2012. http://alexricemath.com/wp-content/uploads/2013/06/AlexThesis.pdf
17. A. RICE, *Sárközy's theorem for $\mathcal{P}$ -intersective polynomials*, Acta Arith. 157 (2013), no. 1, 69–89
18. I. RUZSA, *Difference sets without squares*, Period. Math. Hungar. 15 (1984), 205–209
19. I. RUZSA, T. SANDERS, *Difference sets and the primes*, Acta. Arith. 131, no. 3 (2008), 281–301
20. A. SÁRKÖZY, *On difference sets of sequences of integers I*, Acta. Math. Hungar. 31(1–2) (1978), 125–149
21. A. SÁRKÖZY, *On difference sets of sequences of integers III*, Acta. Math. Hungar. 31(3–4) (1978), 355–386
22. S. SLIJEPCEVIĆ, *A polynomial Sárközy-Furstenberg theorem with upper bounds*, Acta Math. Hungar. 98 (2003), 275–280

Egyptian Fractions, Nonstandard Extensions of $\mathbb{R}$, and Some Diophantine Equations Without Many Solutions

David A. Ross

Abstract Non-Archimedean extensions of $\mathbb{R}$ are used to simplify and extend results related to the study of Egyptian fractions.

1 Introduction

Egyptian fractions are numbers a which can be expressed as the sum of reciprocals, e.g. $a = 1/a_1 + 1/a_2 + \cdots + 1/a_s$ with $a_i \in \mathbb{N}$. Such fractions have been a fruitful source for interesting mathematical problems since Fibonacci's *Liber Abaci*. For example, in 1921 Kellogg [7] conjectured a bound (later proved by Curtiss [3]) on the number of positive integer solutions for the Diophantine equation

$$1/x_1 + 1/x_2 + \cdots + 1/x_s = 1.$$

The question is equivalent to asking for the number of ways 1 can be expressed as an Egyptian fraction with s terms. Such questions about the structure of solutions to equations like the Kellogg equation have drawn the attention of Erdos [4], Graham [5], Sierpinski [10], and more recently Nathanson [9].

We show that many such results can be proved quite easily with the help of sufficiently saturated elementary extensions of $\mathbb{R}$ as an ordered field. The approach makes it possible not only to give new proofs of known results, but also to extend them in significant ways. For example, the result noted in Sect. 2.3 (and proved in Sect. 3.2) that the set A_s is not only nowhere dense, but in fact compact, was missed by Sierpinski. Similarly, our proof of Lagarias's Theorem in Sect. 3.3 gives extended information about large solutions of the equation.

For convenience we use the language of nonstandard analysis, though it is easy to see that we mainly use a few straightforward consequences of compactness. We review notation and results we need in Appendix 4.

D. A. Ross (✉)
Department of Mathematics, University of Hawaii, Honolulu, HI 96822, USA
e-mail: ross@math.hawaii.edu

M. B. Nathanson (ed.), *Combinatorial and Additive Number Theory III*,
Springer Proceedings in Mathematics & Statistics 297,
https://doi.org/10.1007/978-3-030-31106-3_15

2 The Structure of Some Sets of Egyptian Fractions

This section is motivated by the results of Sierpinski [10]. We begin with some notation.

$\mathbb{N}$ does not contain 0. $\mathbb{Z}^\# = \mathbb{Z} \setminus \{0\}$.

For $c \in \mathbb{R}$, $\text{monad}(c) = \{x \in {}^*\mathbb{R} : x \approx c\} = \bigcap_{n\in\mathbb{N}} {}^*(c - 1/n, c + 1/n)$.

For $s \in \mathbb{N}$,

$$A_s = \{1/n_1 + 1/n_2 + \cdots + 1/n_s \ : \ n_i \in \mathbb{Z}^\#\}$$

$$B_s = \{1/n_1 + 1/n_2 + \cdots + 1/n_s \ : \ n_i \in \mathbb{N}\}$$

Note $B_s \subseteq A_s \subseteq [-s, s]$ and $A_s \subseteq A_{s+1}$, $B_s \subseteq B_{s+1}$ (since $1/n = 1/2n + 1/2n$).

2.1 *Number of A_3-Representations*

Theorem 2.1 ([10], Théorème 1) *Let $0 \neq a \in A_3 \setminus A_1$. Then a has only finitely many representations as $a = 1/n_1 + 1/n_2 + 1/n_3$*

Proof Else by overflow $a = 1/n_1 + 1/n_2 + 1/n_3$ where at least one of n_1, n_2, n_3 is infinite. There are three cases:

1. Only one (say n_1) is finite. Then $a = {}^\circ a = {}^\circ(1/n_1 + 1/n_2 + 1/n_3) = 1/n_1$, so $a \in A_1$, a contradiction.
2. Only one (say n_1) is infinite. Then $1/n_1 = a - 1/n_2 - 1/n_3$ is standard, a contradiction.
3. n_1, n_2 and n_3 are all infinite. Then $a = {}^\circ a = {}^\circ(1/n_1 + 1/n_2 + 1/n_3) = 0$, a contradiction.

Since none of these cases is possible, the theorem is proved. □

2.2 *Mycielski's Theorem*

Sierpinski attributes this result to Jan Mycielski.

Theorem 2.2 ([10], Théorème 3) *B_s has no strictly increasing sequences.*

Proof Else let s be least where B_s contains an increasing sequence a_n. Let $M \in {}^*\mathbb{N}$ be infinite, and $x_1, \ldots, x_s \in {}^*\mathbb{N}$ with $a_M = 1/x_1 + 1/x_2 + \cdots + 1/x_s$. We may suppose that x_i is finite for $i \leq r$, x_i infinite otherwise. Consider two cases:

1. $r = s$. Then $a_M = {}^\circ a_M$ is standard, and by transfer $a_n = a_M$ for some standard n, contradicting strict monotonicity.

2. $r < s$. Then by transfer there is a subsequence, which WOLG we can again call a_n, and $b_n \in B_{s-r}$, with $a_n = 1/x_1 + 1/x_2 + \cdots + 1/x_r + b_n$. Then b_n is an increasing sequence in B_{s-r}, contradicting minimality of s. □

2.3 Nowhere Dense Sets

Recall that E is a *nowhere dense* (nwd) subset of $\mathbb{R}$ provided for every $a < b$ there are $a < a' < b' < b$ with $E \cap (a', b') = \varnothing$. Nonstandardly, that means for every open interval (a, b) there is a monad $\mu \subseteq {}^*(a, b)$ with $\mu \cap {}^*E = \varnothing$.

Theorem 2.3 ([10], Théorème 2) *For $s \in \mathbb{N}$, A_s is nwd.*

Proof Let $a < b$, let $c \in (a, b) \setminus A_s$, $c \neq 0$, and $\mu = \text{monad}(c)$.
Claim: $\mu \cap {}^*A_s = \varnothing$. Otherwise, let $x \in \mu \cap {}^*A_s$, in particular $x = 1/x_1 + 1/x_2 + \cdots + 1/x_s$ for some $x_1, \ldots, x_s \in {}^*\mathbb{Z}^\#$. We may suppose that x_i is finite for $i \leq r$, x_i infinite otherwise. Then:

$$c = {}^\circ x = \begin{cases} 0, & \text{if r = 0;} \\ 1/x_1 + 1/x_2 + \cdots + 1/x_r \in A_r \subseteq A_s, & \text{otherwise.} \end{cases}$$

This contradicts the choice of c.

Let v be any $*$-interval in μ, for example $v = (c - \epsilon, c + \epsilon)$ where ϵ is a positive infinitesimal. Then v witnesses "there is a subinterval v of ${}^*(c, d)$ with $v \cap {}^*A_s = \varnothing$," so by transfer there is a subinterval v of (c, d) with $v \cap A_s = \varnothing$, proving the result. □

The proof shows more, that $A_s \cup \{0\}$ is closed (and therefore compact; see Sect. 3.2 for a general result). Since A_s is countable, and closed countable sets are always nwd, the last paragraph of the proof is superfluous if one prefers to cite elementary results about perfect sets.

2.4 Sierpinski's Proof

Sierpinski's proof of Theorem 2.3 used a strong lemma of independent interest. We can prove a generalization of it very like the proof of Theorem 2.3.

Lemma 2.1 *Let $E \subseteq \mathbb{R}$ be nwd, Let $B \subseteq \mathbb{R}$ have no limit points except possibly* 0. *If either E or B is bounded then $H = E + B$ is nwd.* □

Proof Given $a < b$, let $c \in (a, b)$ and $\mu = \text{monad}(c)$, with $\mu \cap {}^*E = \varnothing$. Note that $E + x$ is nwd for each $x \in B$, and so for any finite $I \subseteq B$ so is $E + I$. Put $B_n = \{b \in B : 1/n < |b| < n\}$. Note that B_n is finite.

Claim: For some finite $I \subseteq B$, $\mu \cap {}^*H \subseteq {}^*E + I$.

Suppose not. Then by saturation there is an

$$x \in \bigcap_{n\in\mathbb{N}}{}^*\big((c - 1/n, c + 1/n) \cap (H \setminus (E + B_n)\big),$$

and $x = e + b$ for some $e \in {}^*E$ and $b \in {}^*B$ with b either infinite or infinitesimal. If b is infinite then $e = x - b \approx c - b$ is infinite, which means that B and E are both unbounded. If b is infinitesimal then $e \approx x \approx c$, so $\mu \cap {}^*E \neq \varnothing$. Either way this is a contradiction, proving the claim.

For this finite I, ${}^*E + I$ is *-nwd, so there is a *-interval $v \subseteq \mu \subseteq {}^*(a, b)$ such that $v \cap {}^*E + I = v \cap {}^*H = \varnothing$. By transfer, there is an interval $v \subseteq (a, b)$ with $v \cap H = \varnothing$, witnessing that H is nwd. □

Corollary 2.1 ([10], Lemme 2) *Let E be nwd, then $\bigcup_{n\in\mathbb{Z}^\#} E + 1/n$ is nwd.*

Proof Let $B = \{1/n : n \in \mathbb{Z}^\#\}$ in Lemma 2.1

We used, by the way, the following elementary standard lemma that we will not prove.

Lemma 2.2 ([10], Lemme 1) *The union of two (and therefore finitely many) nwd sets is nwd.*

3 Egyptian-Like Equations

3.1 Kellogg's Equation

Consider a slight generalization of Kellogg's equation:

$$a_1/x_1 + \cdots + a_s/x_s = a \tag{1}$$

where $s \in \mathbb{N}$ and $a, a_1, \ldots, a_s$ are fixed positive real numbers.

Lemma 3.1 *Equation 1 has at most finitely many solutions with $x_i \in \mathbb{N}$.*

The equation where each $a_i = 1$ is discussed in Sierpinski [10].

The following proof does not originate with the author, though we have been unable to discover a reference.

Proof Otherwise by overspill there is a *-solution $x_1, \ldots, x_s \in {}^*\mathbb{N}$ with at least one x_i infinite. We may suppose that x_i is finite for $i \leq r$, x_i infinite otherwise. We have:

$$a - (a_1/x_1 + \cdots + a_r/x_r) = a_{r+1}/x_{r+1} + \cdots + a_s/x_s$$

The left-hand-side of this identity is a standard real number, the right-hand-side is a nonzero infinitesimal, a contradiction.

Štefan Znám has considered a similar equation (see Brenton and Vasiliu [2]) in connection with the problem of finding finite sets of natural numbers such that each is a divisor of the product of the rest, plus one:

$$1/x_1 + \cdots + 1/x_s + 1/x_1 x_2 \cdots x_s = a$$

More generally, consider a *generalized Znám equation*:

$$a = a_1/x_1 + \cdots + a_s/x_s + b/x_1 x_2 \cdots x_s \qquad (2)$$

where $s \in \mathbb{N}$ and $a, b, a_1, \ldots, a_s$ are fixed positive real numbers.

Lemma 3.2 *Equation 2 has at most finitely many solutions with* $x_i \in \mathbb{N}$.

Proof Otherwise as before there is a *-solution $x_1, \ldots, x_n \in {}^*\mathbb{N}$ with at least one x_i infinite. We may suppose that x_i is finite for $i \leq r$, x_i infinite otherwise. We have:

$$a - (a_1/x_1 + \cdots + a_r/x_r) = a_{r+1}/x_{r+1} + \cdots + a_s/x_s + b/x_1 x_2 \cdots x_s$$

The left-hand side of this identity is a standard real number, the right-hand side is a nonzero infinitesimal, a contradiction. □

3.2 *Equations with More Complicated Terms*

These arguments translate fairly easily to equations whose terms are even more complicated. For example, consider the equation

$$\sum_I \frac{a_I}{\prod\limits_{i \in I} x_i} = a_\varnothing \qquad (3)$$

where the sum ranges over nonempty subsets I of $\{1, \ldots, s\}$, and $a_I \in \mathbb{R}$.

Theorem 3.1 *Suppose in Eq. 3 that every* $a_I \geq 0$ *and* $a_\varnothing > 0$. *The following are equivalent:*

1. *For every* $i \leq s$ *there is an* I *with* $i \in I$ *and* $a_I \neq 0$.
2. *Equation 3 has only finitely many solutions in* $\mathbb{N}$.

Proof (2) ⇒ (1) is trivial. For (1) ⇒ (2), suppose there are infinitely many solutions in $\mathbb{N}$. Then there is a solution in ${}^*\mathbb{N}$ with at least one x_i infinite. Let I be a subset of $\{1, \ldots, s\}$ with $a_I > 0$. Then $\frac{a_I}{\prod_{i \in I} x_i}$ is a positive infinitesimal. It follows that $a_\varnothing$ is the sum of one or more positive infinitesimals and zero or more positive real numbers, which is impossible. □

Now, let

$$C_s = \left\{ \sum_I \frac{a_I}{\prod_{i \in I} n_i} : n_i \in \mathbb{N} \right\},$$

the set of real numbers $a_\varnothing$ which can occur in Eq. 3 with positive integer solutions.

For the next result we add the condition that the coefficients a_I only depend on the cardinality of I, so they do not depend on rearrangements of the variables $x_1, \ldots, x_s$.

Theorem 3.2 *The set* $\{0\} \cup C_1 \cup \cdots \cup C_s$ *is compact.*

Proof Let $x = \sum_I \frac{a_I}{\prod_{i \in I} n_i} \in {}^*C_s$. We may suppose that n_i is finite for $i \leq r$, n_i infinite otherwise. Then $c = {}^\circ x = 0$ if $r = 0$, otherwise

$$c = \sum_J \frac{a_J}{\prod_{i \in j} n_i}$$

where the sum ranges over nonempty subsets J of $\{1, \ldots, r\}$, so $c \in C_r$ by the condition on a_I. Either way, $c \in \{0\} \cup C_1 \cup \cdots \cup C_s$, proving the theorem. □

3.3 A Theorem of Lagarias

We now consider the case when solutions are allowed to be negative. Following Lagarias [8] Consider the following special case of the generalized Znám equation,

$$c(1/x_1 + \cdots + 1/x_s) + b/x_1 x_2 \cdots x_s = a \tag{4}$$

where $a, b, c \in \mathbb{Z}^\#$, $c \geq 1$,and $\gcd(b, c) = 1$.

Straightforward examination of Eq. 4 with infinite values for some of the terms x_i will give us conditions under which this equation has infinitely many solution.

So suppose Eq. 4 has a solution $x_1, \ldots, x_s$, $x_i \in {}^*\mathbb{Z}^\#$, with one or more infinite values. We can suppose that x_i is finite for $i \leq r$, x_i infinite otherwise. Since $c(1/x_{r+1} + \cdots + 1/x_s) \approx b/x_1 x_2 \cdots x_s \approx 0$, and $c(1/x_1 + \cdots + 1/x_r)$ is standard, we have

$$\begin{aligned} c(1/x_1 + \cdots + 1/x_r) &= a, \text{ and} \\ c(1/x_{r+1} + \cdots + 1/x_s) &= -\, b/x_1 x_2 \cdots x_s. \end{aligned} \tag{5}$$

From (5), $c(\prod_{i>r} x_i)(\sum_{i>r} \frac{1}{x_i}) = -b/\prod_{i \leq r} x_i$ and $(\prod_{i>r} x_i)(\sum_{i>r} \frac{1}{x_i}) \in {}^*\mathbb{Z}$, so c and $\prod_{i \leq r} x_i$ divide b; therefore,

$$c = 1 \text{ and } b = b' \prod_{i \le r} x_i \text{ for some } b' \in \mathbb{Z}^{\#} \tag{6}$$

If $d = gcd(x_j, x_k)$ is infinite for some $r < j < k \le s$ then

$$0 \approx -b'/d = \Big(\prod_{i>r, i\ne k} x_i\Big)\Big(\frac{x_k}{d}\Big)\Big(\sum_{i>r} \frac{1}{x_i}\Big) \in {}^{*}\mathbb{Z}, \tag{7}$$

but since $b'/x_k \ne 0$, this is a contradiction; therefore,

$$x_j, x_k \text{ have no infinite common divisors, } r < j \ne k \le s \tag{8}$$

In particular, $|x_{r+1}|, \ldots, |x_s|$ are distinct.

Since $|a| = |\sum_{i \le r} \frac{1}{x_i}| \le \sum_{i \le r} |\frac{1}{x_i}| \le r$,

$$\text{If } |a| = r \text{ then } x_1 = x_2 = \cdots = x_r = \text{sign}(a) = \pm 1 \tag{9}$$

If $r = s - 1$, ie only x_s is infinite, then from (5) $1/x_s = -b/\prod_{i \le s} x_i$, or $b = -\prod_{i<s} x_i$. Combined with (9) we get

$$\begin{aligned} &\text{If } |a| = r = s - 1 \text{ then } x_1 = x_2 = \cdots = x_r = \text{sign}(a) = \pm 1 \\ &\text{and } b = -(\text{sign}(a)^{s-1}) \end{aligned}$$

Note that if $|a| = s - 1$ then $s - 1 = |a| \le r \le s - 1$; it follows:

$$\begin{aligned} &\text{If } |a| = s - 1 \text{ then } x_1 = x_2 = \cdots = x_r = \text{sign}(a) = \pm 1 \\ &\text{and } b = -(\text{sign}(a)^{s-1}) \end{aligned} \tag{10}$$

With the aid of transfer and Lemma 4.1 we can combine results (6), (8), and (10) into the following theorem.

Theorem 3.3 *Let $a, b, c \in \mathbb{Z}^{\#}$ with $c \ge 1$ and $gcd(b, c) = 1$. If Eq. 4 has infinitely many integer solutions, then:*

(i) $c = 1$;
(ii) Either (a) $|a| = s - 1$ and $b = -(\text{sign}(a)^{s-1})$, or (b) $|a| < s - 1$ and b is arbitrary;
(iii) There is no sequence $\{(x_1^n, x_2^n, \ldots, x_s^n)\}_n$ of solutions of Eq. 4 such that $\lim_{n\to\infty} |x_i^n| = \infty$ for every i;
(iv) There is no sequence $\{(x_1^n, x_2^n, \ldots, x_s^n)\}_n$ of solutions of Eq. 4 such that for some $j \ne k$, $\lim_{n\to\infty} \gcd(|x_j^n|, |x_k^n|) = \infty$;

(v) *For some $a_1, \ldots, a_r$, $r < s$, and every $N \in \mathbb{N}$, there is a solution of Eq. 4 with $x_i = a_i$ for $i \le r$, and $|x_i| > N$ for $r < i \le s$. For any such $a_1, \ldots, a_r$, $\sum_{i \le r} \frac{1}{a_i} = a$ and $\prod_{i \le r} a_i$ divides b.*

Parts (i) and (ii) of Theorem 3.3 comprise the main result of Lagarias [8]. That paper proves a converse, that under conditions (i) and (ii) there are infinitely many solutions to Eq. 4. This is easy to see by observing that for every a, b, c, and s satisfying (i) and (ii), one of the following is a solution (where $\sigma = \text{sign}(a)$, H is infinite, and M, N, and P are nonzero integers with $b = MN + MP + NP$):

$$(\underbrace{\sigma, \ldots, \sigma}_{s-1}, H)$$

$$(\underbrace{\sigma, \ldots, \sigma}_{s-k}, \underbrace{1, -1, 1, -1, \ldots, 1, -1}_{k-2}, \pm H, \mp(H + b))$$

$$(\underbrace{\sigma, \ldots, \sigma}_{s-k}, \underbrace{1, -1, 1, -1, \ldots, 1, -1}_{k-3}, \pm M, \pm N, \pm P)$$

Acknowledgements This paper grew out of discussions with Melvyn Nathanson which began at an American Institute of Mathematics workshop on *Nonstandard methods in combinatorial number theory*, August 14–18, 2017. The author is grateful to Nathanson for these discussion and for locating the paper [10] of Sierpinski, and to the organizers of the AIM workshop.

4 Appendix: Nonstandard Extensions of $\mathbb{R}$

In this section we review the properties we need for the models used in this paper. Let

$$\mathcal{R} = \langle \mathbb{R}, +, \times, 0, 1, \le, \ldots \rangle$$

be the real numbers considered as a first order structure in a countable language $\mathfrak{L}$ extending the language of ordered fields. Consider a non-Archimedean ordered field extension:

$$^*\mathcal{R} = \langle ^*\mathbb{R}, +, \times, 0, 1, \le, \ldots \rangle$$

(By convention, we don't put stars on the extensions of the usual operation symbols.)

Since $^*\mathbb{R}$ is non-Archimedean, it has a positive *infinitesimal* ε, ie $\varepsilon > 0$ and $-1/n < \varepsilon < 1/n$ for each $n \in \mathbb{N}^+$.

Note $1/\varepsilon$ is larger in absolute value than every positive integer N. Denote by $\text{Fin}(^*\mathbb{R})$ the *finite* elements of $^*\mathbb{R}$, $x \in \text{Fin}(^*\mathbb{R}) \iff (\exists N \in \mathbb{N})[\,-N < x < N]$ For $p, q \in {}^*\mathbb{R}$ write $p \approx q$ if $p - q$ is an infinitesimal.

Here are some useful properties of arithmetic in the ordered field $^*\mathcal{R}$. (The proofs of these, and all other results in this section, can be found in any basic introduction to

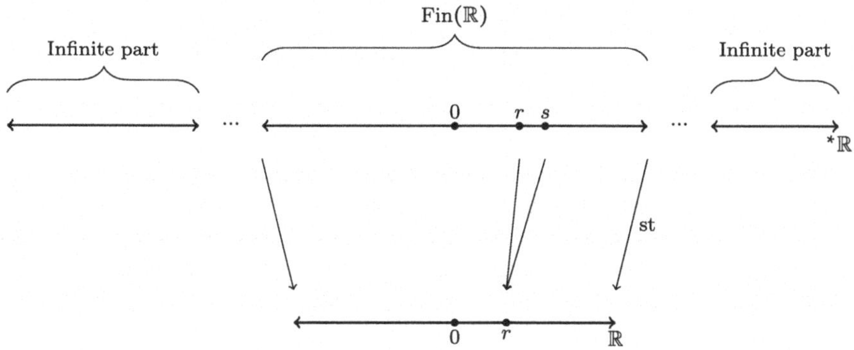

Fig. 1 The standard part map

nonstandard analysis, but are also very easy exercises not requiring any mathematical logic.)

1. $\approx$ is an equivalence relation on $\mathbb{R}$.
2. The finite and infinitesimal elements of $^*\mathbb{R}$ are subrings of $^*\mathbb{R}$.
3. The infinitesimals form a maximal multiplicative ideal in $\mathrm{Fin}(^*\mathbb{R})$.
4. Denote by $\mathrm{st}(x)$ or $^\circ x$ the quotient map from $\mathrm{Fin}(^*\mathbb{R})$ onto $\mathbb{R}$ (sometimes called the *standard part* map).
5. For $x \in \mathbb{R}$, $\mathrm{st}^{-1}(x) = \mathrm{monad}(x)$ (see Sect. 2). $\mathrm{Fin}(^*\mathbb{R}) = \bigcup_{x \in \mathbb{R}} \mathrm{monad}(x)$.
6. st is a ring homomorphism, $^\circ x \approx x$ for all $x \in \mathrm{Fin}(^*\mathbb{R})$, and $^\circ x = x$ for all $x \in \mathbb{R}$.
7. The order does not *strictly* respect $\approx$, since if $x' \approx x < y \approx y'$ it might happen that $x' \not< y'$ (even $x' > y'$), but it *is* the case that if $x \le y$ then $x' < y'$ *or* $x' \approx y'$, which we write $x' \lesssim y'$.
8. If $x \lesssim y$, $x \approx x'$, and $y \approx y'$ then $x' \lesssim y'$.
9. If $x \lnsim y$, $x \approx x'$, and $y \approx y'$ then $x' \lnsim y'$.

The standard part map is illustrated in Fig. 1 (where in the picture r is standard and $s \approx r$).

The properties above apply to *any* non-Archimedean ordered field extension of $\mathbb{R}$. It will be convenient for this paper to assume that our extension satisfies some additional properties.

First, the structure $\mathcal{R}$ should contain enough additional predicates so that all the sets of interest to us are *first-order definable*. To that end we will assume that the language $\mathcal{L}$ contains unary predicate symbols for $\mathbb{N}$, $\mathbb{Z}^\#$, A_r, B_r as well as symbols for any other mathematical elements that appear in the paper. In particular, when proving a result like Lemma 3.1 we assume that there are symbols a_i for the corresponding real constants appearing in the statement of the lemma. More subtly, when proving Theorem 2.2 we assume that there is a function symbol in the language for the *function* $a(i) = a_i$ from $\mathbb{N}$ to $\mathbb{R}$.

Elements of $\mathbb{R}$, as well as any constant, set, or function which is definable in $\mathcal{R}$, will be called *standard*.

If E is a definable set in $\mathcal{R}$, write $^*E$ for the corresponding set in $^*\mathcal{R}$, e.g.:

$$^*A_s = \left\{ \frac{1}{x_1} + \cdots + \frac{1}{x_s} \ : \ x_i \in {^*\mathbb{Z}^\#} \right\}$$

Next, we assume that $^*\mathcal{R}$ is an elementary extension of $\mathcal{R}$. *Elementary extension* means that any first-order statement about elements of $\mathcal{R}$ is true in $^*\mathcal{R}$ if and only if it is true in $\mathcal{R}$; in nonstandard analysis this is usually called the *transfer property*. In particular, $^*\mathcal{R}$ is an ordered field, and finite Boolean equations and inequalities that hold for definable subsets of $\mathcal{R}$ hold as well for the stars of these sets. Moreover, if A is a *finite* definable subset of $\mathbb{R}$ then $^*A = A$.

Finally, it will be convenient to assume that the extension $^*\mathcal{R}$ satisfies properties such as the following: if $A \subseteq \mathbb{N}^n$ is definable in $\mathcal{R}$ then

$$A \text{ is infinite} \iff {^*A} \text{ has elements with infinite components.}$$

To that end we assume that the structure $^*\mathcal{R}$ is an $\aleph_1$-saturated elementary extension of $\mathcal{R}$ (see, for example, [1] or [6]). It is a basic and straightforward result in Model Theory that such $\aleph_1$-saturated elementary extensions exist.

$\aleph_1$-saturation of $^*\mathcal{R}$ and transfer together imply the following useful property:

Lemma 4.1 *1. If a_n is a definable standard sequence then $\lim_{n\to\infty} a_n = \infty$ if and only if for any infinite M, a_M is infinite.*
2. *A definable set $A \subseteq \mathbb{R}$ contains arbitrarily large numbers if and only if $^*A$ contains an infinite number.*
3. *A definable subset $A \subseteq \mathbb{N}$ is infinite if and only if $^*A$ contains an infinite number.*
4. *A definable subset $A \subseteq \mathbb{N}^n$ is infinite if and only if for some $\langle a_1, \ldots, a_n\rangle \in {^*A^n}$ and $i \leq n$, a_i is infinite.*

In the framework of nonstandard analysis the second property in Lemma 4.1, together with its generalizations, is called *overflow*.

We conclude this appendix with some basic topology.

Lemma 4.2 *Let $U \subset \mathbb{R}$ be definable. The following are equivalent:*

1. *U is open.*
2. *For every $x \in U$, $\operatorname{monad}(x) \subseteq {^*U}$.*
3. *$\operatorname{st}^{-1}[U] \subseteq {^*U}$.*

Lemma 4.3 *Let $K \subset \mathbb{R}$ be definable. The following are equivalent:*

1. *K is compact.*
2. *For every $x \in {^*K}$ there is a $y \in K$ with $^\circ x = y$.*
3. *$^*K \subseteq \operatorname{st}^{-1}[K]$.*

References

1. Arkeryd, L.O., Cutland, N.J., Henson, C.W. (Eds.), *Nonstandard Analysis: Theory and Applications*, Springer Netherlands, Nato Science Series C, **493**, 1997.
2. Brenton, L., and Vasiliu, A., *Znam's Problem*, Mathematics Magazine, 2002, 3–11.
3. Curtiss, D. R., *On Kellogg's Diophantine Problem*, Amer. Math. Monthly, **29**, 1922, 380–387.
4. Erdös, P., *On a Diophantine equation*, Matematikai Lapok., 1950, 192–210.
5. Graham, R. L., *On finite sums of unit fractions*, Proc. London Math. Soc., 1964, 193–207.
6. Keisler, H.J., *Foundations of infinitesimal calculus*, Prindle Weber & Schmidt 1976, 214pp. Available online at http://www.math.wisc.edu/~keisler/foundations.pdf.
7. Kellogg, O. D., *On a Diophantine problem*, Amer. Math. Monthly **28**, 1921, 300–303.
8. Lagarias, J. C., *Cyclic systems of simultaneous congruences*, Int. J. Number Theory **6**, 2010, no. 2, 219–245.
9. Nathanson, M. B., *Weighted real Egyptian numbers*, Functiones et Approximatio Commentarii Mathematici, 2018.
10. Sierpinski, W., *Sur les décompositions de nombres rationnels en fractions primaires*, 1956.

A Dual-Radix Approach to Steiner's 1-Cycle Theorem

Andrey Rukhin

Abstract This article presents three algebraic proofs of Steiner's *1-Cycle Theorem* [14] within the context of the (accelerated) $3x + 1$ dynamical system. Furthermore, under an assumption of an exponential upper-bound on the iterates, the article demonstrates that the only 1-cycles in the (accelerated) $3x - 1$ dynamical system are (1) and $(5, 7)$.

1 Introduction

Within the context of the $3x + 1$ Problem, Steiner's *1-cycle Theorem* [14] is a result pertaining to the non-existence of *1-cycles* (or *circuits*): for all $a, b \in \mathbb{N}$, Steiner shows that a rational expression of the form

$$\frac{2^a - 1}{2^{a+b} - 3^b} \tag{1}$$

does not assume a positive integer value except in the case where $a = b = 1$. In the proof, the author appeals to the continued fraction expansion of $\log_2 3$, transcendental number theory, and extensive numerical computation (see [13]). This argument serves as the basis for demonstrating the non-existence of 2-cycles in [12], and the non-existence of m-cycles in [13] where $m \leq 68$.

The result has been strengthened in [4] as follows: Let C denote a cycle in the (accelerated) $3x + 1$ dynamical system $T : 2\mathbb{Z} + 1 \to 2\mathbb{Z} + 1$, defined by the mapping

$$T(x) = \frac{3x + 1}{2^{e(x)}}$$

This work was supported by the Naval Surface Warfare Center Dahlgren Division's In-House Laboratory Independent Research Program.

A. Rukhin (✉)
Naval Surface Warfare Center, Dahlgren, VA 22448, USA
e-mail: andrey.rukhin@navy.mil

M. B. Nathanson (ed.), *Combinatorial and Additive Number Theory III*,
Springer Proceedings in Mathematics & Statistics 297,
https://doi.org/10.1007/978-3-030-31106-3_16

where $e(x)$ is the 2-adic valuation of the quantity $3x + 1$. If $e(x) \geq 2$, the element x is said to be a *descending element* in C, and we define $\delta(C)$ to be the number of descending elements in C. Theorem 1.1 in [4] demonstrates that the number of cycles satisfying the inequality $\delta(C) < 2\log(|C|)$ is finite; Steiner's result addresses the case where $d(C) = 1$ by showing that the only (accelerated) cycle with a single descending element is the cycle including 1.

However, the author in [9] declares that the "most remarkable thing about [Steiner's theorem] is the weakness of its conclusion compared to the strength of the methods used in its proof." This article offers alternative proofs of this theorem by demonstrating the non-integrality of the maximal element of a 1-cycle

$$\frac{(2^{a+1}+1)3^{b-1} - 2^{a+b}}{2^{a+b} - 3^b} = 2 \cdot 3^{b-1}\left(\frac{2^a - 1}{2^{a+b} - 3^b}\right) - 1$$

within a variety of algebraic settings. Assuming the upper bound on periodic iterates established in [2], these proofs exploit the fact that the denominator in the above expression is coprime to both 2 and 3. Based on the results in [11], the first proof appeals to elementary modular arithmetic, the second proof exploits identities on weighted binomial coefficients and the Fibonacci numbers, and the third proof analyzes the 2-adic and 3-adic digits of the values in a 1-cycle.

The article concludes with a similiar analyses of the existence of 1-cycles within the (accelerated) $3x - 1$ dynamical system: we will demonstrate that, under the assumption of an exponential upper bound on the iterate values of a periodic orbit, the only 1-cycles are (1) and (5, 7).

2 Overview

2.1 Notation

This manuscript inherits all of the notation and definitions established in [11], which we summarize here. Let $\tau \in \mathbb{N}$, and let $\mathbf{e}, \mathbf{f} \in \mathbb{N}^\tau$ where $\mathbf{e} = (e_0, \ldots, e_{\tau-1})$ and $\mathbf{f} = (f_0, \ldots, f_{\tau-1})$. For each $u \in \mathbb{Z}$, define $E_u = \sum_{0 \leq w < u} e_{w \bmod \tau}$ and $\overline{E}_u = \sum_{0 \leq w < u} e_{(\tau-1-w) \bmod \tau}$; we will define F_u and $\overline{F}_u$ in an analogous manner with the elements of $\mathbf{f}$.

For a positive integer b, we will write $[b] = \{1, \ldots, b\}$ and $[b) = \{1, \ldots, b-1\}$; furthermore, we will write $[b]_0 = [b] \cup \{0\}$ and $[b)_0 = [b) \cup \{0\}$.

For any integer a and positive base b ($b \geq 1$), let $[a]_b$ denote the element[1] of $[b)_0$ that satisfies the equivalence $[a]_b \equiv a \bmod b$. We will also write $[a]_b^{-1}$ to denote the element in $[b)_0$ that satisfies the equivalence $[a]_b\,[a]_b^{-1} \underset{b}{\equiv} 1$.

---

[1]This element is also known as the *standard* (or *canonical*) *representative* of the equivalence class $\overline{a} \bmod b$.

For the maximal iterate value $n_{\max}$ within a 1-cycle, we will define $\mu_\tau = n_{\max} \bmod 3^\tau$ and $\lambda_\tau = n_{\max} \bmod 2^{e+\tau-1}$ for $e, \tau \in \mathbb{N}$

We will write $(-)^u$ to denote the quantity $(-1)^u$ for each $u \in \mathbb{N}_0$.

2.2 Argument Overview

The dual-radix approach to the non-existence of circuits is based upon the following premises:

i. We will establish an upper bound of 3^τ for a potential, periodic iterate value over $\mathbb{N}$ for the (accelerated) $3x+1$ Problem. In this context, the authors in [2] have demonstrated that the maximal iterate $n_{\max}$ within a periodic orbit admits the upper bound

$$n_{\max} < \frac{\left(\frac{3}{2}\right)^{\tau-1}}{1 - \frac{3^\tau}{2^{\overline{E}_\tau}}} \leq \tau^C \left(\frac{3}{2}\right)^{\tau-1} = o\left(3^{\tau-1}\right) \tag{2}$$

for some effectively computable constant C (by applying the result in [1]). A recent upper bound on C is available in [10], in which the author establishes the inequality[2]

$$\left|-\overline{E}_\tau \log 2 + \tau \log 3\right| \geq \overline{E}_\tau^{-13.3}; \tag{3}$$

consequently, assuming $2^{\overline{E}_\tau} > 3^\tau$, we can bound[3] the denominator in (2) from below $1 - \frac{3^\tau}{2^{\overline{E}_\tau}} \geq \frac{\overline{E}_\tau^{-13.3}}{2}$. According to [5], for a periodic orbit over $\mathbb{N}$ of length $\overline{E}_\tau$, the ratio $\frac{\overline{E}_\tau}{\tau}$ satisfies the inequality

$$\frac{\overline{E}_\tau}{\tau} \leq \lg\left(3 + \frac{1}{n_{\min}}\right) \leq 2;$$

numerical computation yields $n_{\max} < \left(\frac{3}{2}\right)^{\tau-1} 2 \cdot (2\tau)^{13.3} < 3^\tau$ when $\tau \geq 103$. Thus, if $n_{\max} > 3^\tau$ and $n_{\max} \in \mathbb{N}$, then $\tau < 103$. However, the author in [7] demonstrates that the length of a non-trivial periodic orbit (excluding 1) over $\mathbb{N}$ must satisfy the inequality $2\tau \geq \overline{E}_\tau \geq 35,400$.

Thus, if $n_{\max} \in \mathbb{N}$, then $n_{\max} < 3^\tau < 2^{\overline{E}_\tau}$, and the equalities $n_{\max} = \mu_\tau = \lambda_\tau$ must hold.

ii. Within a circuit of order τ in the (accelerated) $3x+1$ dynamical system, the maximal element equals

---

[2]In their notation, we set $u_0 = 0$, $u_1 = -\overline{E}_\tau$, and $u_2 = \tau$.

[3]We can shed the logarithms: when $|w| < 1$, the power series expansion of $\log(1+w) = \sum_{u\geq 1}(-1)^{u-1}\frac{w^u}{u}$ yields $|\log(1+w)| \leq 2|w|$ when $|w| \leq \frac{1}{2}$. See [6] (Corollary 1.6).

$$\frac{(2^e+1)3^{\tau-1}-2^{e+\tau-1}}{2^{e+\tau-1}-3^\tau}=2\cdot 3^{\tau-1}\left(\frac{2^{e-1}-1}{2^{e+\tau-1}-3^\tau}\right)-1$$

for some $e \in \mathbb{N}$ (see [3]).
When $\tau = 1$, the left-hand side the equality above satisfies the inequality $\frac{1}{2^e-3} \leq 1$; the maximal iterate equals 1 when $e = 2$, and the ratio in (1) equals 1.
When $\tau > 1$, we will analyze the difference of canonical residues

$$\mu_\tau = \left[(2^e+1)3^{\tau-1}-2^{e+\tau-1}\right][2^{e+\tau-1}]^{-1} \bmod 3^\tau$$

and

$$\lambda_\tau = \left[(2^e+1)3^{\tau-1}-2^{e+\tau-1}\right][-3^\tau]^{-1} \bmod 2^{\overline{E}_\tau};$$

we will demonstrate the inequality $\mu_\tau \neq \lambda_\tau$ (contradicting the assumption that $n_{\max} = \mu_\tau = \lambda_\tau$ as per above).
We will also perform similar analyses on the maximal element of a circuit within the (accelerated) $3x - 1$ dynamical system; we will show that, assuming[4] the inequality $n_{\max} < 2^{\overline{E}_\tau}$, a circuit over $\mathbb{N}$ exists if and only if either $e = 1$, or $\tau = e = 2$.

3 Circuits with the $3x + 1$ Dynamical System

Throughout the remainder of the manuscript, unless otherwise stated, we assume that

i. $\tau \in \mathbb{N}$ with $\tau \geq 2$;
ii. $\mathbf{f} = (1, \ldots, 1) \in \mathbb{N}^\tau$;
iii. $\mathbf{e} = (\underbrace{1, \ldots, 1}_{\tau-1}, e)$ for some $e \in \mathbb{N}$; and
iv. $\mathbf{a} = (a_0, \ldots, a_{\tau-1}) \in \{-1, +1\}^\tau$.

We begin with the following assumptions.

Assumption 3.1 Assume 3.1 and 3.3 from [11], and let $\mathbf{a} = \mathbf{1}^\tau$. Let $N = (2^e + 1)3^{\tau-1} - 2^{e+\tau-1}$, and let $D = 2^{e+\tau-1} - 3^\tau$ where $D > 0$.

Assume that

$$n_{\max} = \frac{N}{D} < \min\left(3^\tau, 2^{\overline{E}_\tau}\right),$$

let $\mu_\tau = n_{\max} \bmod 3^\tau$, and let $\lambda_\tau = n_{\max} \bmod 2^{e+\tau-1}$.

Under these assumptions, if $n_{\max} \in \mathbb{N}$, then the chain of equalities $n_{\max} = \mu_\tau = \lambda_\tau$ holds.

---

[4] Appealing to a similar argument outlined above, this condition holds for finitely many τ for each fixed $e \in \mathbb{N}$.

Our goal for the remainder of this subsection is to prove the following theorem.

Theorem 1 *Assume 3.1.*
We have the equalities

$$\mu_\tau = \begin{cases} 3^{\tau-1} - 1 & e \underset{2}{\equiv} 0 \\ 3^\tau - 1 & e \underset{2}{\equiv} 1 \end{cases}$$

when $\tau \underset{2}{\equiv} 0$, *and*

$$\mu_\tau = \begin{cases} 2 \cdot 3^{\tau-1} - 1 & e \underset{2}{\equiv} 0 \\ 3^\tau - 1 & e \underset{2}{\equiv} 1 \end{cases}$$

when $\tau \underset{2}{\equiv} 1$.
Furthermore, when $\tau \underset{2}{\equiv} 1 \underset{2}{\equiv} e - 1$, *then*

$$\lambda_\tau = 2^e \left(\frac{2^{\tau-1} - 1}{3} \right) + \frac{2^{e+\tau-1} - 1}{3} = \frac{(2^\tau - 1)2^e - 1}{3}.$$

For completeness, we have

$$\lambda_\tau = \begin{cases} \frac{(2^{\tau-1}-1)2^e-1}{3} & e \underset{2}{\equiv} 0 \\ 2^{e+\tau-1} - \frac{2^e+1}{3} & e \underset{2}{\equiv} 1 \end{cases}$$

when $\tau \underset{2}{\equiv} 0$, and

$$\lambda_\tau = \begin{cases} \frac{(2^{\tau}-1)2^e-1}{3} & e \underset{2}{\equiv} 0 \\ 2^{e+\tau-1} - \frac{2^e+1}{3} & e \underset{2}{\equiv} 1 \end{cases}$$

when $\tau \underset{2}{\equiv} 1$. However, in order to expedite the proofs, we exclude three out of the four cases when the corresponding canonical 3-residue μ_τ is even (assuring the inequality $\mu_\tau \neq \lambda_\tau$). We exclude the remaining case with the following lemma.

Lemma 1 *Assume that* $\tau \underset{2}{\equiv} 1 \underset{2}{\equiv} e - 1$; *furthermore, let* $\mu_\tau = 2 \cdot 3^{\tau-1} - 1$, *and* $\lambda_\tau = \frac{(2^\tau-1)2^e-1}{3}$. *Then, the inequality* $\mu_\tau \neq \lambda_\tau$ *holds.*

Proof By way of contradiction, assume that the natural number e satisfies the equality $2 \cdot 3^{\tau-1} - 1 = \frac{(2^\tau-1)2^e-1}{3}$; equivalently, we require that the equality $2\,(3^\tau - 1) = (2^\tau - 1)2^e$ holds. However, we have that

$$2^{e-2}\,(2^\tau - 1) = \frac{3^\tau - 1}{2} \underset{2}{\equiv} \sum_{0 \leq w < \tau} 3^w \underset{2}{\equiv} 1$$

for all odd, positive τ. When $e = 2$, the value of τ must satisfy the equality $2 - \frac{1}{2^\tau} = \left(\frac{3}{2}\right)^\tau$; however, this equality fails to hold for $\tau > 1$. □

Lemma 1, Assumptions 3.1, and Theorem 1, along with the bounds provided in [5, 7, 13], demonstrate the non-existence of circuits in the $3x + 1$ dynamical system.

3.1 Elementary Modular Arithmetic

Our first proof of Theorem 1 appeals to elementary modular arithmetic.

Proof We will write

$$\mu_\tau \underset{3^\tau}{\equiv} ND^{-1} \underset{3^\tau}{\equiv} \left[(2^e + 1)3^{\tau-1} - 2^{e+\tau-1}\right]\left[2^{e+\tau-1}\right]^{-1} \underset{3^\tau}{\equiv} \left[\left[2^{\tau-1}\right]_{3^1}^{-1} + \left[2^{e+\tau-1}\right]_{3^1}^{-1}\right] 3^{\tau-1} - 1.$$

It follows that $\mu_\tau \underset{3^\tau}{\equiv} 3^{\tau-1}(-)^{\tau-1}\left[1 + (-)^e\right] - 1$. Thus, when $e \underset{2}{\equiv} 1$, we have $\mu_\tau = 3^\tau - 1 \underset{2}{\equiv} 0$. Similarly, when $e \underset{2}{\equiv} 0$ and $\tau \underset{2}{\equiv} 0$, we have $\mu_\tau = 3^{\tau-1} - 1 \underset{2}{\equiv} 0$. When $\tau \underset{2}{\equiv} 1 \underset{2}{\equiv} e - 1$, we arrive at the equality $\mu_\tau = 2 \cdot 3^{\tau-1} - 1$.

For the **2**-remainder, we begin by writing

$$\lambda_\tau \underset{2^{e+\tau-1}}{\equiv} ND^{-1} \underset{2^{e+\tau-1}}{\equiv} \left[(2^e + 1)3^{\tau-1} - 2^{e+\tau-1}\right]\left[-3^\tau\right]^{-1} \underset{2^{e+\tau-1}}{\equiv} 2^e\,[-3]_{2^{\tau-1}}^{-1} + [-3]^{-1}.$$

When $\tau \underset{2}{\equiv} 1 \underset{2}{\equiv} e - 1$, we have $\left[-3^1\right]_{2^{\tau-1}}^{-1} = \frac{2^{\tau-1}-1}{3}$ and $\left[-3^1\right]_{2^{e+\tau-1}}^{-1} = \frac{2^{e+\tau-1}-1}{3}$.

As

$$2^e\left(\frac{2^{\tau-1} - 1}{3}\right) + \frac{2^{e+\tau-1} - 1}{3} = \frac{2\left(2^{e+\tau-1}\right) - 2^e - 1}{3} < 2^{e+\tau-1},$$

we arrive at the chain of equalities $\lambda_\tau = 2^e\left(\frac{2^{\tau-1}-1}{3}\right) + \frac{2^{e+\tau-1}-1}{3} = \frac{(2^\tau-1)2^e-1}{3}$. □

3.2 Weighted Binomial Coefficients

The previous approach is apparently limited; it is unclear to the author how to extrapolate this approach to admissible sequences of order τ with an arbitrary **2**-grading $(e_0, \ldots, e_{\tau-1})$. In this subsection, we introduce a more robust approach to identifying the 3-residues and **2**-remainders of the iterates of an admissible cycle in a $(3, 2)$-system. Moreover, we do so by connecting the residues of $(3, 2)$-systems to the well-known *Fibonacci sequence* by way of elementary equivalence identities, which we establish first.

Lemma 2 *For $a, b, z \in \mathbb{N}$, the equivalence*

$$\left(\sum_{0 \le w < b} z^w\right)^a \underset{z^b}{\equiv} \sum_{0 \le w < b} \binom{a-1+w}{w} z^w$$

holds.

Proof Define $S_b(z) = \sum_{0 \le w < b} z^w$, and define $T_{a,b}(z) = \sum_{0 \le w < b} \binom{a-1+w}{w} z^w$. The proof is by induction on b.

When $b = 1$, we arrive at the equivalence $1^a \underset{z}{\equiv} \binom{a-1}{0}$ for all $a, z \in \mathbb{N}$.

Assume the claim holds for $b \in \mathbb{N}$. The identity $S_{b+1}(z) = zS_b(z) + 1$ allows the chain of equivalences

$$\left[S_{b+1}(z)\right]^a \underset{z^{b+1}}{\equiv} \sum_{0 \le y < b+1} \binom{a}{y} z^y \left[S_b(z)\right]^y \underset{z^{b+1}}{\equiv} \binom{a}{0} z^0 + \sum_{1 \le y < b+1} \binom{a}{y} z^y T_{y,b}(z).$$

We will recast the coefficient of z^0 as $\binom{a-1}{0}$, and we will write

$$\sum_{1 \le y < b+1} \binom{a}{y} z^y T_{y,b}(z) = \sum_{1 \le y < b+1} \sum_{0 \le u < b} z^{u+y} \binom{a}{y} \binom{y-1+u}{u}.$$

For each $w \in [b+1)$, the coefficient of z^w is $\sum_{1 \le y \le w} \binom{a}{y}\binom{w-1}{w-y} = \sum_{0 \le y < w} \binom{a}{w-y}$ $\binom{w-1}{y}$, which equals $\binom{a-1+w}{w}$ as per the *Vandermonde–Chu* identity. □

Identity *(Fibonacci Identity)* Let $F_0 = 0$, $F_1 = 1$, and $F_n = F_{n-1} + F_{n-2}$ for $n \ge 2$. The equality $F_n = \sum_{0 \le k < n} \binom{n-1-k}{k}$ holds.

We will use these identities to establish the *remainder approximation functions.*

Lemma 3 *Define the map $M_\tau : \mathbb{N}^\tau \times \mathbb{N}^\tau \to \mathbb{Z}$ to be*

$$M_\tau = M_\tau(\mathbf{e}, \mathbf{a}) = \sum_{0 \le w < u} (-)^{E_{w+1}} 3^w a_w \sum_{0 \le y < \tau - w} \binom{E_{w+1} - 1 + y}{y} 3^y,$$

and define the map $\Lambda_\tau : \mathbb{N}^\tau \times \mathbb{N}^\tau \to \mathbb{Z}$ to be

$$\Lambda_\tau = \Lambda_\tau(\mathbf{e}, \mathbf{a}) = \sum_{0 \le w < \tau} (-)^w 2^{\overline{E}_w} a_{\tau-1-w} \sum_{0 \le y < \eta_w} \binom{w+y}{y} 4^y,$$

where $\eta_w = \left\lceil \frac{E_{\tau-w}}{2} \right\rceil$.

Then, the equivalences $M_\tau \underset{3^\tau}{\equiv} \mu_\tau$ and $\Lambda_\tau \underset{2^{\overline{E}_\tau}}{\equiv} \lambda_\tau$ hold.

Proof We will make use of the following elementary identities involving *Euler's totient function* ϕ: we have $3^{\phi(2)} - 1 = 2$ and $2^{\phi(3)} - 1 = 3$. In light of these identities, we will appeal to Lemma 2: for $a, b \in \mathbb{N}$, we will write

$$[2^a]^{-1} \underset{3^b}{\equiv} \left(\frac{1 - 3^{\phi(2)\left\lceil \frac{b}{\phi(2)} \right\rceil}}{2} \right)^a \underset{3^b}{\equiv} (-)^a \left(\sum_{0 \le y < b} 3^y \right)^a \underset{3^b}{\equiv} (-)^a \sum_{0 \le y < b} \binom{a - 1 + y}{y} 3^y,$$

and

$$\left[3^b\right]^{-1} \underset{2^a}{\equiv} \left(\frac{1 - 2^{\phi(3)\left\lceil \frac{a}{\phi(3)} \right\rceil}}{3} \right)^b \underset{2^a}{\equiv} (-)^b \left(\sum_{0 \le y < \lceil \frac{a}{2} \rceil} 4^y \right)^b \underset{2^a}{\equiv} (-)^b \sum_{0 \le y < \lceil \frac{a}{2} \rceil} \binom{b - 1 + y}{y} 4^y.$$

We derive the 3-remainder approximation function as follows:

$$\begin{aligned} \mu_\tau \underset{3^\tau}{\equiv} \left[ND^{-1}\right]_{3^\tau} &\underset{3^\tau}{\equiv} \sum_{0 \le w < \tau} 3^w 2^{\overline{E}_{\tau-1-w}} a_w \left[2^{\overline{E}_\tau}\right]^{-1} \\ &\underset{3^\tau}{\equiv} \sum_{0 \le w < \tau} (-)^{E_{w+1}} 3^w a_w \sum_{0 \le y < \tau - w} \binom{E_{w+1} - 1 + y}{y} 3^y. \end{aligned}$$

We derive the **2**-remainder approximation function analogously:

$$\lambda_\tau \underset{2^{\overline{E}_\tau}}{\equiv} \sum_{0 \le w < \tau} 3^w 2^{\overline{E}_{\tau-1-w}} a_w \left[-3^\tau\right]^{-1} \underset{2^{\overline{E}_\tau}}{\equiv} \sum_{0 \le w < \tau} (-)^w 2^{\overline{E}_w} a_{\tau-1-w} \sum_{0 \le y < \eta_w} \binom{w + y}{y} 4^y.$$

□

It will prove useful to re-index these double-sums: for example, in the 3-residue approximation, for each fixed $w \in [\tau)_0$ the coefficient of 3^w is

$$S_w = \sum_{0 \le y \le w} (-)^{E_{y+1}} \binom{E_{y+1} - 1 + w - y}{w - y} a_y;$$

thus, we can write $M_\tau = \sum_{0 \le w < \tau} 3^w S_w$.

The following example illustrates the connection between an orbit over $\mathbb{N}$ within the $3x + 1$ dynamical system and the Fibonacci Sequence.

3.2.1 Example: The (1, 4, 2)-Orbit in the 3x + 1 Dynamical System

For this example, define $e_y = 2$ and $a_y = 1$ for each $y \in [\tau)_0$; thus, the sum $E_{y+1} = 2(y + 1) \underset{2}{\equiv} 0$. We can express the 3-remainder approximation as $M_\tau =$

$\sum_{0 \le w < \tau} 3^w S_w$, where

$$S_w := \sum_{0 \le y \le w} (-)^{2(y+1)} \binom{2(y+1) - 1 + w - y}{w - y} = \sum_{0 \le y \le w} \binom{2w + 1 - y}{y}.$$

The sequence $(S_w)_{w \ge 0}$ is the even-indexed bisection of the Fibonacci sequence $(F_w)_{w \ge 0}$ as per Identity 1.3.1; we have $S_w = F_{2(w+1)}$ for $w \ge 0$. It is known[5] that this bisection satisfies the recurrence[6] $F_{2w} = 3F_{2(w-1)} - F_{2(w-2)}$ for $w \ge 0$; thus, we will write $M_\tau = \sum_{0 \le w < \tau} 3^w S_w = \sum_{0 \le w < \tau} 3^w F_{2(w+1)}$, and we continue by writing

$$\begin{aligned}\sum_{0 \le w < \tau} 3^w \left[3F_{2w} - F_{2(w-1)}\right] = & \sum_{0 \le w < \tau - 1} 3^{w+1} F_{2w} + 3^\tau F_{2(\tau-1)} - F_{-2} \\ & - \sum_{1 \le w < \tau} 3^w F_{2(w-1)} = 3^\tau F_{2(\tau-1)} + 1.\end{aligned}$$

For the **2**-remainder approximation, we have the equalities $\Lambda_\tau = \sum_{0 \le w < \tau} 4^w \sum_{0 \le y \le w} \binom{w}{y}(-1)^y = \sum_{0 \le w < \tau} 4^w (1-1)^w = 1$ for $\tau \in \mathbb{N}$.

The Fibonacci sequence appears within the **2**-remainder approximation for the following proof of Theorem 1. In order to expedite the derivation of this **2**-remainder, we will first prove the following lemma.

Lemma 4 *For $a \in \mathbb{N}_0$, let F_a denote the ath Fibonacci number; furthermore, for $k \in \mathbb{N}_0$, define $\sigma(a, k) = 2\binom{a+1}{k} - \binom{a}{k}$, and define $\mathcal{S}(k) = \sum_{0 \le i < k} \sigma(2k - i, i + 1)$.*

For $k \in \mathbb{N}_0$, the equality $\mathcal{S}(k) = F_{2k+2} + 2F_{2k+1} - 3$ holds.

Proof Assume the conditions within the statement of the lemma. For $k = 0$, we have $\mathcal{S}(k) = 0 = F_2 + 2F_1 - 3$. When $k > 0$, we will write

$$\begin{aligned}\mathcal{S}(k) &= \sum_{0 \le i < k} \left[2\binom{2k - i + 1}{i + 1} - \binom{2k - i}{i + 1}\right] \\ &= \sum_{1 \le i < k+1} \left[2\binom{2k + 2 - i}{i} - \binom{2k + 1 - i}{i}\right] \\ &= 2\left[F_{2k+3} - \binom{2k+2}{0} - \binom{k+1}{k+1}\right] - \left[F_{2k+2} - \binom{2k+1}{0}\right] \\ &= F_{2k+2} + 2F_{2k+1} - 3.\end{aligned}$$

□

We proceed with the proof of the theorem.

[5]OEIS:A001906

[6]We assume the standard definition $F_{-u} = (-)^{u-1} F_u$ for $u \in \mathbb{N}$.

Proof First, we will demonstrate the equality $M_\tau = -1 + 3^{\tau-1}(-)^{\tau-1}\left[1+(-)^e\right]$; afterwards, when assuming $\tau \underset{2}{\equiv} 1 \underset{2}{\equiv} e-1$, we will show that

$$\Lambda_\tau = 2^e\left(\frac{2^{\tau-1}-1}{3}\right) + \frac{2^{e+\tau-1}-1}{3} + 2^{e+\tau-1}\left(F_{\tau-2}-1\right).$$

In circuits, we have

$$E_w = \begin{cases} w & w < \tau \\ e+\tau-1 & w = \tau, \end{cases}$$

for $w \in [\tau)$. Thus, when $w < \tau - 1$, we have

$$\begin{aligned} S_w &= \sum_{0\le y\le w} (-)^{E_{y+1}}\binom{E_{y+1}-1+w-y}{w-y} \\ &= \sum_{0\le y\le w} (-)^{y+1}\binom{w}{w-y} \\ &= -\sum_{0\le y\le w} (-)^{w-y}\binom{w}{y} \\ &= -(1-1)^w \\ &= \begin{cases} 0 & w > 0 \\ -1 & w = 0. \end{cases}; \end{aligned}$$

when $w = \tau - 1 \ge 1$, we have

$$\begin{aligned} S_{\tau-1} &= \sum_{0\le y\le \tau-1} (-)^{E_{y+1}}\binom{E_{y+1}-1+\tau-1-y}{\tau-1-y} \\ &= \sum_{0\le y\le \tau-2} (-)^{y+1}\binom{\tau-1}{\tau-1-y} + (-)^{e+\tau-1}\binom{e+\tau-2}{0} \\ &= -(1-1)^{\tau-1} + (-)^{\tau-1}\binom{\tau-1}{\tau-1} + (-)^{e+\tau-1}\binom{e+\tau-2}{0} \\ &= (-)^{\tau-1}\left[1+(-)^e\right]. \end{aligned}$$

It follows that $M_\tau = -1 + 3^{\tau-1}(-)^{\tau-1}\left[1+(-)^e\right]$. Thus, when $e \underset{2}{\equiv} 1$, we have $\mu_\tau = 3^\tau - 1$. Similarly, when $e \underset{2}{\equiv} 0$ and $\tau \underset{2}{\equiv} 0$, we have $\mu_\tau = 3^{\tau-1} - 1$.

When $\tau \underset{2}{\equiv} 1 \underset{2}{\equiv} e-1$, we arrive at the equality $\mu_\tau = 2\cdot 3^{\tau-1} - 1$. Continuing with these parity conditions, we let T_w denote the sum $\sum_{0\le y<\left\lceil \frac{E_{\tau-w}}{2}\right\rceil}\binom{w+y}{y}4^y$. We write

$$\begin{aligned}\Lambda_\tau &= \sum_{0\le w<\tau} (-)^w 2^{\overline{E}_w} T_w \\ &= T_0 + \sum_{1\le w<\tau} (-)^w 2^{\overline{E}_w} T_w \\ &= \sum_{0\le y<\frac{e+\tau-1}{2}} \binom{y}{y} 4^y + \sum_{1\le w<\tau} (-)^w 2^{\overline{E}_w} \binom{w}{0} + \sum_{1\le w<\tau} (-)^w 2^{\overline{E}_w} \left[T_w - \binom{w}{0}\right].\end{aligned}$$

We proceed with the first two sums in the final expression. When $e+\tau-1 \underset{2}{\equiv} 0$, we will write

$$T_0 = \sum_{0\le y<\frac{e+\tau-1}{2}} \binom{y}{y} 4^y = \frac{2^{e+\tau-1}-1}{3}.$$

In circuits, we have $\overline{E}_w = e+w-1$ for $w \in [\tau)$; thus, when $\tau-1 \underset{2}{\equiv} 0$, we will also write

$$\begin{aligned}\sum_{1\le w<\tau} (-)^w 2^{\overline{E}_w} \binom{w}{0} &\underset{2^{e+\tau-1}}{\equiv} 2^e \sum_{0\le w<\tau-1} (-)^{w+1} 2^w \\ &\underset{2^{e+\tau-1}}{\equiv} 2^e \sum_{0\le w<\frac{\tau-1}{2}} \left[2^{2w+1} - 2^{2w}\right] \\ &\underset{2^{e+\tau-1}}{\equiv} 2^e \sum_{0\le w<\frac{\tau-1}{2}} 4^w \\ &\underset{2^{e+\tau-1}}{\equiv} 2^e \left(\frac{2^{\tau-1}-1}{3}\right).\end{aligned}$$

What remains to be shown is that $\sum_{1\le w<\tau}(-)^w 2^{\overline{E}_w}\left[T_w - \binom{w}{0}\right] \underset{2^{e+\tau-1}}{\equiv} 0$. To this end, for each $k \in \mathbb{N}_0$, we will define

$$\widehat{\Lambda}_{2k+1} = \sum_{1\le w<2k-1} (-)^w 2^{w-1} \sum_{1\le y<\left\lceil\frac{2k+1-w}{2}\right\rceil} \binom{w+y}{y} 4^y;$$

we will show that

$$\sum_{1\le w<\tau} (-)^w 2^{\overline{E}_w} \left[T_w - \binom{w}{0}\right] = 2^e \widehat{\Lambda}_\tau = 2^{e+\tau-1}\left(F_{\tau-2}-1\right).$$

Assume the notation from the statement of Lemma 4. We will demonstrate the chain of equalities

$$\widehat{\Lambda}_{2k+1} = \widehat{\Lambda}_{2k-1} + 4^{k-1}\mathcal{S}(k-1) = 4^k\left(F_{2k-1}-1\right)$$

inductively for $k \in \mathbb{N}$. Firstly, we have $\widehat{\Lambda}_3 = 0 + 4^0 \mathcal{S}(0) = 4^0 (F_1 - 1) = 0$ for $k = 1$. Assuming the inductive claim, we proceed with the chain of equalities for $k \geq 2$:

$$\widehat{\Lambda}_{2k+1} = \sum_{1 \leq w < 2k-1} (-)^w 2^{w-1} \sum_{1 \leq y < \lceil \frac{2k+1-w}{2} \rceil} \binom{w+y}{y} 4^y = \widehat{\Lambda}_{2k-1} + A_k,$$

where

$$A_k = \sum_{1 \leq w < 2k-1} (-)^w 2^{w-1} \binom{w + \lceil \frac{2k-1-w}{2} \rceil}{\lceil \frac{2k-1-w}{2} \rceil} 4^{\lceil \frac{2k-1-w}{2} \rceil}.$$

The sum

$$\begin{aligned}
A_k &= \sum_{1 \leq w < 2k-1} (-)^w 2^{w-1} \binom{k + w + \lceil \frac{-1-w}{2} \rceil}{k + \lceil \frac{-1-w}{2} \rceil} 4^{k + \lceil \frac{-1-w}{2} \rceil} \\
&= \sum_{1 \leq w < \frac{2k-1}{2}} \left[2^{2w-1} \binom{k+w}{k-w} - 2^{2w-2} \binom{k-1+w}{k-w} \right] 4^{k-w} \\
&= 4^{k-1} \sum_{1 \leq w < k} \left[2 \binom{k+w}{k-w} - \binom{k-1+w}{k-w} \right] \\
&= 4^{k-1} \sum_{1 \leq w < k} \left[2 \binom{2k-w}{w} - \binom{2k-1-w}{w} \right] \\
&= 4^{k-1} \sum_{0 \leq w < k-1} \left[2 \binom{2k-1-w}{w+1} - \binom{2k-2-w}{w+1} \right] \\
&= 4^{k-1} \mathcal{S}(k-1).
\end{aligned}$$

Thus, with Lemma 4 and the inductive hypothesis, we can write

$$\widehat{\Lambda}_{2k+1} = \widehat{\Lambda}_{2k-1} + 4^{k-1} \mathcal{S}(k-1) = 4^{k-1} \left[F_{2k-3} + F_{2k-2} + 3F_{2k-1} - 4 \right] = 4^k \left[F_{2k-1} - 1 \right]$$

as required. Consequently, when $\tau \underset{2}{\equiv} 1 \underset{2}{\equiv} e - 1$, the **2**-remainder approximation

$$\Lambda_\tau = 2^e \left(\frac{2^{\tau-1} - 1}{3} \right) + \frac{2^{e+\tau-1} - 1}{3} + 2^{e+\tau-1} (F_{\tau-2} - 1) \underset{2^{e+\tau-1}}{\equiv} 2^e \left(\frac{2^{\tau-1} - 1}{3} \right) + \frac{2^{e+\tau-1} - 1}{3}.$$

□

Note that the approach within this subsection exploits the serendipitous pair of identities $3^{\phi(2)} - 1 = 2$ and $2^{\phi(3)} - 1 = 3$. In general, Euler's Theorem allows one to write $m^{\phi(l)} - 1 = [-l]^{-1}_{m^{\phi(l)}} l$, and $l^{\phi(m)} - 1 = [-m]^{-1}_{l^{\phi(m)}} m$; however, for arbitrary, coprime m and l exceeding 1, the terms $[-l]^{-1}_{m^{\phi(l)}}$ and $[-m]^{-1}_{l^{\phi(m)}}$ may prevent one from executing the approach above in an analogous manner.

3.3 Dual-Radix Modular Division

The approach in this section, based on the work in [11], demonstrates a different method of proving Theorem 1 using *dual-radix modular division.*

Proof Under the assumption that

$$e_w = \begin{cases} 1 & w \in [\tau - 1)_0 \\ e & w = \tau - 1, \end{cases}$$

we have the following initial conditions for the recurrence in Theorem 4.4 in [11]. For $v \in [\tau)_0$, the 3-adic digit $d_{v,0} \underset{3}{\equiv} [2^{e_v}]^{-1}$; thus, we have

$$d_{v,0} = \begin{cases} 2 & v \in [\tau - 1)_0 \\ 1 + e \bmod 2 & v = \tau - 1; \end{cases}$$

furthermore, the **2**-adic digit $b_{v,0} \underset{2^{e_{v-1}}}{\equiv} [-3]^{-1}$; thus, we have

$$b_{v,0} = \begin{cases} \frac{2^{2\lceil \frac{e}{2} \rceil} - 1}{3} & v = 0 \\ 1 & v \in [\tau - 1]. \end{cases}$$

For $u > 0$, the equivalences

$$d_{v,u} \underset{3}{\equiv} \left[2^{e_v}\right]^{-1} \left[d_{v+1,u-1} - b_{v+u,u-1}\right]$$

and

$$b_{v,u} \underset{2^{e_{v-1-u}}}{\equiv} [-3]^{-1} \left[d_{v-u,u-1} - b_{v-1,u-1}\right]$$

yield, by induction on u, the equalities $d_{v,u} = 2[2 - 1] = 2$ for $v < \tau - 1 - u$, and $b_{v,u} = 1[2 - 1] = 1$ for $v > u$.

Firstly, we will identify the 3-adic digits of the 3-remainder of $n_0 = n_{\max}$. When $e \underset{2}{\equiv} 1$, we have the initial condition $d_{\tau-1,0} = 2$. Thus, for $u \in [\tau)$, the digit $d_{\tau-1-u,u} \underset{3}{\equiv} [2^{e_{\tau-1-u}}]^{-1} \left[d_{\tau-u,u-1} - b_{\tau-1,u-1}\right] \underset{3}{\equiv} 2\,[2 - 1] \underset{3}{\equiv} 2$, and thus we have $d_{0,\tau-1} = 2$. Consequently, we have $\mu_\tau = \sum_{0 \le w < \tau} 3^w d_{0,w} = 3^\tau - 1$.

When $e \underset{2}{\equiv} 0$, we have the initial condition $d_{\tau-1,0} = 1$, and $d_{\tau-2,1} \underset{3}{\equiv} \left[2^1\right]^{-1} \left[d_{\tau-1,0} - b_{\tau-1,0}\right] \underset{3}{\equiv} \left[2^1\right]^{-1} [1 - 1] \underset{3}{\equiv} 0$. By induction, for $u \in [\tau)$ where $u \underset{2}{\equiv} 0$, the digit

$$d_{\tau-1-u,u} \underset{3}{\equiv} \left[2^{e_{\tau-1-u}}\right]^{-1} \left[d_{\tau-u,u-1} - b_{\tau-1,u-1}\right] \underset{3}{\equiv} 2\,[0 - 1] \underset{3}{\equiv} 1.$$

For $u \underset{2}{\equiv} 1$, the digit $d_{\tau-1-u,u} \underset{3}{\equiv} [2^{e_{\tau-1-u}}]^{-1} \left[d_{\tau-u,u-1} - b_{\tau-1,u-1}\right] \underset{3}{\equiv} 2[1-1] \underset{3}{\equiv} 0$. Thus, the digit $d_{0,\tau-1} = \tau \bmod 2$. Thus, when $\tau \underset{2}{\equiv} 0$, the 3-adic remainder $\mu_\tau = \sum_{0 \le w < \tau-1} 3^w(2) + 3^{\tau-1}(0) = 3^{\tau-1} - 1$; and, when $\tau \underset{2}{\equiv} 1$, the 3-adic residue $\mu_\tau = \sum_{0 \le w < \tau-1} 3^w(2) + 3^{\tau-1}(1) = 2 \cdot 3^{\tau-1} - 1$.

We will now determine the **2**-adic digits of n when $\tau \underset{2}{\equiv} 1 \underset{2}{\equiv} e-1$: the initial **2**-adic digit $b_{0,0} = \frac{2^e-1}{3}$, and the digit $b_{0,1} \underset{2^{e_{\tau-2}}}{\equiv} [-3]^{-1} \left[d_{\tau-1,0} - b_{\tau-1,0}\right] \underset{2^1}{\equiv} (1) \cdot [1-1] \underset{2^1}{\equiv} 0$. For $u \in [\tau)$ where $u \underset{2}{\equiv} 0$, we have $b_{0,u} \underset{2^{e_{\tau-1-u}}}{\equiv} [-3]^{-1} \left[d_{\tau-u,u-1} - b_{\tau-1,u-1}\right] \underset{2^1}{\equiv} (1) \cdot [0-1] \underset{2^1}{\equiv} 1$, and, when $u \underset{2}{\equiv} 1$, we have $b_{0,u} \underset{2^{e_{\tau-1-u}}}{\equiv} [-3]^{-1} \left[d_{\tau-u,u-1} - b_{\tau-1,u-1}\right] \underset{2^1}{\equiv} (1) \cdot [1-1] \underset{2^1}{\equiv} 0$. Thus, when $\tau \underset{2}{\equiv} 1 \underset{2}{\equiv} e-1$, the **2**-adic remainder

$$\begin{aligned}
\lambda_\tau &= b_{0,0} + \sum_{1 \le u < \tau} 2^{\overline{E}_u} b_{0,u} \\
&= \frac{2^e-1}{3} + 2^e \sum_{2 \le u < \tau} 2^{u-1}[u \underset{2}{\equiv} 0] \\
&= \frac{2^e-1}{3} + 2^{e+1} \sum_{0 \le u < \tau-2} 2^u [u \underset{2}{\equiv} 0] \\
&= \frac{2^e-1}{3} + 2^{e+1} \sum_{0 \le u \le \frac{\tau-3}{2}} 4^u \\
&= \frac{2^e-1}{3} + 2^{e+1} \left(\frac{4^{\frac{\tau-1}{2}} - 1}{3}\right) \\
&= 2^e \left(\frac{2^{\tau-1}-1}{3}\right) + \frac{2^{e+\tau-1}-1}{3}.
\end{aligned}$$

□

3.4 Circuits in the $3x - 1$ Dynamical System

We conclude this article by applying the previous analyses to the $3x - 1$ dynamical system; now, we will consider the case where $a_w = -1$ for all $w \in [\tau)_0$.

We will extend the argument in [2] to the case where $3^\tau > 2^{\overline{E}_\tau}$: the magnitude of the numerator of a maximal iterate in a periodic orbit can be bound from above as follows:

$$\left|\left(2^e+1\right)3^{\tau-1} - 2^{\overline{E}_\tau}\right| = 3^\tau \left[\frac{2^e+1}{3} - \frac{2^{\overline{E}_\tau}}{3^\tau}\right] < 3^{\tau-1}\left(2^e+1\right).$$

We can bound the denominator $3^{\tau} - 2^{\overline{E}_{\tau}}$ from below by appealing to the inequality (3) once again to conclude that the maximal iterate $n_{\max}$ within a periodic orbit in the $3x - 1$ dynamical system satisfies the inequality

$$n_{\max} < \frac{\frac{2^e+1}{3}}{1 - \frac{2^{e+\tau-1}}{3^{\tau}}} < \left(\frac{2^e+1}{3}\right) 2\,(e+\tau-1)^{13.3} = o(2^{e+\tau-1})$$

for any fixed $e \in \mathbb{N}$. Thus, we will reuse the notation of the previous section and begin with the following assumptions.

Assumption 3.2 Assume 3.1, except that now we assume that $N = 2^{e+\tau-1} - (2^e+1)3^{\tau-1}$, and $D = 2^{e+\tau-1} - 3^{\tau} < 0$.

As before, define $\mu_{\tau} = ND^{-1} \bmod 3^{\tau}$ and $\lambda_{\tau} = ND^{-1} \bmod 2^{e+\tau-1}$.

Our goal for the remainder of this subsection is to prove the following theorem.

Theorem 2 *Assume (3.2).*

The 3*-remainder*

$$\mu_{\tau} = \begin{cases} 2 \cdot 3^{\tau-1} + 1 & e \underset{2}{\equiv} 0 \\ 1 & e \underset{2}{\equiv} 1 \end{cases}$$

when $\tau \underset{2}{\equiv} 0$*, and*

$$\mu_{\tau} = \begin{cases} 3^{\tau-1} + 1 & e \underset{2}{\equiv} 0 \\ 1 & e \underset{2}{\equiv} 1 \end{cases}$$

when $\tau \underset{2}{\equiv} 1$.

The **2***-remainder*

$$\lambda_{\tau} = \begin{cases} \frac{2^e(2^{\tau}+1)+1}{3} & e \underset{2}{\equiv} 0 \\ \frac{2^e+1}{3} & e \underset{2}{\equiv} 1 \end{cases}$$

when $\tau \underset{2}{\equiv} 0$*, and*

$$\lambda_{\tau} = \begin{cases} \frac{2^e(2^{\tau-1}+1)+1}{3} & e \underset{2}{\equiv} 0 \\ \frac{2^e+1}{3} & e \underset{2}{\equiv} 1 \end{cases}$$

when $\tau \underset{2}{\equiv} 1$.

Analogous to Lemma 1, the following lemma will aid in identifying circuits within the $3x - 1$ Dynamical System.

Lemma 5 *Assume that the* 3*-remainder is*

$$\mu_\tau = \begin{cases} 2 \cdot 3^{\tau-1} + 1 & e \underset{2}{\equiv} 0 \\ 1 & e \underset{2}{\equiv} 1 \end{cases}$$

when $\tau \underset{2}{\equiv} 0$*, and*

$$\mu_\tau = \begin{cases} 3^{\tau-1} + 1 & e \underset{2}{\equiv} 0 \\ 1 & e \underset{2}{\equiv} 1 \end{cases}$$

when $\tau \underset{2}{\equiv} 1$*. Moreover, assume that the* **2***-remainder is*

$$\lambda_\tau = \begin{cases} \frac{2^e(2^\tau+1)+1}{3} & e \underset{2}{\equiv} 0 \\ \frac{2^e+1}{3} & e \underset{2}{\equiv} 1 \end{cases}$$

when $\tau \underset{2}{\equiv} 0$*, and*

$$\lambda_\tau = \begin{cases} \frac{2^e(2^{\tau-1}+1)+1}{3} & e \underset{2}{\equiv} 0 \\ \frac{2^e+1}{3} & e \underset{2}{\equiv} 1 \end{cases}$$

when $\tau \underset{2}{\equiv} 1$*.*

The equality $\mu_\tau = \lambda_\tau$ *holds if and only if either i.* $e = 1$ *or ii.* $e = \tau = 2$*.*

Proof When $e \underset{2}{\equiv} 1$, we require that the equality $\frac{2^e+1}{3} = 1$ holds; consequently, we require that $e = 1$ (irrespective of the parity of τ).

When $e \underset{2}{\equiv} 0$ and $\tau \underset{2}{\equiv} 0$, we require that the equality $2 \cdot 3^{\tau-1} + 1 = \frac{2^e(2^\tau+1)+1}{3}$ holds. Equivalently, we require that $2 \cdot 3^\tau + 3 = 2^e\,(2^\tau + 1) + 1$; after simplifying, we require that $\frac{3^\tau+1}{2^{e-1}} = 2^\tau + 1$. When $\tau \underset{2}{\equiv} 0$, the numerator on the left-hand side $9^{\frac{\tau}{2}} + 1 \underset{4}{\equiv} 2$; thus, it follows that we require that $e = 2$. The equality $3^\tau = 2^{\tau+1} + 1$ holds only when $\tau = 2$ as per a result of Gersonides[7] on *harmonic numbers*.

When $e \underset{2}{\equiv} 0$ and $\tau \underset{2}{\equiv} 1$, we have $\mu_\tau \underset{2}{\equiv} 0$ and $\lambda_\tau \underset{2}{\equiv} 1$. □

Proof (Theorem 2) We can write

$$\mu_\tau \underset{3^\tau}{\equiv} N\left[2^{e+\tau-1} - 3^\tau\right]^{-1} \underset{3^\tau}{\equiv} \left[2^{e+\tau-1} - (2^e + 1)3^{\tau-1}\right]\left[2^{e+\tau-1}\right]^{-1}$$
$$\underset{3^\tau}{\equiv} 1 - \left[\left[2^{\tau-1}\right]_{3^1}^{-1} + \left[2^{e+\tau-1}\right]_{3^1}^{-1}\right]3^{\tau-1}.$$

[7] Levi Ben Gerson, 1342 AD. See [8].

As $[2^u]_{3^1}^{-1} \underset{3}{\equiv} (-)^u$ for $u \in \mathbb{N}$, it follows that $\mu_\tau \underset{3^\tau}{\equiv} 1 + 3^{\tau-1}(-)^\tau\left[1 + (-)^e\right]$. For the **2**-remainder, we begin by writing

$$\lambda_\tau \underset{2^{e+\tau-1}}{\equiv} N\left[2^{\overline{E}_\tau} - 3^\tau\right]^{-1} \underset{2^{e+\tau-1}}{\equiv} [2^{e+\tau-1} - (2^e+1)3^{\tau-1}][-3^\tau]^{-1} \underset{2^{e+\tau-1}}{\equiv} 2^e\,[3]_{2^{\tau-1}}^{-1} + [3]_{2^{e+\tau-1}}^{-1}.$$

We will write $[3]_{2^{\tau-1}}^{-1} = \frac{2^{\tau-(\tau-1) \bmod 2}+1}{3}$, and $[3]_{2^{e+\tau-1}}^{-1} = \frac{2^{e+\tau-(e+\tau-1) \bmod 2}+1}{3}$, and we will complete the proof by cases.

i. $(e \underset{2}{\equiv} 0, \tau \underset{2}{\equiv} 0)$ $\mu_\tau = 2\cdot 3^{\tau-1}+1$, and $\lambda_\tau = \left[2^e\left(\frac{2^{\tau-1}+1}{3}\right) + \frac{2^{e+\tau-1}+1}{3}\right] \bmod 2^{e+\tau-1} = \frac{2^{e+\tau}+2^e+1}{3}$

ii. $(e \underset{2}{\equiv} 0, \tau \underset{2}{\equiv} 1)$ $\mu_\tau = 3^{\tau-1}+1$, and $\lambda_\tau = \left[2^e\left(\frac{2^{\tau}+1}{3}\right) + \frac{2^{e+\tau}+1}{3}\right] \bmod 2^{e+\tau-1} = \frac{2^{e+\tau-1}+2^e+1}{3}$.

iii. $(e \underset{2}{\equiv} 1, \tau \underset{2}{\equiv} 0)$ $\mu_\tau = 1$, and $\lambda_\tau = \left[2^e\left(\frac{2^{\tau-1}+1}{3}\right) + \frac{2^{e+\tau}+1}{3}\right] \bmod 2^{e+\tau-1} = \frac{2^e+1}{3}$.

iv. $(e \underset{2}{\equiv} 1, \tau \underset{2}{\equiv} 1)$ $\mu_\tau = 1$, and $\lambda_\tau = \left[2^e\left(\frac{2^{\tau}+1}{3}\right) + \frac{2^{e+\tau-1}+1}{3}\right] \bmod 2^{e+\tau-1} = \frac{2^e+1}{3}$. □

Thus, under the assumption that $n < 2^{e+\tau-1}$, the only circuits within the $3x-1$ dynamical system are (1) and $(5, 7)$.

References

1. A. Baker, G. Wüstholz, Logarithmic forms and group varieties. Journal für die reine und angewandte Mathematik **442**, 19–62 (1993)
2. E.G. Belaga, M. Mignotte, Embedding the 3x + 1 conjecture in a 3x + d context. Exp. Math. **7**(2), 145–151 (1998)
3. C. Böhm, G. Sontacchi, On the Existence of Cycles of Given Length in Integer Sequences Like $x_{n+1} = x_n/2$ if x_n Even, and $x_{n+1} = 3x_n + 1$ Otherwise. Atti Accad. Naz. Lincei Rend. Cl. Sci. Fis. Mat. Natur. (8) **64**(3), 260–264 (1978)
4. T. Brox, Collatz cycles with few descents. Acta Arith. **92**(2), 181–188 (2000)
5. S. Eliahou, The 3x+1 problem: new lower bounds on nontrivial cycle lengths. Discret. Math. **118**(1), 45–56 (1993)
6. J. Evertse, Linear forms in logarithms (2011). http://www.math.leidenuniv.nl/~evertse/dio2011-linforms.pdf
7. L.E. Garner, On the collatz $3n + 1$ algorithm. Proc. Am. Math. Soc. **82**(1), 19–22 (1981)
8. I. Grattan-Guinness, *The Norton History of the Mathematical Sciences: The Rainbow of Mathematics* (W.W. Norton, New York, 1998)
9. J. Lagarias, The 3x + 1 problem and its generalizations. Am. Math. Mon. **92**(1), 3–23 (1985)
10. G. Rhin, Approximants de Padé et mesures effectives d'irrationalité, in *Seminaire de Théorie des Nombres*, Paris 1985–1986, ed. by Goldstein C. Progress in Mathematics, vol. 71 (Birkhäuser Boston, Boston, 1987)
11. A. Rukhin, A dual-radix division algorithm for computing periodic orbits within syracuse dynamical systems (2018) (preprint, submitted)
12. J. Simons, On the nonexistence of 2-cycles for the problem. Math. Comput. **74**, 1565–1572 (2005)

13. J. Simons, B. de Weger, Theoretical and computational bounds for m-cycles of the 3n+1-problem. Acta Arith. **117**(1), 51–70 (2005)
14. R. Steiner, A theorem on the syracuse problem, in *Proceedings of the Seventh Manitoba Conference on Numerical Mathematics and Computing* (1977)

Potentially Stably Rational Del Pezzo Surfaces over Nonclosed Fields

Yuri Tschinkel and Kaiqi Yang

1 Introduction

A geometrically rational surface S over a nonclosed field k is k-birational to either a del Pezzo surface of degree $n \in [1, \ldots, 9]$ or a conic bundle (see [6]). Throughout, we assume that $S(k) \neq \emptyset$. This implies k-rationality of S when $n \in [5, \ldots, 9]$ or when the number of degenerate fibers of the conic bundle is at most 3.

Let G_k be the absolute Galois group of k, it acts on exceptional curves and on the geometric Picard group $\mathrm{Pic}(\bar{S})$ of S. The surface S is called *split* over k if all exceptional curves are defined over k, and *minimal* if no blow-downs are possible over k, i.e., there are no G_k-orbits consisting of pairwise disjoint exceptional curves. A minimal del Pezzo surface of degree ≤ 4 over k is not rational (see, e.g., [10, Theorem 3.3.1]). A surface S is called *stably rational* over k if $S \times \mathbb{P}^m$ is birational to $\mathbb{P}^{m+2}$, over k. A necessary condition for stable rationality of S over k is

Condition (H1)

$$\mathrm{H}^1(G_{k'}, \mathrm{Pic}(\bar{S})) = 0, \quad \text{for all finite extensions} \quad k'/k.$$

As a special case of a general conjecture of Colliot-Thélène and Sansuc one expects that this is also sufficient:

Y. Tschinkel (✉)
Simons Foundation, New York, NY 10012, USA
e-mail: tschinkel@cims.nyu.edu

Y. Tschinkel · K. Yang
Courant Institute, NYU, New York, NY 10012, USA
e-mail: ky994@nyu.edu

M. B. Nathanson (ed.), *Combinatorial and Additive Number Theory III*,
Springer Proceedings in Mathematics & Statistics 297,
https://doi.org/10.1007/978-3-030-31106-3_17

Conjecture 1.1 *If S satisfies* (H1) *then S is stably rational over k.*

Only one example of a minimal, and thus nonrational, but stably rational del Pezzo surface of degree ≤ 4 is known at present [2, 4, 5]; in this case, the Galois group acts via the symmetric group $\mathfrak{S}_3$, the smallest nonabelian group (see Sect. 2 for a description of this action). Finding another example is a major open problem. There are however examples of minimal del Pezzo surfaces of degrees $1 \leq n \leq 4$ and of conic bundles with at least 4 degenerate fibers, *failing* (H1) and thus stable rationality over k.

For $n = 3, 2$, and 1, the Galois group G_k acts on the primitive Picard group of S (the orthogonal complement of the canonical class in $\mathrm{Pic}(S)$) through the Weyl group $W(\mathsf{E}_{9-n})$; for $n = 4$ and conic bundles with $n + 1$ degenerate fibers through $W(\mathsf{D}_{n+1})$. These actions have been extensively studied, in connection with arithmetic applications and rationality questions, e.g., the Hasse Principle and Weak Approximation, when k is a number field (see e.g., [1, 7–9, 11, 12]).

This note is inspired by a recent result of Colliot-Thélène concerning stable rationality of geometrically rational surfaces over quasi-finite k, i.e., perfect fields with procyclic absolute Galois groups [3]. The main result of [3] is that over such fields, stably rational surfaces are actually rational. This follows from:

Theorem 1.2 ([3], Theorem 4.1) *Let S be a surface over k, geometrically rational with $S(k) \neq \emptyset$. If S is split by a cyclic extension and is not k-rational then there exists a finite separable extension k'/k such that*

$$\mathrm{H}^1(G_{k'}, \mathrm{Pic}(\bar{S})) \neq 0.$$

The proof proceeds via a case-by-case analysis of actions of (conjugacy classes of) elements of the corresponding Weyl groups, investigated in connection with the study of the Hasse-Weil zeta function of del Pezzo surfaces. For $n = 4$ this is due to [9, 11] and also follows from [7]; for $n = 3$ this goes back to Trepalin.

For general k, it is of interest to identify Galois actions potentially giving rise to minimal, stably rational surfaces, i.e., those satisfying (H1). This has been done in [7] for del Pezzo surfaces of degree 4. Our main result is a classification of the relevant actions in degrees 3, 2, and 1. In particular, this immediately gives an alternative proof of Theorem 1.2 for del Pezzo surfaces; there are simply no cyclic groups on the list of actions in Sects. 3 and 4.

The computation is organized as follows: the `magma` program produces a list of subgroups (modulo conjugation); then, starting with small groups, computes first cohomology groups. When it finds a group with nontrivial first cohomology, it eliminates all groups containing it. In this way, the poset of subgroups is rapidly exhausted. After that, minimality and presence of conic bundles are easily checked. The code and lists of orbit decompositions for subgroups satisfying (H1) are available at:

`cims.nyu.edu/~tschinke/papers/yuri/18h1dp/magma/`

2 Degree 4 and 3

We use the following notation:

- $\mathfrak{C}_n$–cyclic group of order n
- $\mathfrak{D}_n$–dihedral group of order $2n$
- $\mathfrak{F}_n$–Frobenius group of order $n(n-1)$
- $\mathfrak{S}_n$–symmetric group of order $n!$

Let S be a minimal del Pezzo surface of degree 4, satisfying Condition (H1). We recall Theorems E and F from [7]:

- If S admits a conic bundle structure then S is k-birational to
$$x^2 - ay^2 = f_3(t), \quad \deg(f_3) = 3,$$
where $a = \mathrm{disc}(f_3)$. The Galois group of the splitting field is $\mathfrak{S}_3$. One of the degenerate fibers, over ∞, is defined over k, the other three, corresponding to roots of f_3, are permuted by the $\mathfrak{S}_3$ action, the components of all singular fibers are exchanged the Galois action of the discriminant quadratic extension. A surface S of this type is not rational but stably rational over k.
- Assume that S does not admit a conic bundle structure over k. Let $\tilde{S} \to S$ be a blowup, with center in a suitable k-rational point; $\tilde{S}$ is a smooth (nonminimal) cubic surface admitting a conic bundle with 5 degenerate fibers. Then $\tilde{S}$ is of type I_1, I_2, or I_3 listed in [7, Theorem 6.15]. The Galois groups of corresponding splitting fields are $\mathfrak{S}_2 \times \mathfrak{S}_3$ in the first case, a nontrivial extension of $\mathfrak{S}_3$ by $\mathfrak{S}_2$ in the second case, and a nontrivial central extension of $\mathfrak{S}_2 \times \mathfrak{S}_3$ by $\mathfrak{S}_2$ in the third case. In Case 1, there are two degenerate fibers defined over k, with nontrivial Galois action on the components of the fibers, and three Galois conjugated degenerate fibers. In the Cases 2 and 3, the Galois-action has two orbits on the set of degenerate fibers, of length 2 and 3.

Our first result is:

Proposition 2.1 *There are no minimal cubic surfaces satisfying Condition* (H1). *In particular, a k-minimal cubic surface is not stably rational over k.*

Proof Direct calculation with `magma`. □

3 Degree 2

In the description below we encode the Galois action on the set of exceptional curves as follows: we write $\{v_1^{r_1}, \dots, v_m^{r_m}\}$ for the decomposition into orbits, where v_j are dual intersection graphs, enumerated below, and r_j are their multiplicities. For minimal del Pezzo surfaces of degree 2 we find unique orbit types with cardinality 4, 8, 18,

24, 30, 42, two types of cardinality 2 and 12, and three types of cardinality 6 and 10. The occurring graphs for each orbit are symmetrical: each vertex has the same number of outgoing edges (with multiplicities). We write

$$(n)[s_1^{t_1}, \ldots, s_d^{t_d}]$$

for a graph with n vertices, where each vertex has t_j outgoing edges of multiplicity s_j (equal to the intersection number between the two exceptional curves connected by this edge). The corresponding graphs are listed below:

- $2_c := (2)[1]\ \bullet - \bullet$, $2 := (2)[2]\ \bullet = \bullet$
- $4 := (4)[1^3]$
- $6_1 := (6)[1^2, 2]$, $6_2 := (6)[1^4]$, $6_c = (6)[1]$ conic bundle
- $8 := (8)[1^3, 2]$
- $10_1 := (10)[1^4, 2]$, $10_2 := (10)[1^6]$, $10_c = (10)[1]$ conic bundle
- $12 := (12)[1^5, 2]$, $12_c = (12)[1]$, conic bundle
- $14 := (14)[1^6, 2]$
- $18 := (18)[1^8, 2]$
- $24 := (24)[1^{11}, 2]$
- $30 := (30)[1^{14}, 2]$
- $42 := (42)[1^{20}, 2]$

In the following propositions we list the structure of Galois groups of splitting fields, the structure or orbits on the set of exceptional curves, and the stabilizers for each orbit.

Proposition 3.1 *Assume that S is a minimal degree 2 del Pezzo surface over k satisfying Condition* (H1). *Then S either admits a conic bundle structure over k or is one of the following types, each corresponding to a conjugacy class of subgroups in $W(\mathsf{E}_7)$:*

dP2(1) $\mathfrak{D}_7$: $\{14^4\}$, *trivial stabilizer*
dP2(2) $\mathfrak{F}_7$: $\{14, 42\}$, *specializes to* dP2(1), *when restricted to* $\mathfrak{D}_7 \subset \mathfrak{F}_7$.
dP2(3) $\mathfrak{D}_{15}$: $\{6_1, 10_1^2, 30\}$, *stabilizers* $\{\mathfrak{C}_5, \mathfrak{C}_3, 1\}$.
dP2(4) $\mathfrak{C}_3 \rtimes \mathfrak{F}_5$: $\{6_1, 10_2^2, 30\}$, *stabilizers* $\{\mathfrak{D}_5, \mathfrak{C}_6, \mathfrak{C}_2\}$, *with* $\mathfrak{C}_2$ *not normal.*

Below we list all possible conic bundle types. Each X admits two conic bundle structures over k, with isomorphic Galois actions on the set of exceptional fibers of the corresponding conic bundle. We organize by cardinalities of orbits on these sets, and by the orbit structure on the set of exceptional curves of X.

3 + 3:

D6(1) $\mathfrak{S}_3$: $\{2, 6_1^3, 6_2^2, 6_c^4\}$, stabilizers $\{\mathfrak{C}_3, 1, 1, 1\}$
D6(2) $\mathfrak{C}_3 \rtimes \mathfrak{S}_3$: $\{2, 6_1^2, 6_c^4, 18\}$, stabilizer $\{\mathfrak{C}_3^2, \mathfrak{C}_3, \mathfrak{C}_3, 1\}$

5 + 1:

D6(3) $\mathfrak{D}_5$: $\{2_c^2, 2, 10_1^3, 10_c^2\}$, stabilizer $\{\mathfrak{C}_5, \mathfrak{C}_5, 1, 1\}$

D6(4) $\mathfrak{F}_5$: $\{2_c^2, 2, 10_1, 10_2^2, 10_c^2\}$, stabilizer $\{\mathfrak{D}_5, \mathfrak{D}_5, \mathfrak{C}_2, \mathfrak{C}_2, \mathfrak{C}_2\}$; $\mathfrak{C}_2$ is not normal

6:

D6(5) $\mathfrak{D}_6$: $\{2, 6_1, 12^2, 12_c^2\}$, stabilizer $\{\mathfrak{C}_6, \mathfrak{C}_2, 1, 1\}$.
D6(6) $\mathfrak{D}_6$: $\{2, 6_1, 6_2^2, 12, 12_c^2\}$, stabilizer $\{\mathfrak{S}_3, \mathfrak{C}_2, \mathfrak{C}_2, 1, 1\}$.
D6(7) $\mathfrak{S}_4$: $\{2, 6_1, 12_c^2, 24\}$, stabilizer $\{\mathfrak{A}_4, \mathfrak{C}_2^2, \mathfrak{C}_2, 1\}$.
D6(8) $\mathfrak{S}_4$: $\{4^2, 6_2^2, 12, 12_c^2\}$, stabilizer $\{\mathfrak{S}_3, \mathfrak{C}_2^2, \mathfrak{C}_2, \mathfrak{C}_2\}$.
D6(9) $\mathfrak{S}_4$: $\{6_2^2, 8, 12, 12_c^2\}$, stabilizer $\{\mathfrak{C}_4, \mathfrak{C}_3, \mathfrak{C}_2, \mathfrak{C}_2\}$.
D6(10) $\mathfrak{S}_3^2$: $\{2, 12, 12_c^2, 18\}$, stabilizer $\{\mathfrak{C}_3 \times \mathfrak{S}_3, \mathfrak{C}_3, \mathfrak{C}_3, \mathfrak{C}_2\}$.
D6(11) $\mathfrak{C}_2 \times \mathfrak{S}_4$: $\{2, 6_1, 12_c^2, 24\}$, stabilizer $\{\mathfrak{C}_2 \times \mathfrak{A}_4, \mathfrak{C}_2^3, \mathfrak{C}_2^2, \mathfrak{C}_2\}$, the stabilizer $\mathfrak{C}_2$ is not normal, and this case does not reduce to D6(7), with $\mathfrak{S}_4$-action.
D6(12) $\mathfrak{C}_2 \times \mathfrak{S}_4$: $\{6_2^2, 8, 12, 12_c^2\}$, stabilizer $\{\mathfrak{D}_4, \mathfrak{S}_3, \mathfrak{C}_2^2, \mathfrak{C}_2^2\}$.
D6(13) $\mathfrak{S}_5$: $\{2, 12_c^2, 30\}$, stabilizer $\{\mathfrak{A}_5, \mathfrak{D}_5, \mathfrak{C}_2^2\}$.
D6(14) $\mathfrak{S}_5$: $\{10_2^2, 12, 12_c^2\}$, stabilizer $\{\mathfrak{D}_6, \mathfrak{D}_5, \mathfrak{D}_5\}$.

Some types above are specializations of other types, by restriction to subgroups:

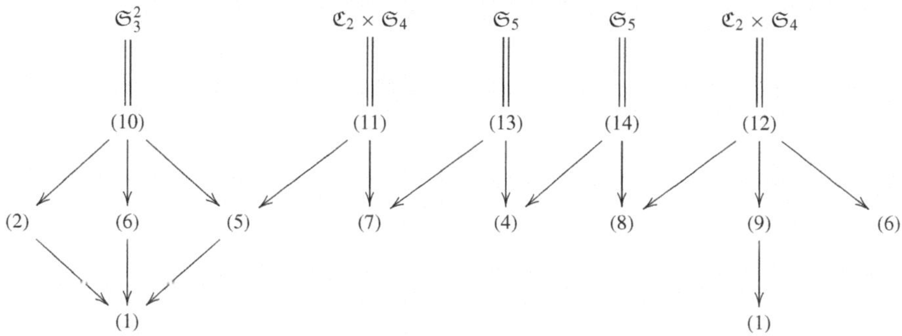

4 Degree 1

Proposition 4.1 *If S is a minimal degree 1 del Pezzo surface satisfying Condition* (H1) *then S is a conic bundle over k.*

As Galois orbits we have unions of degenerate fibers of conic bundles (4_c, 6_c, 8_c, 10_c) and several new orbit types:

- $3 := (3)[2^2]$
- $4_1 := (4)[2^2]$, $4_2 := (4)[1^2, 2]$, $4_3 = (4)[1^2, 3]$,
- $5 := (5)[1^2, 2^2]$.
- $6_3 := (6)[2^2, 3]$, $6_4 := (6)[1^3, 2^2]$
- $10_3 := (10)[1^3, 2^4]$, $10_4 := (10)[1^4, 2^2, 3]$,
- $12_1 := (12)[1, 2^6]$ $12_2 := (12)[1^4, 2^3]$, $12_3 := (12)[1^2, 2^4, 3]$, $12_4 := (12)[1^8, 2]$, $12_5 := (12)[1^6, 2^2, 3]$

- $20_1 := (20)[1^2, 2^8, 3]$, $20_2 := (20)[1^8, 2^4]$, $20_3 := (20)[1^6, 2^6, 3]$, $20_4 := (20)[1^{12}, 2^2]$, $20_5 := (20)[1^9, 2^6]$
- $24_1 := (24)[1^2, 2^{10}, 3]$, $24_2 := (24)[1^{13}, 2^3]$
- $36_1 := (36)[1^{18}, 2^5]$, $36_2 := (36)[1^{18}, 2^8, 3]$.
- $40 := (40)[1^{18}, 2^{10}, 3]$

The types of occurring conic bundles are listed below, each corresponding to a conjugacy class of subgroups in $W(\mathsf{E}_8)$:
1 + 3 + 3:

D7(1) $\mathfrak{S}_3^2$: $\{2_c^2, 3^4, 4_2^2, 6_3^2, 6_c^4, 12_2^4, 12_3^2, 36_1^2, 36_2\}$,
stabilizer $\{\mathfrak{C}_3 \rtimes \mathfrak{S}_3, \mathfrak{D}_6, \mathfrak{C}_3^2, \mathfrak{S}_3, \mathfrak{S}_3, \mathfrak{C}_3, \mathfrak{C}_3, 1, 1\}$

1 + 1 + 5:

D7(2) $\mathfrak{D}_{10}$: $\{2_c^4, 4_1^2, 4_3, 5^4, 10_2^2, 10_c^2, 20_1^2, 20_2^4, 20_4^2\}$, stabilizer $\{\mathfrak{D}_5, \mathfrak{C}_5, \mathfrak{C}_5, \mathfrak{C}_2^2, \mathfrak{C}_2, \mathfrak{C}_2, 1, 1, 1\}$.
D7(3) $\mathfrak{C}_2 \times \mathfrak{F}_5$: $\{2_c^4, 4_1^2, 4_3, 10_1^2, 10_c^2, 20_1^2, 20_3, 20_4^6\}$, stabilizer $\{\mathfrak{F}_5, \mathfrak{D}_5, \mathfrak{D}_5, \mathfrak{C}_2^2, \mathfrak{C}_2^2, \mathfrak{C}_2, \mathfrak{C}_2, \mathfrak{C}_2\}$.

2 + 5:

D7(4) $\mathfrak{C}_5 \rtimes \mathfrak{C}_4$: $\{4_1^2, 4_3, 4_c^2, 5^4, 10_2^2, 10_c^2, 20_2^4, 20_4^2, 20_5^2\}$, stabilizer $\{\mathfrak{C}_5, \mathfrak{C}_5, \mathfrak{C}_5, \mathfrak{C}_4, \mathfrak{C}_2, \mathfrak{C}_2, 1, 1, 1\}$
D7(5) $\mathfrak{F}_5$: $\{4_1^2, 4_3, 4_c^2, 10_1^2, 10_c^2, 20_3, 20_4^6, 20_5^2\}$, stabilizer $\{\mathfrak{C}_5, \mathfrak{C}_5, \mathfrak{C}_5, \mathfrak{C}_2, \mathfrak{C}_2, 1, 1, 1\}$
D7(6) $\mathfrak{C}_5 \rtimes \mathfrak{D}_4$: $\{4_1^2, 4_3, 4_c^2, 5^4, 10_2^2, 10_c^2, 20_2^4, 20_4^2, 40\}$, stabilizer $\{\mathfrak{C}_{10}, \mathfrak{C}_{10}, \mathfrak{D}_5, \mathfrak{D}_4, \mathfrak{C}_2^2, \mathfrak{C}_2^2, \mathfrak{C}_2, \mathfrak{C}_2, 1\}$
D7(7) $\mathfrak{C}_2 \times \mathfrak{F}_5$: $\{4_1^2, 4_3, 4_c^2, 10_1^2, 10_c^2, 20_3, 20_4^6, 20_5^2\}$, stabilizer $\{\mathfrak{D}_5, \mathfrak{D}_5, \mathfrak{D}_5, \mathfrak{C}_2^2, \mathfrak{C}_2^2, \mathfrak{C}_2, \mathfrak{C}_2, \mathfrak{C}_2\}$; the stabilizer $\mathfrak{C}_2$ is not normal and we cannot reduce to D7(5) $= \mathfrak{F}_5$
D7(8) $\mathfrak{C}_2^2 \rtimes \mathfrak{F}_5$: $\{4_1^2, 4_3, 4_c^2, 10_1^2, 10_c^2, 20_3, 20_4^6, 40\}$, stabilizer $\{\mathfrak{D}_{10}, \mathfrak{D}_{10}, \mathfrak{F}_5, \mathfrak{C}_2^3, \mathfrak{C}_2^3, \mathfrak{C}_2^2, \mathfrak{C}_2^2, \mathfrak{C}_2\}$

1 + 6:

D7(9) $(\mathfrak{C}_3 \rtimes \mathfrak{S}_3) \rtimes \mathfrak{C}_2$: $\{2_c^2, 4_2^2, 6_2^2, 12_1^2, 12_4^4, 12_5, 12_c^2, 36_1^2, 36_2\}$, stabilizer $\{\mathfrak{C}_3 \rtimes \mathfrak{S}_3, \mathfrak{C}_3^2, \mathfrak{S}_3, \mathfrak{C}_3, \mathfrak{C}_3, \mathfrak{C}_3, \mathfrak{C}_3, 1, 1\}$
D7(10) $\mathfrak{S}_3 \wr \mathfrak{C}_2$: $\{2_2^2, 4_2^2, 6_2^2, 12_5, 12_c^2, 24_1, 24_2^2, 36_1^2, 36_2\}$, stabilizer $\{\mathfrak{S}_3^2, \mathfrak{C}_3 \times \mathfrak{S}_3, \mathfrak{D}_6, \mathfrak{S}_3, \mathfrak{S}_3, \mathfrak{C}_3, \mathfrak{C}_3, \mathfrak{C}_2, \mathfrak{C}_2\}$

Again, some types are specializations, by restriction to subgroups:

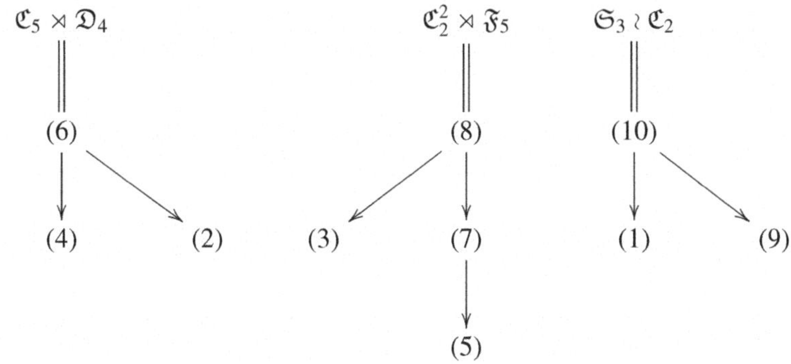

Acknowledgements We are grateful to J.-L. Colliot-Thélène for helpful comments and suggestions. The first author was partially supported by NSF grant 1601912.

References

1. Banwait, B., Fité, F., Loughran, D.: Del Pezzo surfaces over finite fields and their Frobenius traces (2016). arXiv:1606.00300.
2. Beauville, A., Colliot-Thélène, J.L., Sansuc, J.J., Swinnerton-Dyer, P.: Variétés stablement rationnelles non rationnelles. Ann. of Math. (2) **121**(2), 283–318 (1985).
3. Colliot-Thélène, J.L.: Surfaces stablement rationnelles sur un corps quasi-fini (2017). arXiv:1711.09595.
4. Colliot-Thélène, J.L., Sansuc, J.J., Swinnerton-Dyer, P.: Intersections of two quadrics and Châtelet surfaces. I. J. Reine Angew. Math. **373**, 37–107 (1987).
5. Colliot-Thélène, J.L., Sansuc, J.J., Swinnerton-Dyer, P.: Intersections of two quadrics and Châtelet surfaces. II. J. Reine Angew. Math. **374**, 72–168 (1987).
6. Iskovskih, V.A.: Minimal models of rational surfaces over arbitrary fields. Izv. Akad. Nauk SSSR Ser. Mat. **43**(1), 19–43, 237 (1979).
7. Kunyavskiĭ, B.E., Skorobogatov, A.N., Tsfasman, M.A.: del Pezzo surfaces of degree four. Mém. Soc. Math. France (N.S.) (37), 113 (1989).
8. Li, S.: Rational points on del Pezzo surfaces of degree 1 and 2. arxiv.org/0904.3555.
9. Manin, Y.I.: Rational surfaces over perfect fields. II. Mat. Sb. (N.S.) **72 (114)**, 161–192 (1967).
10. Manin, Y.I., Tsfasman, M.A.: Rational varieties: algebra, geometry, arithmetic. Uspekhi Mat. Nauk **41**(2(248)), 43–94 (1986).
11. Swinnerton-Dyer, P.: The zeta function of a cubic surface over a finite field. Proc. Cambridge Philos. Soc. **63**, 55–71 (1967).
12. Urabe, T.: Calculation of Manin's invariant for Del Pezzo surfaces. Math. Comp. **65**(213), 247–258, S15–S23 (1996).

The manufacturer's authorised representative in the EU is Springer Nature Customer Service Centre GmbH, Europaplatz 3, 69115 Heidelberg, Germany. If you have any concerns regarding our products, please contact ProductSafety@springernature.com

Printed and bound by CPI Group (UK) Ltd, Croydon, CR0 4YY

24/07/2026

02175907-0005